Phosphate in Soils

Interaction with Micronutrients, Radionuclides and Heavy Metals

ADVANCES IN TRACE ELEMENTS IN THE ENVIRONMENT

Series Editor: H. Magdi Selim

Louisiana State University, Baton Rouge, USA

Permeable Reactive Barrier: Sustainable Groundwater Remediation
edited by Ravi Naidu and Volker Birke

Phosphate in Soils: Interaction with Micronutrients, Radionuclides and Heavy Metals
edited by H. Magdi Selim

Phosphate in Soils

Interaction with Micronutrients, Radionuclides and Heavy Metals

Edited by
H. Magdi Selim

CRC Press
Taylor & Francis Group
Boca Raton London New York

CRC Press is an imprint of the
Taylor & Francis Group, an **informa** business

CRC Press
Taylor & Francis Group
6000 Broken Sound Parkway NW, Suite 300
Boca Raton, FL 33487-2742

First issued in paperback 2017

ISBN-13: 978-1-4822-3679-8 (hbk)
ISBN-13: 978-1-138-80318-3 (pbk)

Visit the Taylor & Francis Web site at
http://www.taylorandfrancis.com

and the CRC Press Web site at
http://www.crcpress.com

To the memory of my brother

Sami

Contents

Preface

Phosphorus (P) in the form of phosphate is a commonly used fertilizer worldwide. Phosphorus is a macronutrient essential for plant growth and crop yields. Due to its complexity, phosphorous interactions in soils continue to be extensively investigated. Concurrently, the presence of phosphate in soils influences the bioavailability, retention, and mobility of heavy metals in various ways. This book provides a focused effort on recent research findings of the influence of phosphorus on trace elements, metal(loid)s, and radio nuclides in soils.

Trace elements such as copper (Cu) and zinc (Zn) along with P are often incorporated in soils as fertilizer amendments (as micronutrients) to increase crop yields. When this occurs, phosphates can affect trace element dynamics through pH changes or through modifications operating on the electric charge of the colloids, increasing the negative charge and then conditioning sorption of new adsorbates. Chapter 1 provides an overview of the reactions of metal(loid)s and common P compounds that are used as fertilizer in soils, and then focuses on the mechanisms for the (im)mobilization of metal(loid)s by the various P forms. Implications of P forms on the transformation of metal(loid)s are presented in relation to sequestration and phytoavailability of metal(loid)s in soils.

In Chapter 2, the effect of phosphorus on copper and zinc adsorption in acid soils is emphasized. The chapter clearly illustrates that the presence of phosphates in solution increases the retention of both Cu and Zn on acid soils and details the possible mechanisms involved in such enhancement of sorption. Increases of Cu adsorption take place mainly via slow reaction sites, whereas in soils containing high amounts of iron (Fe) and aluminum (Al) oxyhydroxides, the Cu adsorption increase is in relation to both fast and slow reaction sites.

In Chapter 3, phosphate-containing materials are also evaluated as remediating agents for the *in situ* immobilization of radiological soil contaminants, including uranium (U) and other actinides. An extensive discussion is presented on the dynamic processes controlling the interaction of U and P in contaminated soils. This information is a prerequisite to proper remediation management efforts that will enable protection of the soil–water environment. In Chapter 4, the influence of the presence of phosphate on the movement of three heavy metals—zinc, cadmium (Cd), and vanadium (V)—is presented in detail. Phosphate proved to be highly competitive with all three heavy metals to various degrees. As well, detailed descriptions of the ways to describe the competitive effect of phosphate during transport are presented.

Phosphorus fertilizers contain Cd as a contaminant at levels varying from trace amounts to as much as 300 mg Cd kg^{-1} of dry product and therefore can be a major source of Cd input to agricultural systems (Chapter 5). As both total Cd input into the soil and fertilizer-induced changes in soil properties are a function of the rate of fertilizer application, management practices that improve fertilizer use efficiency should be adopted to minimize fertilizer input while maintaining crop yield potential.

In Chapter 6, mathematical models of competitive arsenic (As)–P sorption and transport in geological porous media are reviewed. First, surface complexation models commonly used to describe the chemical mechanisms of the competitive reaction between PO_4^{3-} and AsO_4^{3-} are examined, and then equilibrium and kinetic multicomponent models based on the empirical Freundlich or Langmuir equations are presented. This is followed by findings on the influence of phosphate compounds on speciation, mobility, and bioavailability of heavy metals in soils (Chapter 7). The role of phosphates on *in situ* and phytoremediation of heavy metals for contaminated soils is discussed.

Chapter 8 emphasizes on the influence of phosphate addition on various heavy metals species in soils, and their solubility/mobility and availability. The focus is also on immobilization of heavy metals in soils by phosphate materials (natural and synthetic) and chemical speciation and ionic interaction between phosphate and heavy metals. In Chapter 9, a comprehensive literature search provided extensive information on testing various high phosphate materials for remediation of trace element contaminated sites. Several materials have been applied to reduce Pb solubility in soil and Pb phytoavailability, which are directly related to Pb extractability and Pb uptake by plants. The use of less soluble rock fertilizers is feasible for practical implementation due to lower cost and lower risk of eutrophication. In Chapter 10, the reactivity of trace elements in several soils is presented. The focus is on trace element behavior when soils receive long-term applications of high rates of sewage sludge to land cropped with maize.

In Chapter 11, a case study illustrates various remediation efforts of acidic soils and remediation of Cu, Zn, and lead (Pb) contaminated soils around nonferrous industrial plants. The influence of applications of organomineral mixtures containing lime, phosphorite–glauconite, peat, coal, and Fe hydroxides on the solubility, speciation, and bioavailability of Cu, Zn, and Pb to ryegrass plants is presented. Chapter 12 emphasizes the significance of common ions (cations and anions) on phosphate mobility and sorption in soils. Such ions are encountered as part of phosphate fertilizers applied to soils.

I wish to sincerely thank all contributors who made this book possible in a timely manner. Special thanks also go to the editorial staff of Taylor & Francis/CRC Press, in particular Ms. Irma Britton, for her help and support throughout the publication of this book.

H. Magdi Selim

Editor

H. Magdi Selim is a professor of soil physics, A. George and Mildred L. Caldwell Endowed Professor, School of Plant, Environmental and Soil Sciences at Louisiana State University at Baton Rouge. He received his BS degree in soil science from Alexandria University, Alexandria, Egypt, and his MS and PhD degrees in soil physics from Iowa State University, Ames, Iowa. He is internationally recognized for his research in the areas of kinetics of reactive chemicals in heterogeneous porous media and transport modeling of dissolved chemicals in water-saturated and -unsaturated soils. He is the original developer of several concept models for describing the retention processes of dissolved chemicals in soils and natural materials in porous media. Pioneering work also includes multistep/multireaction and nonlinear kinetic models for trace elements, heavy metals, radionuclides, explosive contaminants, and phosphorus and pesticides in soils and geological media.

Professor Selim is the author or coauthor of numerous scientific publications in several journals. He is also the editor and coauthor of several books. Dr. Selim is the recipient of several professional awards and honors. He is a member of the American Society of Agronomy, Soil Science Society of America, International Society of Soil Science, International Society of Trace Element Biogeochemistry, Louisiana Association of Agronomy, and American Society of Sugarcane Technology. Dr. Selim was elected chair of the Soil Physics Division of the Soil Science Society of America. He served on numerous committees of the Soil Science Society of America, American Society of Agronomy, and the International Society of Trace Element Biogeochemistry. He also served as an associate editor of *Water Resources Research* and *Soil Science Society of America Journal* and as a technical editor of *Journal of Environmental Quality*.

Professor Selim is a fellow of the American Society of Agronomy and the Soil Science Society of America. He has received numerous awards including the Phi Kappa Phi Research Award, the Gamma Sigma Delta Award for Research, the Joe Sedberry Graduate Teaching Award, the First Mississippi Research Award, the Doyle Chambers Achievements Award, and the EPA Environmental Excellence Award. Recent awards include the Soil Science Society of America Soil Science Research Award and the International Union of Soil Science von Liebeg Award.

Contributors

Manuel Arias-Estévez
Area de Edafoloxía e Química Agrícola
Departamento de Bioloxía Vexetal e Ciencia do Solo
Universidade de Vigo
Ourense, Spain

Irena Atanassova
Nikola Poushkarov Institute of Soil Science
Agrotechnologies and Plant Protection
Sofia, Bulgaria

Maya Benkova
Nikola Poushkarov Institute of Soil Science
Agrotechnologies and Plant Protection
Sofia, Bulgaria

Nanthi S. Bolan
Centre for Environmental Risk Assessment and Remediation
University of South Australia
Mawson Lakes, Australia

and

Cooperative Research Centre for Contamination
Assessment and Remediation of the Environment
Adelaide, Australia

Shiv S. Bolan
Centre for Environmental Risk Assessment and Remediation
University of South Australia
Mawson Lakes, Australia

Shea W. Buettner
Savannah River Ecology Laboratory
Aiken, South Carolina

Hyun-shik Chang
Savannah River Ecology Laboratory
Aiken, South Carolina

Shibao Chen
Institute of Agricultural Resources and Regional Planning
Chinese Academy of Agricultural Sciences
Beijing, China

David Fernández-Calviño
Department of Plant and Environmental Sciences
University of Copenhagen
Frederiksberg, Denmark

Cynthia A. Grant
Agriculture and Agri-Food Canada
Brandon Research Centre
Brandon, Manitoba, Canada

Petra Kidd
Instituto de Investigaciones Agrobiológicas de Galicia (IIAG)
Consejo Superior de Investigaciones Cientificas (CSIC)
Santiago de Compostela, Spain

Anitha Kunhikrishnan
Chemical Safety Division
Department of Agro-Food Safety
National Academy of Agricultural Science
Wanju-gun, Jeollabuk-do, Republic of Korea

Gabriel Maurício Peruca de Melo
Universidade Camilo Castelo Branco
Descalvado, São Paulo, Brazil

Valéria Peruca de Melo
Universidade Camilo Castelo Branco
Descalvado, São Paulo, Brazil

Wanderley José de Melo
Technology Department
Faculdade de Ciências Agrárias e Veterinárias
Universidade Estadual Paulista
Campus de Jaboticabal, São Paulo, Brazil

Ravi Naidu
Centre for Environmental Risk Assessment and Remediation
University of South Australia
Mawson Lakes, Australia

and

Cooperative Research Centre for Contamination
Assessment and Remediation of the Environment
Adelaide, Australia

Jinhee Park
Korea Institute of Geoscience and Mineral Resources
Daejeon, Republic of Korea

Cristina Pérez-Novo
Tecnopole (Parque Tecnolóxico de Galicia)
CACTI
Ourense, Spain

Jörg Rinklebe
University of Wuppertal
Department D, Soil- and Groundwater-Management
Wuppertal, Germany

John C. Seaman
Savannah River Ecology Laboratory
Aiken, South Carolina

H. Magdi Selim
Louisiana State University
Baton Rouge, Louisiana

Sabry M. Shaheen
University of Kafrelsheikh
Faculty of Agriculture
Department of Soil and Water Sciences
Kafr El-Sheikh, Egypt

Grzegorz Siebielec
Institute of Soil Science and Plant Cultivation (IUNG)
Pulawy, Poland

Zhengguo Song
Agro-Environmental Protection Institute
Ministry of Agriculture
Tianjin, China

Christos D. Tsadilas
National Agricultural Research Foundation
Institute of Soil Mapping and Classification
Larissa, Greece

Aleksandra Ukalska-Jaruga
Institute of Soil Science and Plant Cultivation (IUNG)
Pulawy, Poland

Minggang Xu
Institute of Agricultural Resources and Regional Planning
Chinese Academy of Agricultural Sciences
Beijing, China

Hua Zhang
Key Laboratory of Coastal Zone
Environmental Processes and Ecological Remediation
Yantai Institute of Coastal Zone Research
Chinese Academy of Sciences
Yantai, China

Shiwei Zhou
Yantai Institute of Coastal Zone Research
Chinese Academy of Sciences
Yantai, China

1

Phosphorus-Induced (Im)mobilization of Heavy Metal(loid)s in Soils

Anitha Kunhikrishnan, Jinhee Park, Shiv S. Bolan, Ravi Naidu, and Nanthi S. Bolan

CONTENTS

The term *heavy metal(loid)* in general includes elements (both metals and metalloids) with an atomic density greater than 6 g cm^{-3} (with the exception of arsenic [As], boron [B], and selenium [Se]). This group includes both biologically essential (e.g., cobalt [Co], copper [Cu], chromium [Cr], manganese [Mn], and zinc [Zn]) and nonessential (e.g., cadmium [Cd], lead [Pb], nickel [Ni], and mercury [Hg]) elements. The essential elements (for plant, animal, or human nutrition) are required in low concentrations and hence are known as trace elements or micronutrients. The nonessential metal(loid)s are phytotoxic and/or zootoxic and are widely known as toxic elements. Both groups are toxic to plants, animals, and/or humans at exorbitant concentrations.

As land treatment becomes one of the important waste management practices, soil is increasingly being seen as a major source of metal(loid)s reaching the food chain, mainly through plant uptake and animal transfer. Indiscriminate waste disposal practices have led to significant buildup of a wide range of metal(loid)s, such as As, Cr, Cu, Ni, Pb, Cd, Hg, Se, and Zn. Unlike organic contaminants, most metal(loid)s do not undergo microbial or chemical degradation, and the total concentration of these metal(loid)s in soils persists for a long time after their introduction (Adriano et al. 2004). The mobilization of metal(loid)s in soils for plant uptake and leaching to groundwater can, however, be minimized by reducing the bioavailability of metal(loid)s through chemical and biological immobilization (Harmsen and Naidu 2013). Recently there has been increasing interest in the immobilization of metal(loid)s using a range of inorganic compounds, such as lime, phosphate (P) compounds (e.g., apatite rocks), and alkaline waste materials, and organic compounds, such as exceptional quality biosolid (Park et al. 2011a; Zhou and Haynes 2010).

Regular application of P fertilizers has been identified as the main source of heavy metal(loid) contamination of soils in some countries (Loganathan et al. 2008; van Kauwenbergh 2002). Some of these P fertilizers, which act as a source of heavy metal(loid) contamination of agricultural soils, have also been found to act as a sink for the immobilization of these metal(loid)s (Bolan et al. 2003a; Miretzky and Fernandez-Cirelli 2008). P amendment has often been proposed as a practical remediation option for sites with Pb-contaminated soils (Freeman 2012; Ma et al. 2008). Application of P compounds to soils, however, can impact both mobilization and immobilization of metal(loid)s, the effect being dependent on the nature of the P compound, soil type, and metal(loid) species (Sanderson et al. 2014).

Following a brief overview of the reactions of metal(loid)s and common P compounds that are used as fertilizer in soils, this chapter focuses on the mechanisms for the (im)mobilization of metal(loid)s by P compounds. The practical implications of P compounds on the transformation of metal(loid)s are discussed in relation to sequestration and phytoavailability of metal(loid)s in soils.

Sources of Heavy Metal(loid)s in Soil Environment

In terrestrial ecosystems, the soil is the main repository of chemical contaminants. Heavy metal(loid)s reach the soil environment through both pedogenic and anthropogenic processes. Most heavy metal(loid)s occur naturally in soil parent materials, chiefly in forms that are not readily bioavailable for uptake by higher plants. Unlike pedogenic inputs, heavy metal(loid)s added through anthropogenic activities typically have a high bioavailability (Lamb et al. 2009; Naidu and Bolan 2008; Naidu et al. 1996).

Geogenic

Most of the heavy metal(loid)s occur in nature, the major source of which is weathering of soil parent materials including igneous and sedimentary rocks and coal. For example, coal is estimated to release 45,000 tonnes of As annually, while human activities release approximately 50,000 tonnes (Ferguson and Gavis 1972; Mahimairaja et al. 2005). The As content of igneous rocks varies widely (up to 100 mg kg^{-1}); the average content is 2 to 3 mg kg^{-1}. Sedimentary rocks also vary in their As content, from small amounts in limestone and sandstone up to 15,000 mg kg^{-1} in some Mn ores (Yan-Chu 1994). Although the anthropogenic As source is becoming increasingly important, extensive As contamination of groundwaters in many countries including Bangladesh, India, China, and Mexico is of geological origin, transported by rivers from sedimentary rocks in the Himalayas over tens of thousands of years (Armienta et al. 1997; Del Razo et al. 1990; Mahimairaja et al. 2005; Yu et al. 2007).

In nature, Cd occurs mainly in association with Zn ores, of which sphalerite (zinc sulfide) forms the main commercial source of Cd. Although Cd occurs in most soil parent materials, its concentration in common soil-forming rocks such as igneous rocks, sandstones, and limestone is generally low (Bramley 1990; Loganathan et al. 2012; MacDonald et al. 2005); however, the concentration of Cd in rocks derived from lake sediments and marine black shales is considered to be high.

Similarly, seleniferous soils found in the areas of Se-rich rocks, such as black shales, carbonaceous limestones, carbonaceous cherts, mudstones, and seleniferous coal, are a major source of Se input to soil. Also, irrigation of Se-rich groundwater has shown to contaminate irrigation drainwater and surface waters in the San Joaquin Valley in California, United States, and in Punjab, India (Dhillon and Dhillon 2014; Kaush and Pallud 2013). Volcanic and geological activities mobilize natural Hg from deep reservoirs in the earth to the atmosphere. Annual emission of Hg from global mercuriferous belts, the zone along plate tectonic boundaries including western North America, central Europe, and southern China, was estimated up to 500 Mg $year^{-1}$ (Lindqvist et al. 1991; Selin 2009).

Anthropogenic

Anthropogenic activities, primarily associated with industrial processes, manufacturing, and the disposal of domestic and industrial waste materials, are the major source of metal(loid) enrichment in soils. In urbanized areas, heavy metal(loid)s can be accumulated in soil from vehicle emissions and industrial discharges (Li et al. 2001; Rakib et al. 2014). Cd, Cu, Pb, and Zn can be found in soils contaminated by vehicle emissions because they are included in gasoline, car components, oil lubricant, and industrial incinerators (Adriano 2001). Atmospheric pollution from Pb-based gasoline was a major issue in many countries where there was no constraint on the usage of leaded gasoline (Fenger 2009). In agricultural lands, soil irrigation with wastewater accumulates heavy metal(loid)s such as Cd, Zn, Cr, Ni, Pb, and Mn in surface soil (Kunhikrishnan et al. 2012; Sharma et al. 2007). While biosolids is the major source of metal(loid) inputs in Europe and North America, P fertilizers are considered to be the major source of heavy metal(loid) input, especially Cd, in Australia and New Zealand (Bolan et al. 2003a; Haynes et al. 2009; Loganathan et al. 2008; McLaughlin et al. 1996). High heavy metal(loid) concentrations such as Cu, Zn, Pb, Co, Ni, Cd, and As have been reported in soils close to mine sites (Bech et al. 1997; Kabir et al. 2012).

Large quantities of Cu are used in agriculture, horticulture, and animal industries as a trace element; in many formulations of Cu containing fungicides, such as copper oxychloride and Bordeaux mixture; and as a growth promoter in piggery and poultry units (Bolan et al. 2003b; Wighwick et al. 2013). Accumulation of Cu in agricultural soils resulting from continuous use of Cu fungicides and biosolids application has been reported in many countries (Chopin et al. 2008; Heemsbergen et al. 2010; Hildebrandt et al. 2008; Wightwick et al. 2013). Chromium is used as Cr(III) in the tannery industry and as Cr(VI) in the timber treatment industry. Chromate is highly toxic and carcinogenic even when present in very low concentrations in water. Large-scale use of chromated copper arsenate (CCA)-treated timber as fence post and in vineyards can also result in the release of Cu, Cr, and As to soil environment (Robinson et al. 2006; Schwer and McNear 2011).

Reactions of Metal(loid)s in Soils

Metal(loid) ions can be retained in the soil by sorption, precipitation, and complexation reactions, and are removed from soil through plant uptake, leaching, and volatilization (Figure 1.1). Although most metal(loid)s are not subject to volatilization losses, some metal(loid)s such as As, Hg, and

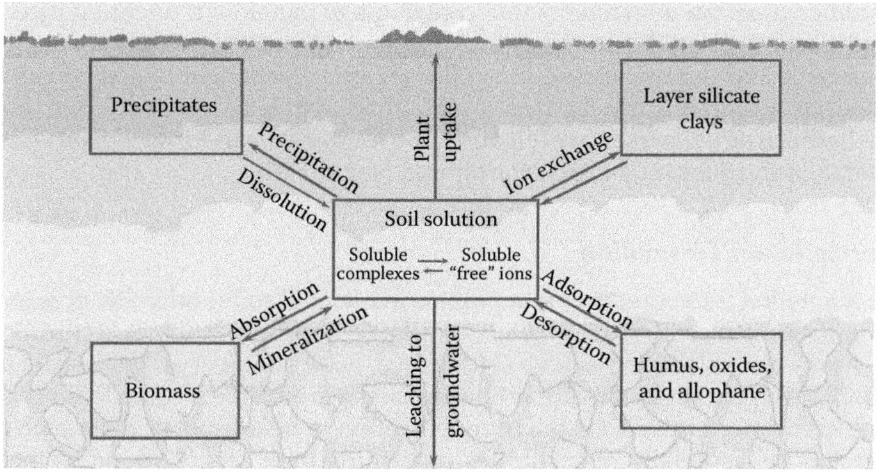

FIGURE 1.1
Reactions of metal(loid)s in soils. (With kind permission from Springer Science+Business Media: *Trace Elements in Terrestrial Environments: Biogeochemistry, Bioavailability and Risks of Metals,* Second Edition, 2001, Adriano, D.C.)

Se tend to form gaseous compounds (Bolan et al. 2013a; Mahimairaja et al. 2005).

Sorption/Desorption Process

Both soil properties and soil solution composition determine the dynamic equilibrium between metal(loid)s in solution and the soil solid phase. The concentration of metal(loid)s in soil solution is influenced by the nature of both organic and inorganic ligand ions, and soil pH through their influence on metal(loid) sorption processes (Bolan et al. 2003c; Harter and Naidu 1995). Two reasons have been given for the effect of inorganic anions on the sorption of metal(loid) cations such as Pb and Cd (Hong et al. 2008, 2010; Lackovic et al. 2003; Naidu et al. 1994). Hong et al. (2008) indicated that inorganic anions form ion pair complexes with metal(loid)s, thereby reducing their sorption. Naidu et al. (1994) indicated that the specific sorption of ligand anions is likely to increase the negative charge on soil particles, thereby increasing the sorption of Cd. The effect of pH values > 6 in lowering free metal(loid) ion activities in soils has been attributed to the increase in pH-dependent surface charge on oxides of Fe, Al, and Mn, chelation by organic matter, or precipitation of metal(loid) hydroxides (e.g., $Pb(OH)_3$) (Mouta et al. 2008; Stahl and James 1991).

Other chemical interactions that contribute to metal(loid) retention by colloid particles include complexation reaction between metal(loid)s and the inorganic and organic ligand ions. The organic component of soil constituents has a high affinity for heavy metal(loid) cations such as Cu, Cd, and Pb because of the presence of ligands or groups that can form chelates with metal(loid)s (Bolan et al. 2011; Harter and Naidu 1995).

Precipitation/Dissolution

Precipitation is the predominant process in high pH soils and in the presence of anions such as SO_4^{2-}, CO_3^{2-}, OH^-, and HPO_4^{2-}, and when the concentration of heavy metal(loid) ion is high (Hong et al. 2007; Naidu et al. 1997; Ok et al. 2010). Precipitation of metal(loid) phosphates/carbonates is considered to be one of the mechanisms for the immobilization of heavy metal(loid)s such as Cu and Pb, especially in substrates containing high concentration of these metal(loid)s. Similarly, liming typically enhances the retention of metal(loid)s (Adriano 2001; Karalić et al. 2013).

Coprecipitation of metal(loid)s especially in the presence of iron oxyhydroxides has also been reported, and often such interactions lead to significant changes in the surface chemical properties of the substrate. For example, arsenate (As(V)) sorption onto ferrihydrite and Ni(II) and trivalent chromium (Cr(III)) sorption onto hydrous iron oxides showed that coprecipitation was a more efficient process than sorption for metal(loid) removal from aqueous solutions. Violante et al. (2007) noticed that As(V) was desorbed by P from a ferrihydrite on which As(V) was added from a Fe–As(V) coprecipitate.

Oxidation/Reduction

Metal(loid)s, including As, Cr, Hg, and Se, are most commonly subjected to microbial oxidation/reduction reactions, thereby influencing their speciation and mobility (Table 1.1). Arsenic in soils and sediments can be oxidized to As(V) by bacteria (Bachate et al. 2012; Battaglia-Brunet et al. 2002). Since As(V) is strongly retained by inorganic soil components, microbial oxidation results in the immobilization of As. Under well-drained conditions, As would be present as As(V), whereas under reduced conditions, arsenite (As(III)) dominates in soils, but elemental arsenic [As(0)] and arsine (H_2As) can also be present. Oxidation of Cr(III) to chromate (Cr(VI)) can enhance the mobilization and bioavailability of Cr. It is primarily mediated abiotically through oxidizing agents such as Mn(IV), whereas reduction of Cr(VI) to Cr(III) is mediated through both abiotic and biotic processes (Choppala et al. 2013).

Chromate can be reduced to Cr(III) in environments where a ready source of electrons (Fe(II)) is available and microbial Cr(VI) reduction occurs in the presence of organic matter as an electron donor (Choppala et al. 2013; Hsu et al.

TABLE 1.1

Equilibrium Dissolution Reaction and the Solubility of Common Crystalline Phosphate Compounds in Soil

Phosphate Compound	Chemical Formula	Equilibrium Dissolution Reaction	Log K_{sp}	Solubility (g 100 g^{-1})
Calcium dihydrogen phosphate	$Ca(H_2PO_4)_2$	$Ca(H_2PO_4)_2(s) \leftrightarrow Ca^{2+} + 2H_2PO_4^-$	−1.14	18
Calcium monohydrogen phosphate	$CaHPO_4$	$CaHPO_4(s) \leftrightarrow Ca^{2+} + HPO_4^-$	−6.6	0.14
Tricalcium phosphate	$Ca_3(PO_4)_2$	$Ca_3(PO_4)_2(s) \leftrightarrow Ca^{2+} + 2PO_4^{3-}$	−24.0	0.02
Hydroxy apatite	$Ca_{10}(PO_4)_6(OH)_2$	$Ca_{10}(PO_4)_6(OH)_2(s) \leftrightarrow 10Ca^{2+} + 6PO_4^{3-} + OH^-$	−55.9	Insoluble
Carbonate apatite	$Ca_{10}(PO_4)_6CO_3$	$Ca_{10}(PO_4)_6(CO)_3(s) \leftrightarrow 10Ca^{2+} + 6PO_4^{3-} + CO_3^{2-}$	−108.3	Insoluble
Fluroapatite	$Ca_{10}(PO_4)_6F_2$	$Ca_{10}(PO_4)_6F_2(s) \leftrightarrow 10Ca^{2+} + 6PO_4^{3-} + 2F^-$	−110.2	Insoluble
Variscite	$AlPO_4{\cdot}2H_2O$	$AlPO_4(s) \leftrightarrow Al^{3+} + PO_4^{3-} + 2H_2O$	−21.0	Insoluble
Strengite	$FePO_4{\cdot}2H_2O$	$FePO_4 2H_2O(s) \leftrightarrow Fe^{3+} + PO_4^{3-} + 2H_2O$	−26.0	Insoluble
Vivianite	$Fe_3(PO_4)_2{\cdot}8H_2O$	$Fe_3(PO_4)_2 8H_2O(s) \leftrightarrow Fe^{2+} + 2PO_4^{3-} + 8H_2O$	−3.11	Insoluble

2009). In living systems, Se tends to be reduced rather than oxidized, and reduction occurs under both aerobic and anaerobic conditions. Dissimilatory selenite (Se(IV)) reduction to Se(0) and its subsequent precipitation is the major biological transformation for remediation of Se oxyanions in anoxic sediments (Yee et al. 2007; Zhang and Frankenberger 2003). In the case of Hg, microorganisms, particularly bacteria, play a major role in reducing reactive Hg(II) to nonreactive Hg(0), which may be subjected to volatilization losses (Wiatrowski et al. 2006).

Methylation/Demethylation

Biomethylation may result in metal(loid) detoxification, since methylated derivatives may be excreted readily from cells and are often volatile and may be less toxic (e.g., organoarsenicals). Microorganisms in soils and sediments act as biologically active methylators (Frankenberger and Arshad 2001; Loseto et al. 2004). Organic matter provides the methyl-donor source for methylation in soils and sediments. Methylation of Hg is controlled by low molecular weight fractions of fulvic acid in soils (Feyte et al. 2010; Ravichandran 2004). Biomethylation is effective in forming volatile compounds of As such as alkylarsines, which could easily be lost to the atmosphere (Lehr 2003; Yin et al. 2011). Methylated As species could result from direct excretion by algae or microbes or from degradation of the excreted arsenicals or more complex cellular organoarsenicals (Li et al. 2009). Methylation of Hg occurs under both aerobic and anaerobic conditions (Randall and Chattopadhyay 2013; Rodríguez Martín-Doimeadios et al. 2004). Under anaerobic conditions, Hg(II) ions can be biologically methylated to form either monomethyl or dimethyl Hg, which are highly toxic and more biologically mobile than the other forms (Adriano et al. 2004; Naja and Volesky 2009). Selenium biomethylation is of interest because it represents a potential mechanism for removing Se from contaminated environments, as methylated compounds such as dimethyl selenide (DMSe) are less toxic than dissolved Se oxyanions.

Reactions of Phosphate Compounds in Soils

The P-induced (im)mobilization of metal(loid)s depends on the form of P compounds used and their reactions in soils. Phosphate compounds that are used as a fertilizer source are broadly grouped into water-soluble (fast-release) and water-insoluble (slow-release) fertilizers (Bolan et al. 1993). The important water-soluble P fertilizers include single superphosphate (SSP), triple superphosphate (TSP), monoammonium phosphate (MAP), and diammonium phosphate (DAP). The important water-insoluble P fertilizers include phosphate rocks (PRs) and basic slag. Monocalcium phosphate (MCP) and ammonium phosphate (AMP) are the principal P components present in superphosphates (SSP and TSP) and ammonium phosphates (MAP and DAP), respectively.

Water-Soluble Compounds

When superphosphate fertilizers are added to soils, the dissolution of MCP results in the formation of slowly soluble dicalcium phosphate (DCP) with a release of phosphoric acid close to the fertilizer granules (Equation 1.1). Phosphoric acid subsequently dissociates into dihydrogen phosphate ($H_2PO_4^-$) and hydrogen ions (protons–H^+). The protons reduce the pH around the fertilizer granules to a very low level (<2 pH). When AMP fertilizers are added to soil, they dissociate into ammonium $\left(NH_4^+\right)$ and $H_2PO_4^-$ ions. The subsequent oxidation of NH_4^+ to nitrate $\left(NO_3^-\right)$ results in the release of protons (Equation 1.2).

$$Ca(H_2PO_4)_2 + H_2O \rightarrow CaHPO_4{\cdot}H_2O + H_3PO_4 \quad (1.1)$$

$$NH_4^+ + 2O_2 \rightarrow NO_3^- + 2H^+ + H_2O \quad (1.2)$$

The acidic solution around the fertilizer granules dissolves the Fe and Al compounds in the soil, resulting in the adsorption and precipitation of P. The retention of P by soils decreases the amount of P available for both plant uptake and leaching to groundwater. The acidity generated can also have important implications to the mobilization of metal(loid)s.

Water-Insoluble Compounds

The solubility of common P compounds that are added to soils and formed as reaction products is given in Table 1.2. When insoluble P fertilizers such as PRs are added to soil, they need to be dissolved in soils for the P to become plant available. In soils, PRs dissolve by using the acid produced in the soils (Equation 1.3). This is a major reason why PRs are very effective as a nutrient source mainly in acid soils (pH < 6.5) (Bolan et al. 1990; Danso et al. 2010) and as a metal(loid) immobilizing agent in acid mine drainage (Evangelou and Zhang 1995; Oliva et al. 2012). Once the PR is dissolved, the P released undergoes similar adsorption and precipitation reactions as in the case of soluble P fertilizers.

$$Ca_{10}(PO_4)_6F_2 + 12H^+ \rightarrow 10Ca^{2+} + 6H_2PO_4^- + 2F^- \quad (1.3)$$

TABLE 1.2

Selected References on the Mobilization of Heavy Metal(loid)s by Water-Soluble and Water-Insoluble Phosphate Compounds

P Compound	Heavy Metal(loid)	Method of Investigation[a]	Proposed Primary Mechanism	Reference
Water-Soluble Phosphate Compounds				
SSP	Cd	PB	Cd source: dissolution	Gray et al. (1999)
TSP	As	PB	Competitive adsorption	Cao and Ma (2004)
TSP	Se	PB	Acidification	Mora et al. (2008)
TSP	As, Sb	TL, PBET	Desorption	Kilgour et al. (2008)
MCP	As	PB, TL	Competitive adsorption	Xenidis et al. (2010)
MAP, DAP	Cd	CF	Acidification	Lambert et al. (2007); Cakmak et al. (2010)
KH_2PO_4	Mo	AD, TL, PB	Desorption	Neunhäuserer et al. (2001)
KH_2PO_4	Se	PB	Competitive root absorption	Li et al. (2008)
KH_2PO_4, $Ca(H_2PO_4)_2$	As, Sb	TL	Desorption	Harris and Lottermoser (2006a,b)
KH_2PO_4	Se, Sb	PB	Competitive adsorption	Nakamaru and Sekine (2008)
KH_2PO_4	Mo	AD	Competitive adsorption	Vistoso et al. (2012)
KH_2PO_4	As	PB	Competitive adsorption	Bolan et al. (2013c)

K_2HPO_4	As	PB	Competitive adsorption	Ultra et al. (2007)
Na_2HPO_4	Se	AD	Desorption	Nakamaru et al. (2006)
Soil P	Se	PB	Desorption	Eich-Greatorex et al. (2010)
Water-Insoluble Phosphate Compounds				
Hydroxyapatite	As	TCLP, PBET	Desorption	Zupančič et al. (2012)
PR	As	CF, PB	Competitive adsorption	Fayiga and Ma (2006); Yan et al. (2012)
PR	As	PB	Competitive adsorption	Lessl and Ma (2013)
PR	Cd	PB	Cd source: dissolution	Mendes et al. (2006)
PR	Cd	CF	Cd source: dissolution	Kuo et al. (2007)
Bone meal, granular superphosphate	As, Sb	TL	Phosphate ligand exchange	Munksgaard and Lottermoser (2013)
Cow bone (hydroxyapatite)	Sb	PB, CF, SEM, EXAFS, XRD	Competitive adsorption	Ahmad et al. (2014)

Note: DAP = diammonium phosphate, MAP = monoammonium phosphate, MCP = monocalcium phosphate, PR = phosphate rock, SSP = single superphosphate, TSP = triple superphosphate.

[a] Method of investigation:

1. Chemical—adsorption/desorption (AD), chemical fractionation (CF), transport/leaching (TL), toxicity characteristics leaching procedure (TCLP)
2. Biological—phytoavailability bioassay (PB), physiologically-based extraction test (PBET)
3. Minerological—x-ray diffraction (XRD), extended x-ray absorption fine structure spectroscopy (EXAFS), scanning electron microscope (SEM)

Mechanisms for (Im)mobilization of Heavy Metal(loid)s by Phosphate Compounds

Phosphate compounds affect the (im)mobilization of metal(loid)s in the soil–plant system through various processes (Figure 1.2). These include phosphate compounds as a metal(loid) source; P-micronutrient imbalance in plant nutrition; direct metal(loid) adsorption by P compounds; anion-induced metal(loid) adsorption and desorption; direct precipitation of metal(loid)s with solution P as metal(loid) phosphates; precipitation through the liming action of PRs; and rhizosphere modification through acidification and mycorrhizal association. Both mobilization and immobilization of metal(loid)s in soils treated with a range of P compounds have been reported, and the probable mechanisms for the P-induced (im)mobilization of metal(loid)s reported in the literature are given in Tables 1.3 and 1.4.

Phosphate Compounds as a Metal(loid) Source

Phosphate compounds contain a range of metal(loid)s (McBride and Spiers 2001; McLaughlin et al. 1996; Molina et al. 2009) (Table 1.5). Addition of P compounds to soils not only helps to overcome the deficiency of some of the essential trace elements, such as molybdenum (Mo), but also introduces toxic metal(loid)s, such as Cd (Bolan et al. 2013b; Loganathan et al. 2012; McLaughlin et al. 1996). In this regard, Cd contamination of agricultural soils is of particular concern because this metal(loid) reaches the food chain

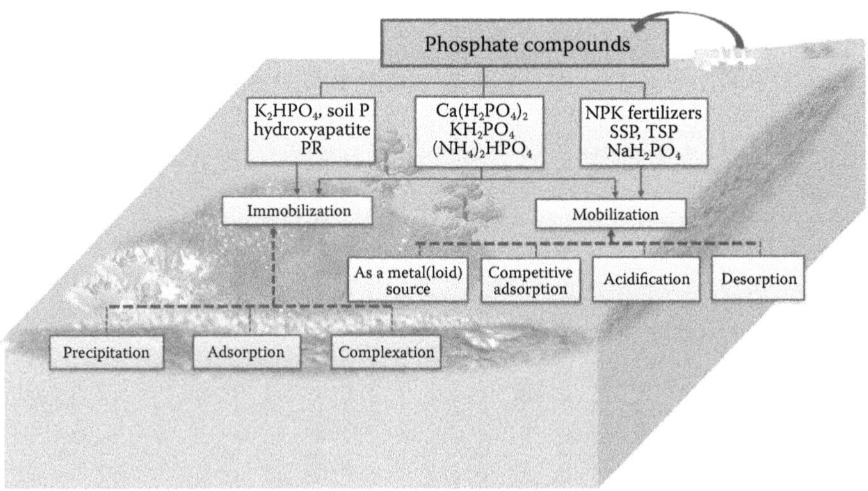

FIGURE 1.2
Plausible mechanisms by which phosphate compounds enhance the (im)mobilization of metal(loid)s in soils.

TABLE 1.3

Selected References on the Immobilization of Heavy Metal(loid)s by Water-Soluble and Water-Insoluble Phosphate Compounds

P Compound	Heavy Metal(loid)	Method of Investigation[a]	Proposed Primary Mechanism	Reference
Water-Soluble Phosphate Compounds				
$Ca(H_2PO_4)_2$	Pb	CF, PB, XRD	Precipitation as metal(loid) phosphates	Hettiarachchi et al. (2000)
$Ca(H_2PO_4)_2H_2O$	Zn	AD	Precipitation as metal(loid) phosphates	Zahedifar et al. (2010)
K_2HPO_4	Cd	AD	Adsorption	Hong et al. (2010)
K_2HPO_4	Cd	PB	Phosphate-induced adsorption	Bolan et al. (2003c)
KH_2PO_4	Cd	PB	Precipitation as metal(loid) phosphates and phosphate-induced adsorption	Dheri et al. (2007)
KH_2PO_4	Cd	AD	Precipitation as metal(loid) phosphates	Ugbaje and Abdu (2013)
$(NH_4)_2HPO_4$	Cd, Pb, Zn	PBET, TL	Precipitation as metal(loid) phosphates	McGowen et al. (2001)
$(NH_4)_2HPO_4$	Pb	PB, CF	Precipitation as metal(loid) phosphates	Khan and Jones (2009); Khan et al. (2012)
Na_2HPO_4	Zn	AD	Metal(loid) phosphate complex	Pérez-Novo et al. (2011)
$NaH_2(PO_4)_3$	Pb	PBET, XRD, XRF, SEM	Precipitation as metal(loid) phosphates	Bosso et al. (2008)
Calcium magnesium phosphate	Pb, Zn	PB, TCLP	Precipitation as metal(loid) phosphates	Wang et al. (2008)
Calcium magnesium phosphate	Cd, Cu, Pb, Zn	CF, TL	Precipitation as metal(loid) phosphates	Zhang and Pu (2011)
Phosphoric acid	Pb	PB, TL, CF	Precipitation as metal(loid) phosphates	Tang et al. (2009)
Calcium-phosphate (phosphoric acid)	As	AD, TL, XRD	Precipitation as metal(loid) phosphates	Nuepane and Donahoe (2013)
Water-Insoluble Phosphate Compounds				
Hydroxyapatite	Cd, Pb, Zn	AD, TCLP	Adsorption/precipitation	Seaman et al. (2001)
Hydroxyapatite	Cu, Cd, Pb, Zn	AD	Ion exchange/surface complexation/precipitation	Chen et al. (2010)

(Continued)

TABLE 1.3 (CONTINUED)

Selected References on the Immobilization of Heavy Metal(loid)s by Water-Soluble and Water-Insoluble Phosphate Compounds

P Compound	Heavy Metal(loid)	Method of Investigation[a]	Proposed Primary Mechanism	Reference
Hydroxyapatite	Pb	PB, CF	Precipitation as metal(loid) phosphates	Waterlot et al. (2011)
Hydroxyapatite	Pb	TCLP, PBET	Precipitation as metal(loid) phosphates	Zupančič et al. (2012)
Bone meal, granular super phosphate	Pb	TL	Precipitation as metal(loid) phosphates	Munksgaard and Lottermoser (2013)
Bone meal	Cd, Zn	TL	Surface complexation/cation exchange	Sneddon et al. (2006)
Hydroxyapatite, PR	Pb, Cd, Zn	PB, CF, EDX, SEM	Precipitation as metal(loid) phosphates	Chen et al. (2007)
Hydroxyapatite, PR	Pb, Cd, Cu, Zn	AD, XRD	Precipitation as metal phosphates	Mignardi et al. (2012)
PR (NCPR)	Pb, Zn, Cd	CF, GI, PBET, PB	Precipitation as metal(loid) phosphates	Basta et al. (2001)
PR	Pb, Cd	PB	Precipitation as metal(loid) phosphates	Fayiga and Ma (2006)
PR	Pb	PBET, XRD, XRF, SEM	Precipitation as metal(loid) phosphates	Bosso et al. (2008)
PR	Cu, Pb	PB	Precipitation as metal(loid) phosphates	Aikpokpodion et al. (2012)
PR with bacteria	Pb	TL, PB	Adsorption/precipitation	Park et al. (2011b, 2012)
PR	Pb	PBET	Adsorption/precipitation	Park et al. (2011c)
P rich biosolid	Pb	PB	Adsorption/precipitation	Siebielec and Chaney (2012)
Biosolid + $(NH_4)_2HPO_4$	Pb	PB	Adsorption/precipitation	Brown et al. (2007)
Biosolid + $Ca(H_2PO_4)_2$	Cu, Cr, Zn	PB, CF	Adsorption/precipitation	Branzini and Zubillaga (2012)

Note: PR = phosphate rock.

[a] Method of investigation:

1. Chemical—adsorption/partitioning (AD), chemical fractionation (CF), transport/leaching (TL), toxicity characteristics leaching procedure (TCLP)
2. Minerological—x-ray diffraction (XRD), x-ray fluorescence (XRF), scanning electron microscope (SEM), energy dispersive x-ray analysis (EDX)
3. Biological—phytoavailability bioassay (PB), physiologically-based extraction test (PBET), microfauna/macrofauna bioassay (MB)

TABLE 1.4

Metal(loid) Concentration in Phosphate Compounds from Various Sources

Phosphate Compound	Concentration (mg kg^{-1})								
	As	Cd	Co	Cu	Zn	Mn	Ni	Pb	Hg
GPR	4	38	3	15	393	7	—	—	—
NFPR	7	3	5	4	57	212	—	—	—
JPR	12	4	<1	8	235	5	—	—	—
NCPR	23	48	2	9	400	7	9–51	<1–51	0.4–2.1
SPR	5	11	3	6	178	91	—	—	—
MPR	3	8	6	4	90	151	—	—	—
NIPR	3	100	6	8	1010	122	—	—	—
APR	7	12	4	12	560	2	—	—	—
MIPR	2	10	<1	6	220	2	—	—	—
CRP	—	2	4	5	95	100	—	—	—
IRP	—	—	109	32	187	975	—	962	—
SSP	—	—	77	15	165	890	—	488	—
TSP	—	—	47	49	418	75	—	238	—
DAP	—	—	16	7.2	112	307	—	195	—

Source: Syers, J.K. et al., *J. Sci. Food Agric.* 37:1057–1064, 1986; McLaughlin, M.J. et al., *Aust. J. Soil Res.* 34:1–54, 1996; With kind permission from Springer Science+Business Media: *Trace Elements in Terrestrial Environments: Biogeochemistry, Bioavailability and Risks of Metals*, Second Edition, 2001, Adriano, D.C.

Note: Phosphate rocks: APR = Arad phosphate rock, CRP = Chatham Rise phosphorite, GPR = Gafsa phosphate rock, IPR = Indian phosphate rock, JPR = Jordan phosphate rock, MIPR = Makatea Island phosphate rock, MPR = Mexican phosphate rock, NCPR = North Carolina phosphate rock, NFPR = North Florida phosphate rock, NIPR = Nauru Island phosphate rock, SPR = Sechura phosphate rock. Phosphate fertilizers: DAP = diammonium phosphate, SSP = single superphosphate, TSP = triple superphosphate.

TABLE 1.5

Solubility of Common Crystalline Metal(loid) Phosphate Compounds

Element	Metal(loid) Phosphate	Chemical Formula	Log K_{sp}
Cd	Cadmium phosphate	$Cd_3(PO_4)_2$	–38.1
Co	Cobalt phosphate	$Co_3(PO_4)_2$	–4.36
		$Co(UO_2)_2(PO_4)_2{\cdot}7H_2O$	–9.9
Cr	Chromium phosphate (green)	$CrPO_4$	–3.07
	Chromium phosphate (purple)	$CrPO_4$	–2.55
Cu	Cornetite	$Cu_3(PO_4)(OH)_3$	–48.0
	Libenthenite	$Cu_2(PO_4)OH$	–28.0
	Torbernite	$Cu(UO_2)_2(PO_4)_2{\cdot}10H_2O$	–41.0
	Metatorbernite	$Cu(UO_2)_2(PO_4)_2{\cdot}8H_2O$	–41.3
	Nissonite	$CuMg(PO_4)(OH){\cdot}2.5H_2O$	–23.6
	Pseudomalachite	$Cu_5(PO_4)_4(OH)_4$	–75.8
	Tagilite	$Cu_2(PO_4)(OH){\cdot}H_2O$	–27.9
	Pyromorphite	$Cu_5(PO_4)_3OH$	–65.6
	Turquois	$CuAl_6(PO_4)_4(OH)_8{\cdot}4H_2O$	–179.0
	Chalcosiderite	$CuFe_6(PO_4)_4(OH)_8{\cdot}4H_2O$	–205.7
	Veszelyite	$CuZn_2PO_4(OH)_3{\cdot}2H_2O$	–45.8
	Liberthenite	Cu_2PO_4OH	—
		$Cu_3(PO_4)_2$	–2.21
		$Cu_3(PO_4)_2{\cdot}3H_2O$	–3.98
		$Cu_5O_2(PO_4)_2{\cdot}6H_2O$	–8.53
Hg	Mercuric phosphate	Hg_2HPO_4	–5.20
Mn	Manganeous phosphate	$MnPO_4{\cdot}1.5H_2O$	–34.82
Ni	Nickel phosphate	$Ni_3(PO_4)_2$	–8.82
		$Ni(UO_2)_2(PO_4)_2{\cdot}7H_2O$	–9.5
Pb	Fluropyromorphite	$Pb_{10}(PO_4)_6(F)_2$	
	Chloropyromorphite	$Pb_{10}(PO_4)_6(Cl)_2$	–25.75
	Hydroxypyromorphite	$Pb_{10}(PO_4)_6(OH)_2$	–18.15
	Pyromorphite	$Pb_3(PO_4)_2$	–4.43
	Plumbogummite	$PbAl_3(PO_4)_2(OH)_5{\cdot}H_2O$	–99.3
	Tsumebite	$Pb_2Cu(PO_4)_2(OH)_3{\cdot}3H_2O$	–51.3
	Dumontite	$Pb_2(UO_2)_4(PO_4)_2(OH)_4$	–91.4
	Corkite	$PbFe_3(PO_4)(OH)_6SO_4$	–112.6
	Hinsdalite	$PbAl_3(PO_4)(OH)_6SO_4$	–99.1
	Parsonsite	$Pb_2UO_2(PO_4)_2{\cdot}2H_2O$	–45.8
	Dewindtite	$Pb(UO_2)_4(PO_4)_2(OH)_4{\cdot}8H_2O$	–92.6
	Renardite	$Pb(UO_2)_4(PO_4)_2(OH)_4{\cdot}7H_2O$	–93.7
	Przhevalskite	$Pb(UO_2)_4(PO_4)_2{\cdot}4H_2O$	–47.4
	Embriyite	$Pb_5(Cr_2O_4)_2(PO_4)_2{\cdot}H_2O$	

TABLE 1.5 (CONTINUED)

Solubility of Common Crystalline Metal(loid) Phosphate Compounds

Element	Metal(loid) Phosphate	Chemical Formula	Log K_{sp}
Zn	Hopeite	$Zn_3(PO_4)_2$	–35.40
	Parahopetite	$Zn_3(PO_4)_2{\cdot}4H_2O$	–17.06
	Phosphophyllite	$Zn_2Fe(PO_4)_2{\cdot}4H_2O$	–35.8
	Scholzite	$Zn_2Ca(PO_4)_2{\cdot}4H_2O$	–34.1
	Spencerite	$Zn_4(PO_4)_2(OH)_2{\cdot}3H_2O$	–52.8
	Turbuttite	$Zn_2(PO_4)(OH)$	–26.6
	Zinc rockbridgeite	$ZnFe_4(PO_4)_3(OH)_5$	–138.6
	Zinc pyromorphite	$Zn_5(PO_4)_3OH$	–63.1
	Fausite	$ZnAl_6(PO_4)_4(OH)_8{\cdot}4H_2O$	–177.7

through regular use of Cd containing fertilizer materials such as SSP and TSP.

Accumulation of Cd in soils through regular fertilizer use has been observed in many countries (Table 1.3). For example, in New Zealand and Australia, most of the Cd accumulation in pasture soils has been derived from the use of P fertilizers containing high Cd concentration. The Cd in most P fertilizers originates mainly from the PRs used for manufacturing P fertilizers. It is important to stress that PR deposits vary in their Cd content and hence manufactured P fertilizers also vary accordingly in their Cd content.

Comparison between virgin or native (i.e., unfertilized) and agricultural (i.e., fertilized) soils has often been used to indicate contamination of soil through agricultural practices. For example, Roberts et al. (1994) obtained evidence for the enrichment of Cd in pastoral soils, and there was a highly significant correlation between total soil P and total soil Cd across all sites, confirming the role of P fertilizers in soil Cd enrichment. This is not surprising considering the long-use history in both New Zealand and Australia of superphosphates manufactured from Nauru Island PR. Nauru superphosphates typically contain 34–69 mg Cd kg^{-1} (Rothbaum et al. 1986; Schipper et al. 2011).

Similarly, Loganathan et al. (1996) examined the movement and distribution of Cd and P in a pastoral soil amended annually for 10 years with four forms of P fertilizers (SSP, DAP, North Carolina PR [NCPR], and Jordan PR [JPR]) that varied in their total Cd content. Both total and plant-available Cd concentrations decreased with soil depth. SSP and NCPR that had higher Cd contents produced higher Cd concentration than DAP, JPR, and control treatments. Approximately 93% of the applied Cd was recovered in the top 120-mm soil, and plant recovery of applied Cd ranged from 1.5% to 4.5%.

Adsorption/Desorption of Metal(loid)s

Depending on the source, application of P compounds can cause direct adsorption of metal(loid)s onto these P compounds, enhance anion-induced metal(loid) adsorption through increased surface charge, decrease metal(loid) adsorption through metal(loid)-complex formation, and induce desorption of metal(loid)s, such as Cr(VI) and As(V), through competition (Table 1.3).

Direct Adsorption by Phosphate Compounds

Adsorption of metal(loid)s by P compounds such as apatite mineral has been observed in a number of studies (Table 1.4), and the order of adsorption by apatite largely depends on pH (Chen et al. 1997a). Adsorption of metal(loid)s onto hydroxyapatite surfaces has been observed for Zn and Cu (Wang and Zhang 2014), Cd (Barka et al. 2012; Middelburg and Comans 1991), strontium (Sr) (Lazic and Vukovic 1991; Rosskopfová et al. 2011), and Ni (Rosskopfová et al. 2013). Metal(loid) adsorption onto apatite is facilitated through the exchange of Ca^{2+} from the apatite particle by the metal(loid) in the soil solution (Zhu et al. 2010). However, Xu et al. (1994) concluded that in the case of Zn and Cd, surface complexation and coprecipitation (Equation 1.4) are the most important mechanisms with ion exchange and solid-state diffusion, also possibly contributing to the overall adsorption process by hydroxyapatite.

$$Ca_{10}(PO_4)_6(OH)_2 + xCd^{2+} \rightarrow (Cd_x,Ca_{10-x})(PO_4)_6(OH)_2 + xCa^{2+} \qquad (1.4)$$

Phosphate-Induced Metal(loid) Adsorption

Anion-induced metal(loid) adsorption has been reported for a number of cations (Bolan et al. 1999a). It has also been shown that Zn^{2+}, Cd^{2+}, or Cu^{2+} adsorption by variable charge components, such as Al and Fe oxides, and by soils can be enhanced by low or moderate enrichment of oxides with P (Ren et al. 2012; Tiberg et al. 2013; Stietiya and Wang 2014; Wang and Xing 2004), causing increases in surface negative charge.

Several mechanisms can be advanced for $H_2PO_4^-$-induced metal(loid) adsorption by soils, including (1) increase in negative charge, (2) cosorption of $H_2PO_4^-$ and metal(loid) as an ion pair, and (3) surface complex formation of metal(loid) on the P compound. While enhanced sorption of a number of metal(loid)s via increased surface charge in the presence of $H_2PO_4^-$ ions has been demonstrated (Bolan et al. 1999b), much work is still needed to provide conclusive evidence for other mechanisms.

The amount of surface charge acquired through specific adsorption of anions (Equation 1.5) depends on the nature of anion adsorbed, pH, and electrolyte concentration of the solute (Bolan et al. 1999b). The most frequently offered causes for P-induced increase in negative charge include (Bolan et al.

2003a; Hingston 1981) (1) a shift in the zero point of charge to lower pH values, (2) neutralization of positive charge, and (3) electrolyte inhibition.

$$FeOOH_{(s)} + HPO_4^{2-} \Leftrightarrow FeOHPO_{4(s)}^{-} + OH^{-} \quad (1.5)$$

Bolan et al. (1999b) obtained a good relationship between the increase in negative charge resulting from $H_2PO_4^-$ enrichment and the increase in Cd^{2+} adsorption. However, only a small fraction of the increased negative charge was occupied by Cd^{2+}. Similarly, Naidu et al. (1994) demonstrated that the effect of ionic strength on Cd^{2+} adsorption was induced through its effect on surface charge. Rietra et al. (2001), Collins et al. (1999), and Helyar et al. (1976) proposed surface complex formation for the increased adsorption of Ca^{2+} and Zn^{2+} onto $H_2PO_4^-$-enriched gibbsite and goethite. In this case, specifically sorbed anions such as $H_2PO_4^-$ form complexes with soil surface so that cations are adsorbed onto the adsorbed anion. Indeed, surface complexation has been suggested as a mechanism for the immobilization of metal(loid)s such as Cd and Zn by hydroxyapatite (Cao et al. 2004; Xu et al. 1994).

Phosphate-Induced Metal(loid) Desorption

Phosphate anions are very effective in desorbing certain oxyanions such as Se(IV), As(V), and Cr(VI) (Aide and Cummings 1997; James et al. 1995). For example, addition of P fertilizers to lead-arsenate-contaminated soils has resulted in an increase in the solubilization and subsequent mobility of As in soils (Peryea 1991; Peryea and Kammereck 1997). This is attributed to increased adsorption competition by $H_2PO_4^-$ ions with the metal(loid) (Davenport and Peryea 1991; Zeng et al. 2012). However, the effect of $H_2PO_4^-$ on metal(loid) desorption depends on the nature of the soil, its sorption capacity, and the extent of saturation with a particular metal(loid) ion (Bolan et al. 2013c; Smith et al. 2002).

Phosphate has also been shown to strongly compete with molybdate anion $\left(MoO_4^{2-}\right)$ for adsorption sites, thereby resulting in increased desorption of the latter (Vistoso et al. 2012; Xie et al. 1993). Neunhäuserer et al. (2001) demonstrated the value of P fertilizers in the phytoremediation of Mo-contaminated soils, where application of soluble P fertilizer greatly enhanced the solubilization of Mo in a soil contaminated with excess Mo, facilitating its removal through plant uptake.

Phosphate-induced mobilization of metal(loid) anions such as As through desorption can facilitate their subsequent plant uptake. However, plant response to As and P interactions in the root zone appears to be system-dependent. In nutrient solution culture, increasing amounts of soluble P at fixed As concentration reduced As phytotoxicity (Lou et al. 2010; Wang and Duan 2009). This was attributed to the inhibition of As uptake by roots in the presence of excessive amounts of P, whereas in soil systems, P fertilizer addition has been shown to enhance the uptake of As due to an increase in the desorption and solubility of As resulting from competitive adsorption of $H_2PO_4^-$ (Cao et al. 2003; Peryea 1991). Thus, the net effect of P on As phytoavailability in soils depends on the extent of P-induced As mobilization in soils and P-induced competition for As uptake by roots (Bolan et al. 2013c; Figures 1.3 and 1.4).

Precipitation of Metal(loid)s

Precipitation as metal(loid) phosphates has been proved to be one of the main mechanisms for the immobilization of metal(loid)s such as Pb and Zn (Table 1.4). These new metal(loid)–P compounds have extremely low solubility over a wide pH range, which makes P application an attractive technology for managing metal-contaminated soils. Advanced studies using x-ray absorption fine structure spectroscopy (XAFS) indicate that formation of surface precipitates may occur even when solution concentration is undersaturated with respect to homogeneous precipitation of pure metal(loid) precipitate phase (Ford and Sparks 2000). The solubility of various metal(loid) phosphates is given in Table 1.5.

As Metal(loid) Phosphates

The P radical can combine with over 30 elements to form P minerals (Clark 1993). The ability of apatite to immobilize dissolved Pb or Pb in contaminated soils through precipitation as Pb phosphate minerals has been well documented. Such precipitates are more commonly manifested as hydroxypyromorphite or as chloropyromorphite (Table 1.5). Two processes for the

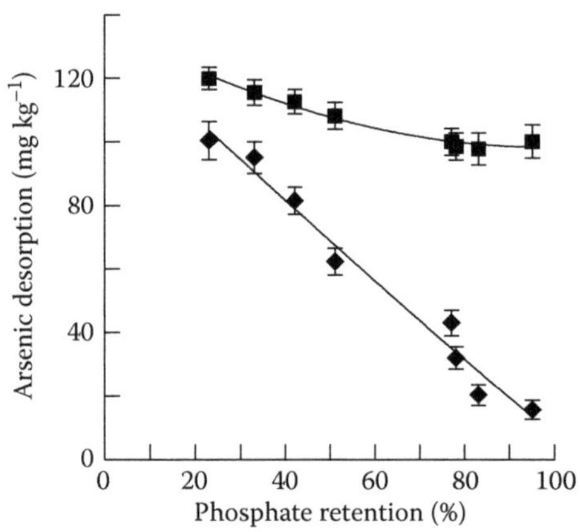

FIGURE 1.3
Relationship between phosphate retention capacity of soils and arsenate (As(V)) desorption in the presence (■) and absence (◆) of phosphate addition. (From Bolan, N.S. et al., *Sci. Total Environ.* 463–464:1154–1162, 2013.)

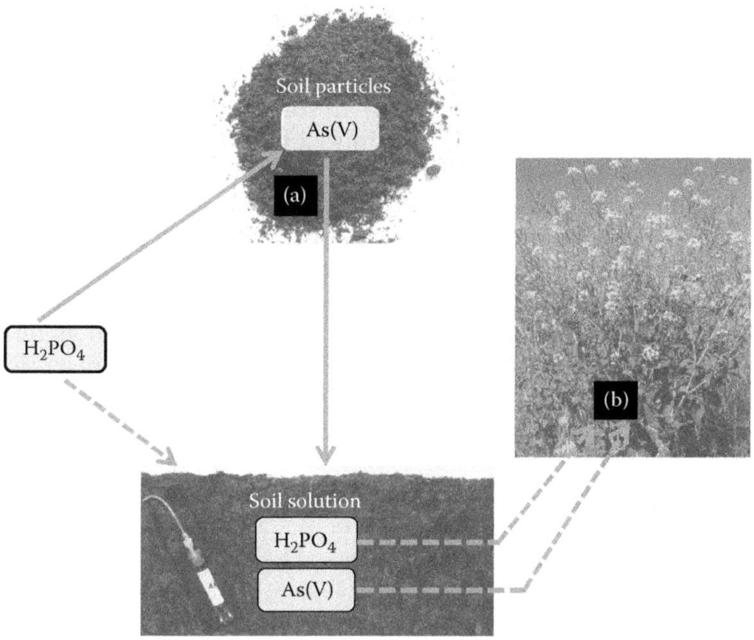

FIGURE 1.4
Phosphate ($H_2PO_4^-$) ions compete for As(V) adsorption by soil particles (a), and for As(V) absorption by plant roots (b). (From Bolan, N.S. et al., *Sci. Total Environ.* 463–464:1154–1162, 2013.)

reaction of dissolved Pb with apatite have been proposed (Laperche and Traina 1998; Miretzky and Fernandez-Cirelli 2008). First, Pb can react with apatite through hydroxyapatite (HA, $Ca_{10}(PO_4)_6(OH)_2$) dissolution (Equation 1.6) followed by precipitation (Equation 1.7) of pure hydroxypyromorphite ($Pb_{10}(PO_4)_6(OH)_2$) as described by Ma et al. (1993) and Mavropoulos et al. (2004). Second, Pb can substitute for Ca in apatite (Equation 1.8). Thus, (Ca, Pb) apatite could be potentially formed by adsorption of Pb or by dissolution of HA followed by coprecipitation of mixed apatites.

$$Ca_{10}(PO_4)_6(OH)_2(s) + 14H^+(aq) \rightarrow 10Ca^{2+}(aq) + 6H_2PO_4^-(aq) + 2H_2O \quad (1.6)$$

$$10Pb^{2+}(aq) + 6H_2PO_4^-(aq) + 2H_2O \rightarrow Pb_{10}(PO_4)_6(OH)_2(s) \quad (1.7)$$

$$Ca_{10}(PO_4)_6(OH)_2(s) + xPb^{2+} \rightarrow (Ca_{10-x}Pb_x)(PO_4)_6(OH)_2(s) + xCa^{2+} \quad (1.8)$$

McGowen et al. (2001) have examined the immobilization of As, Cd, Pb, and Zn in a smelter-contaminated soil using DAP. Application of high levels of DAP at a rate of 2300 mg P kg^{-1} was very effective in immobilizing Cd, Pb, and Zn in the contaminated soil. Others have also shown that $Cd_3(PO_4)_2$

can control Cd solubility in P-sufficient soils or soils amended with P (Hong et al. 2010; Santillan-Medrano and Jurinak 1975; Street et al. 1978). Because the solubility of $Cd_3(PO_4)_2$ is too high to control the concentration of Cd in suspensions involving Fe and Al oxides in soils, it is doubtful that this solid phase can play a significant role in the immobilization of Cd and possibly Zn (Soon 1981; Xiong et al. 1995).

It is important to emphasize that both $H_2PO_4^-$ and metal(loid) ions are readily adsorbed by many arable soils, thereby maintaining typically very low concentrations of these ions in soil solution. Under this condition, precipitation of metal(loid) phosphates in soils is unlikely to occur. Nevertheless, when $H_2PO_4^-$ and metal(loid) ion adsorbing components such as ferromanganese oxyhydroxides undergo dissolution under reducing and acidic conditions, both the metal(loid) and $H_2PO_4^-$ ions are released, elevating their concentrations and subsequently causing precipitation of metal(loid) phosphates. Thus, the synergistic effects of the $H_2PO_4^-$ and metal(loid) adsorbing sediments promote the formation of metal(loid) phosphates in a wide range of geological environments. Park et al. (2011b) suggested the use of P solubilizing bacteria to continuously release $H_2PO_4^-$ from insoluble P sources. Enhanced solubilization of PR by P solubilizing bacteria increased precipitation of Pb, which was evidenced by lowered bioavailable fraction of Pb and plant Pb uptake (Park et al. 2012; Figure 1.5).

Sanderson et al. (2014) investigated the ability of soft PR and commercial phosphate to reduce the bioavailability of Pb, Zn, Ni, Cu, As, and antimony (Sb) in shooting range soils to soil organisms and plants. The effect of P-based amendments varied between soil and P type. They observed that only bioavailability of Pb was reduced for earthworms, and in plants, uptake of only

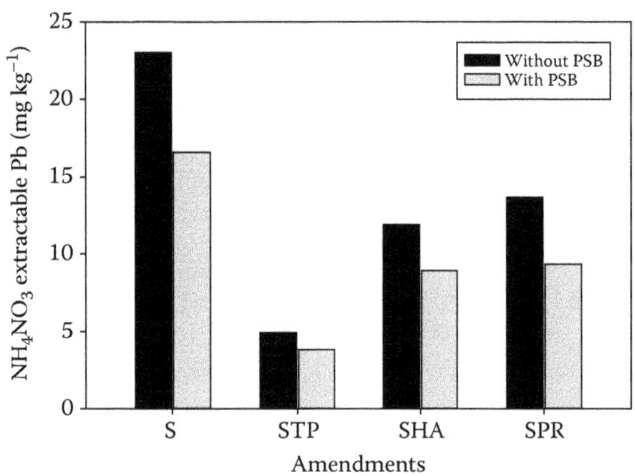

FIGURE 1.5
Immobilization of lead by phosphate amendments and phosphate solubilizing bacteria (PSB). S = soil, STP = soil+tricalcium phosphate, SHA = soil+hydroxyapatite, SPR = soil+rock phosphate. (From Park, J.H. et al., *J. Hazard. Mater.* 185:829–836, 2011.)

Zn, Cu, and Ni was reduced. Microbial activity was increased by greater than 50% with amendments addition. Reduction in bioavailability by P amendments reflects the extent of pyromorphite and metal(loid) P formation.

Through Liming Action of Phosphate Rocks

In addition to the formation of new solid phases discussed above, metal(loid) immobilization may also ensue from the formation of metal(loid) oxyhydroxides due to the buffering capacity (or liming action) of hydroxyapatite (Basta et al. 2001; Chen et al. 1997b; Ma and Rao 1997). Phosphate rocks neutralize acids during the dissolution reactions in soils and thus can also have some liming effect through two processes. First, most PRs contain some free calcium carbonate ($CaCO_3$), which can act as a liming agent itself. Second, the dissolution of the apatite mineral in soils consumes acids, thereby reducing soil acidity. The equivalent liming values of various PRs range from 450 to 560 kg $CaCO_3$ per tonne of PR, with the most liming value coming from the dissolution of apatite mineral. The potential value of PRs as a liming material in mitigating acid mine drainage has been documented both under laboratory and field conditions (Frazer 2001; Mauric and Lottermoser 2011). Treating acid mine drainage with PRs not only neutralizes the acid through their buffering action but also reduces the solution concentration of metal(loid)s through precipitation and ion exchange reactions (Feng et al. 2000; Oliva et al. 2012).

Rhizosphere Modification

Phosphate compounds affect metal(loid) transformation in the rhizosphere primarily through their effects on soil pH (i.e., acidification) and mycorrhizal association.

Acidification

Superphosphates and AMPs are the most common soluble P fertilizers used in agriculture. The dissolution of MCP in soils results in the formation of DCP (Equation 1.1) with a release of phosphoric acid close to the fertilizer granules. Nitrogen-containing P fertilizers such as MAP and DAP cause soil acidification mainly through the natural process of nitrification (Equation 1.2). Acidification caused by P fertilizers induces the mobilization of metal(loid)s. For example, application of DAP has been shown to enhance metal(loid) solubilization through acidification (Basta and McGowen 2004; Khan and Jones 2009). Similarly, Hettiarachchi et al. (2000) have observed that preacidification of Pb-contaminated soils enhances the solubilization of Pb and the subsequent precipitation by P during the *in situ* stabilization of Pb using P fertilizer. Acidification affects the transformation of metal(loid) ions through several ways, foremost of which include (1) modifying the surface charge in variable-charge soils, (2) altering the speciation of metal(loid)s, and (3) influencing the reduction and oxidation reactions of the metal(loid)s (Adriano 2001).

Mycorrhizal Association

Mycorrhizal association is considered to be one of the main factors responsible for metal(loid) tolerance in plants and is likely to have a major influence on the hyperaccumulation of metal(loid)s by certain plant species (Khan et al. 2000; Meier et al. 2012a). The effect of mycorrhizae on P-induced (im)mobilization of metal(loid)s is mediated through two processes. First, enhanced solubilization of P compounds by the mycorrhizal fungi can induce mobilization and the subsequent uptake of metal(loid)s associated with the P compounds. Second, a high level of P in soils has been shown to inhibit mycorrhizal infection (Amijee et al. 1989; Balzergue et al. 2013), which is likely to affect the mobilization of metal(loid)s by the fungi.

The effect of P on mycorrhizal infection and the subsequent uptake of metal(loid)s by plants have been explored (Rabie 2005; Zaefarian et al. 2013). Findings indicate that high levels of bioavailable P in soils inhibited mycorrhizal infection, concomitantly reducing the uptake of metal(loid)s (Mohammad et al. 2003; Rupa et al. 2003). The antagonistic effect of P on mycorrhizal infection is likely to have a strong influence on phytoremediation of metal(loid)s in contaminated soils. There has been some indication that mycorrhizal infection, particularly with the heavy metal(loid) tolerant isolate, could protect plants against heavy metal(loid) phytotoxicity (Hassan et al. 2013; Hildebrandt et al. 2007; Meier et al. 2012a,b; Vogel-Mikuš et al. 2006). These infections might be important in revegetation of polluted sites, and any attempt to immobilize metal(loid)s using high levels of soluble P compounds is likely to affect the mycorrhizae, leading to poor establishment of plant colonies.

Summary and Future Research Needs

Phosphate fertilizers act as an essential nutrient input for sustainable crop production. However, these fertilizers can also act as a source and sink for a number of essential and harmful elements. Among the essential trace elements, reactive PRs are a major source of Mo to legume-based pastures in many countries including Australia and New Zealand. However, toxic metal(loid)s such as Cd and fluorine (F) added through P fertilizers can contaminate soil, thereby creating potential risks to the health of grazing animal and crop and soil quality. Soils that have continuously received P fertilizers over decades or centuries can act as nonpoint sources for P in the environment with detrimental effects on water quality and aquatic biosystems. Thus, the potential risks to environmental quality posed by long-term application of P fertilizers require careful evaluation, and there is an urgent need for the development of management strategies that minimize adverse impacts of such applications to environmental and animal health. Legislative measures have been adopted by regulatory agencies in many countries in recent years to minimize adverse environmental and human health impacts through fertilizer practices. Although such stringent regulatory policies have led to new high-quality fertilizer products containing no or much-reduced levels of metal(loid)s, the cost of such products discourages large-scale applications by the farming community.

Strategies that minimize adverse impacts of long-term P-fertilization are based on the understanding of the dynamics of P–metal(loid) interactions in soils. Phosphate-induced stabilization of metal(loid)s in soils results from adsorption and precipitation reactions. A large number of studies have provided conclusive evidences for the potential value of both water-soluble (e.g., DAP) and water-insoluble (e.g., PRs) P compounds to immobilize metal(loid)s in soils, thereby reducing their mobility and bioavailability for plant uptake. However, P fertilizers mobilize heavy metal(loid) anions such as molybdate and As(V) through competitive desorption, and cations such as Cu^{2+} and Cd^{2+} through rhizosphere acidification and mycorrhizal association. Whenever phytoremediation of contaminated sites is practicable, suitable P compounds can be used to enhance the bioavailability of metal(loid)s for plant uptake. Removal of metal(loid)s through phytoremediation techniques and the subsequent recovery of the metal(loid)s or their safe disposal are attracting research and commercial interests. When it is not possible to remove the metal(loid)s from the contaminated sites by phytoremediation, *in situ* immobilization is considered as an integral part of risk management. However, it is important to recognize that some of these materials contain high levels of metal(loid)s and can act as an agent of metal(loid) introduction to soils. Accordingly, these materials should be inspected prior to their large-scale use as immobilizing agents in contaminated sites.

Although P addition seems to be highly effective in the immobilization of heavy metal(loid)s, excess P in soil and its potential effect on eutrophication of surface water, and the possibility of leaching of certain heavy metal(loid)s such as As(V), remain a concern. It is therefore important that future research should aim to focus on the role of P fertilizers on *in situ* remediation and natural attenuation in multimetal(loid)-contaminated sites with minimum impact on quality of water sources.

Key areas that require further research include

- An investigation of the effect of heavy metal(loid) additions through P fertilizers on soil microbial activity under field conditions
- Influence of soil type on chemodynamics of phosphate–metal(loid) interactions

- The effect of continued application of acidifying P fertilizers on long-term efficacy of P sinks to immobilize heavy metal(loid)s
- The impact of P-rich organic amendments on heavy metal(loid) biogeochemistry in soils
- Application of advanced spectroscopic studies and also *in vivo* animal studies to examine the stability of metal(loid) immobilization
- Long-term stability of P-induced immobilization of heavy metal(loid)s
- Molecular approaches to develop more understanding about the mode of action of P solubilizing bacteria

Acknowledgment

The postdoctoral fellowship program (PJ009828) at the National Academy of Agricultural Science, Rural Development Administration, Republic of Korea, supported Dr. Kunhikrishnan's contribution.

References

Adriano, D.C. 2001. *Trace Elements in Terrestrial Environments: Biogeochemistry, Bioavailability and Risks of Metals*, Second Edition. New York: Springer.

Adriano, D.C., Wenzel, W.W., Vangronsveld, J., and Bolan, N.S. 2004. Role of assisted natural remediation in environmental cleanup. *Geoderma* 122:121–142.

Ahmad, M., Lee, S.S., Lim, J.E., Lee, S., Cho, J.S., Moon, D.H., Hashimoto, Y., and Ok, Y.S. 2014. Speciation and phytoavailability of lead and antimony in a small arms range soil amended with mussel shell, cow bone and biochar: EXAFS spectroscopy and chemical extractions. *Chemosphere* 95:433–441.

Aide, M.T., and Cummings. M.F. 1997. The influence of pH and phosphorus on the adsorption of chromium (VI) on boehmite. *Soil Sci.* 162:599–603.

Aikpokpodion, P., Lajide, L., and Aiyesanmi, A. 2012. In situ remediation activities of rock phosphate in heavy-metal contaminated cocoa plantation soil in Owena, South Western, Nigeria. *Global J. Environ. Res.* 6:51–57.

Amijee, F., Tinker, P.B., and Stribley, D.P. 1989. The development of endomycorrhizal root systems. 7. A detailed study of effects of soil-phosphorus on colonization. *New Phytol.* 111:435–446.

Armienta, M.A., Rodriguez, R., and Cruz, O. 1997. Arsenic content in hair of people exposed to natural arsenic polluted groundwater at Zimapan, Mexico, *Bull. Environ. Contam. Toxicol.* 59:583–589.

Bachate, S.P., Khapare, R.M., Kodam, K.M. 2012. Oxidation of arsenite by two β-proteobacteria isolated from soil. *Appl. Microbiol. Biotechnol.* 93:2135–2145.

Balzergue, C., Chabaud, M., Barker, D.G., Bécard, G., and Rochange, S.F. 2013. High phosphate reduces host ability to develop arbuscular mycorrhizal symbiosis without affecting root calcium spiking responses to the fungus. *Front. Plant Sci.* 4:426.

Barka, N., Ouzaouit, K., Abdennouri, M., El Makhfouk, M., Qourzal, S., Assabbane, A., Ait-Ichou, Y., and Nounah, A. 2012. Kinetics and equilibrium of cadmium removal from aqueous solutions by sorption onto synthesized hydroxyapatite. *Desalin. Water Treat.* 43:8–16.

Basta, N., Gradwohl, R., Snethen, K., and Schroder, J. 2001. Chemical immobilization of lead, zinc, and cadmium in smelter-contaminated soils using biosolids and rock phosphate. *J. Environ. Qual.* 30:1222–1230.

Basta, N., and McGowen, S. 2004. Evaluation of chemical immobilization treatments for reducing heavy metal transport in a smelter-contaminated soil. *Environ. Pollut.* 127:73–82.

Battaglia-Brunet, F., Dictor, M.C., Garrido, F., Crouzet, C., Morin, D., Dekeyser, K., Clarens, M., and Baranger, P. 2002. An arsenic (III)-oxidizing bacterial population: Selection, characterization, and performance in reactors. *J. Appl. Microbiol.* 93:656–667.

Bech, J., Poschenrieder, C., Llugany, M., Barceló, J., Tume, P., Tobias, F., Barranzuela, J., and Vásquez, E. 1997. Arsenic and heavy metal contamination of soil and vegetation around a copper mine in Northern Peru. *Sci. Total Environ.* 203:83–91.

Bolan, N.S., Adriano, D.C., Kunhikrishnan, A., James, T., McDowell, R., and Senesi, N. 2011. Dissolved organic carbon: Biogeochemistry, dynamics and environmental significance in soils. *Adv. Agron.* 110:1–75.

Bolan, N.S., Adriano, D.C., and Naidu, R. 2003a. Role of phosphorus in (im)mobilization and bioavailability of heavy metals in the soil–plant system. *Rev. Environ. Contam. Toxicol.* 177:1–44.

Bolan, N.S., Khan, M.A., Donaldson, J., Adriano, D.C., and Matthew, C. 2003b. Distribution and bioavailability of copper in farm effluent. *Sci. Total Environ.* 309:225–236.

Bolan, N.S., Adriano, D.C., Mani, P., Duraisamy, A., and Arulmozhiselvan, S. 2003c. Immobilization and phytoavailability of cadmium in variable charge soils: I. Effect of phosphate addition. *Plant Soil* 250:83–94.

Bolan, N.S., Choppala, G., Kunhikrishnan, A., Park, J.H., and Naidu, R. 2013a. Microbial transformation of trace elements in soils in relation to bioavailability and remediation. *Rev. Environ. Contam. Toxicol.* 225:1–56.

Bolan, N.S., Makino, T., Kunhikrishnan, A., Kim, P.J., Ishikawa, S., Murakami, M., Naidu, R., and Kirkham, M.B. 2013b. Cadmium contamination and its risk management in rice ecosystems. *Adv. Agron.* 119:183–273.

Bolan, N.S., Mahimairaja, S., Kunhikrishnan, A., and Choppala, G. 2013c. Phosphorus-arsenic interactions in variable-charge soils in relation to arsenic mobility and bioavailability. *Sci. Total Environ.* 463–464:1154–1162.

Bolan, N.S., Hedley, M.J., and Loganathan, P. 1993. Preparation, forms and properties of slow-release phosphate fertilizers. *Fert. Res.* 35:13–24.

Bolan, N.S., Naidu, R., Khan, M.A.R., Tillman, R.W., and Syers, J.K. 1999a. The effects of anion sorption on sorption and leaching of cadmium. *Aust. J. Soil Res.* 37:445–460.

Bolan, N.S., Naidu, R., Syers, J.K., and Tillman, R.W. 1999b. Surface charge and solute interactions in soils. *Adv. Agron.* 67:88–141.

Bolan, N.S., White, R.E., and Hedley, M.J. 1990. A review of the use of phosphate rock as fertilizer for direct application in Australia and New Zealand. *Aust. J. Exp. Agric.* 30:297–313.

Bosso, S.T., Enzweiler, J., and Angélica, R.S. 2008. Lead bioaccessibility in soil and mine wastes after immobilization with phosphate. *Water Air Soil Pollut.* 195:257–273.

Bramley, R.G.V. 1990. Cadmium in New Zealand agriculture. *N. Z. J. Agric. Res.* 33:505–519.

Branzini, A., and Zubillaga, M.S. 2012. Comparative use of soil organic and inorganic amendments in heavy metals stabilization. *Appl. Environ. Soil Sci.* Article ID 721032.

Brown, S., Compton, H., and Basta, N. 2007. Field test of in situ soil amendments at the tar creek national priorities list superfund site. *J. Environ. Qual.* 36:1627–1634.

Cakmak, D., Saljnikov, E., Mrvic, V., Jakovljevic, M., Marjanovic, Z., Sikiric, B., and Maksimovic, S. 2010. Soil properties and trace elements under 40-years of phosphate fertilization. *J. Environ. Qual.* 39:541–547.

Cao, X., and Ma, L.Q. 2004. Effects of compost and phosphate on plant arsenic accumulation from soils near pressure-treated wood. *Environ. Pollut.* 132:435–442.

Cao, X., Ma, L.Q., and Shiralipour, A. 2003. Effects of compost and phosphate amendments on arsenic mobility in soils and arsenic uptake by the hyperaccumulator, *Pteris vittata* L. *Environ. Pollut.* 126:157–167.

Chen, J., Wang, Y., Zhou, D., Cui, Y., Wang, S., and Chen, Y. 2010. Adsorption and desorption of Cu(II), Zn(II), Pb(II), and Cd(II) on the soils amended with nanoscale hydroxyapatite. *Environ. Prog. Sustain. Energy* 29:233–241.

Chen, S.B., Xu, M.G., and Ma, Y.B. 2007. Evaluation of phosphate application on Pb, Cd and Zn bioavailability in metal-contaminated soil. *Ecotox. Environ. Safe.* 67:278–285.

Chen, X.B., Wright, J.V., and Conca, J.L. 1997a. Effects of pH on heavy metal sorption on mineral apatite. *Environ. Sci. Technol.* 31:624–631.

Chen, X.B., Wright, J.V., Conca, J.L., and Peurrung, L.M. 1997b. Evaluation of heavy metal remediation using mineral apatite. *Water Air Soil Pollut.* 98:57–78.

Chopin, E., Marin, B., Mkoungafoko, R., Rigaux, A., Hopgood, M., Delannoy, E., Cances, B., and Laurain, M. 2008. Factors affecting distribution and mobility of trace elements (Cu, Pb, and Zn) in a perennial grapevine (*Vitis vinifera* L.) in the Champagne region of France. *Environ. Pollut.* 156:1092–1098.

Choppala, G.K., Bolan, N.S., Kunhikrishnan, A., Skinner, W., and Seshadri, B. 2013. Concomitant reduction and immobilization of chromium in relation to its bioavailability in soils. *Environ. Sci. Pollut. Res.* http://dx.doi.org/10.1007/s11356-013-1653-6 [epub ahead of print].

Clark, A.M. 1993. *Hey's Mineral Index: Mineral Species, Varieties and Synonyms*, Volume 671. London: Chapman & Hall.

Collins, C.R., Ragnarsdottir, K.V., and Sherman, D.M. 1999. Effect of inorganic and organic ligands on the mechanism of cadmium sorption to goethite. *Geochim. Cosmochim. Acta* 63:2989–3002.

Danso, I., Nuertey, B., Asamoah, P., Tetteh, F., Danso, F., Afari, P., Safo, E., and Logah, V. 2010. The effect of rock phosphate on soil nutrient dynamics, growth, development and yield of oil palm in the semi-deciduous forest zone of Ghana. *J. Sci. Technol.* 30:30–44.

Davenport, J.R., and Peryea, F.J. 1991. Phosphate fertilizers influence leaching of lead and arsenic in a soil contaminated with lead arsenate. *Water Air Soil Pollut.* 57:101–110.

Del Razo, L.M., Arellano, M.A., and Cebrian, M.E. 1990. The oxidation states of arsenic in well-water from a chronic arsenicism area in northern Mexico. *Environ. Pollut.* 64:143–153.
Dheri, G.S., Singh Brar, M., and Malhi, S.S. 2007. Influence of phosphorus application on growth and cadmium uptake of spinach in two cadmium-contaminated soils. *J. Plant Nutr. Soil Sci.* 170:495–499.
Dhillon, K.S., and Dhillon, S.K. 2014. Development and mapping of seleniferous soils in northwestern India. *Chemosphere* 99:56–63.
Eich-Greatorex, S., Krogstad, T., and Sogn, T.A. 2010. Effect of phosphorus status of the soil on selenium availability. *J. Plant Nutri. Soil Sci.* 173:337–344.
Evangelou, V.P., and Zhang, Y.L. 1995. A review: Pyrite oxidation mechanisms and acid mine drainage prevention. CRC *Crit. Rev. Environ. Sci. Technol.* 252:141–199.
Fayiga, A.O., and Ma, L.Q. 2006. Using phosphate rock to immobilize metals in soil and increase arsenic uptake by hyperaccumulator *Pteris Vittata. Sci. Total Environ.* 359:17–25.
Feng, D., Aldrich, C., and Tan, H. 2000. Treatment of acid mine water by use of heavy metal precipitation and ion exchange. *Miner Eng.* 13:623–642.
Fenger, J. 2009. Air pollution in the last 50 years—From local to global. *Atmos. Environ.* 43:13–22.
Ferguson, J.E., and Gavis, J. 1972. A review of the arsenic cycle in natural waters. *Water Res.* 6:1259–1274.
Feyte, S., Tessier, A., Gobeil, C., and Cossa, D. 2010. In situ adsorption of mercury, methylmercury and other elements by iron oxyhydroxides and organic matter in lake sediments. *Appl. Geochem.* 25:984–995.
Ford, R.G., and Sparks, D.L. 2000. The nature of Zn precipitates formed in the presence of pyrophyllite. *Environ. Sci. Technol.* 34:2479–2483.
Frankenberger Jr., W.T., and Arshad, M. 2001. Bioremediation of selenium contaminated sediments and water. *Biofactors* 14:241–254.
Frazer, L. 2001. Probing the depths of a solution for acid mine drainage. *Environ. Health Perspect.* 109:486–489.
Freeman, K.S. 2012. Remediating soil lead with fish bones. *Environ. Health Perspect.* 120:a20–a21.
Gray, C.W., McLaren, R.G., Roberts, A.H.C., and Condron, L.M. 1999. The effect of long-term phosphatic fertiliser applications on the amounts and forms of cadmium in soils under pasture in New Zealand. *Nutr. Cycl. Agroecosyst.* 54:267–277.
Harmsen, J., and Naidu, R. 2013. Bioavailability as a tool in site management. *J. Hazard. Mater.* 261:840–846.
Harris, D.L., and Lottermoser, B.G. 2006a. Evaluation of phosphate fertilizers for ameliorating acid mine waste. *Appl. Geochem.* 21:1216–1225.
Harris, D., and Lottermoser, B. 2006b. Phosphate stabilization of polymineralic mine wastes. *Mineralogical Mag.* 70:1–13.
Harter, R.D.R., and Naidu, R. 1995. Role of metal–organic complexation in metal sorption by soils. *Adv. Agron.* 55:219–264.
Hassan, S.E., Hijri, M., and St-Arnaud, M. 2013. Effect of arbuscular mycorrhizal fungi on trace metal uptake by sunflower plants grown on cadmium contaminated soil. *New Biotechnol.* 30:780–787.
Haynes, R.J., Murtaza, G., and Naidu, R. 2009. Inorganic and organic constituents and contaminants in biosolids: Implications for land application. *Adv. Agron.* 104:165–267.

Heemsbergen, D.A., McLaughlin, M.J., Whatmuff, M., Warne, M.S.J., Broos, K., Bell, M.J., Nash, D., Barry, G., Pritchard, D., and Penney, N. 2010. Bioavailability of zinc and copper in biosolids compared to their soluble salts. *Environ. Pollut.* 158:1907–1915.

Helyar, K.R., Munns, D.N., and Burau, R.G. 1976. Adsorption of phosphate by gibbsite. II. Formation of a surface complex involving divalent cations. *J. Soil Sci.* 27:315–323.

Hettiarachchi, G.M., Pierzynski, G.M., and Ransom, M.D. 2000. In situ stabilization of soil lead using phosphorus and manganese oxide. *Environ Sci. Technol.* 34:4614–4619.

Hildebrandt, A., Guillamon, M., Lacorte, S., Tauler, R., and Barcelo, D. 2008. Impact of pesticides used in agriculture and vineyards to surface and groundwater quality (North Spain). *Water Res.* 42:3315–3326.

Hildebrandt, U., Regvar, M., and Bothe, H. 2007. Arbuscular mycorrhiza and heavy metal tolerance. *Phytochemistry* 68:139–146.

Hingston, F.J. 1981. A review of anion adsorption. In *Adsorption of Inorganics at Solid-Liquid Interfaces*, M.A. Anderson, and A.J. Robin (eds.), Ann Arbor, MI: Ann Arbor Science, 51–90.

Hong, C.O., Chung, D.Y., Lee, D.K., and Kim, P.J. 2010. Comparison of phosphate materials for immobilizing cadmium in soil. *Arch. Environ. Contam. Toxicol.* 58:268–274.

Hong, C.O., Lee, D.K., Chung, D.Y., and Kim, P.J. 2007. Liming effects on cadmium stabilization in upland soil affected by gold mining activity. *Arch. Environ. Contam. Toxicol.* 52:496–502.

Hong, C.O., Lee, D.K., and Kim, P.J. 2008. Feasibility of phosphate fertilizer to immobilize cadmium in a field. *Chemosphere* 70:2009–2015.

Hsu, N.H., Wang, S.L., Lin, Y.C., Sheng, G.D., and Lee, J.F. 2009. Reduction of Cr(VI) by crop residue-derived black carbon. *Environ. Sci. Technol.* 43:8801–8806.

James, B.R., Petura, J.C., Vitale, R.J., and Mussoline, G.R. 1995. Hexavalent chromium extraction from soils: A comparison of five methods. *Environ. Sci. Technol.* 29:2377–2380.

Kabir, E., Ray, S., Kim, K., Yoon, H., Jeon, E., Kim, Y.S., Cho, Y., Yun, S., and Brown, R.J. 2012. Current status of trace metal pollution in soils affected by industrial activities. *The Scientific World Journal.* Article ID 916705.

Karalić, K., Lončarić, Z., Popović, B., Zebec, V., and Kerovec, D. 2013. Liming effect on soil heavy metals availability. *Poljoprivreda* 19:59–64.

Kausch, M.F., and Pallud, C.E. 2013. Science, policy, and management of irrigation-induced selenium contamination in California. *J. Environ. Qual.* 42:1605–1614.

Khan, A.G., Kuek, C., Chaudhry, T.M., Koo, C.S., and Hayes, W.J. 2000. Role of plants, mycorrhizae and phytochelators in heavy metal contaminated land remediation. *Chemosphere* 41:197–207.

Khan, M., and Jones, D. 2009. Effect of composts, lime and diammonium phosphate on the phytoavailability of heavy metals in a copper mine tailing soil. *Pedosphere* 19:631–641.

Khan, M.J., Azeem, M.T., Jan, M.T., and Perveen, S. 2012. Effect of amendments on chemical immobilization of heavy metals in sugar mill contaminated soils. *Soil Environ.* 31:55–56.

Kilgour, D.W., Moseley, R.B., Barnett, M.O., Savage, K.S., and Jardine, P.M. 2008. Potential negative consequences of adding phosphorus-based fertilizers to immobilize lead in soil. *J. Environ. Qual.* 37:1733–1740.

Kunhikrishnan, A., Bolan, N.S., Müller, K., Laurenson, S., Naidu, R., and Kim, W.I. 2012. The influence of wastewater irrigation on the transformation and bioavailability of heavy metal(loid)s in soil. *Adv. Agron.* 115:215–297.

Kuo, S., Huang, B., and Bembenek, R. 2007. Release of cadmium from a triple superphosphate and a phosphate rock in soil. *Soil Sci.* 172:257–265.

Lackovic, K., Angove, M.J., Wells, J.D., and Johnson, B.B. 2003. Modeling the adsorption of Cd(II) onto Muloorina illite and related clay minerals. *J. Colloid Interface Sci.* 257:31–40.

Lamb, D.T., Ming, H., Megharaj, M., and Naidu, R. 2009. Heavy metal (Cu, Zn, Cd, and Pb) partitioning and bioaccessibility in uncontaminated and long-term contaminated soils. *J. Hazard. Mater.* 171:1150–1158.

Lambert, R., Grant, C., and Sauvé, S. 2007. Cadmium and zinc in soil solution extracts following the application of phosphate fertilizers. *Sci. Total Environ.* 378: 293–305.

Laperche, V., and Traina, S.J. 1998. Immobilization of Pb by hydroxyapatite. In *Adsorption of Metals by Geomedia: Variables, Mechanisms, and Model Applications,* J.A. Everett (ed.), Orlando, FL: Academic Press, 225–276.

Lazic, S., and Vukovic, Z. 1991. Ion-exchange of strontium on synthetic hydroxyapatite. *J. Radioanal Nucl. Chem.* 19:161–168.

Lehr, C.R. 2003. Microbial methylation and volatilization of arsenic. PhD thesis, Department of Chemistry, The University of British Columbia, Canada.

Lessl, J.T., and Ma, L.Q. 2013. Sparingly-soluble phosphate rock induced significant plant growth and arsenic uptake by pteris vittata from three contaminated soils. *Environ. Sci. Technol.* 47:5311–5318.

Li, H., McGrath, S.P., and Zhao, F. 2008. Selenium uptake, translocation and speciation in wheat supplied with selenate or selenite. *New Phytol.* 178:92–102.

Li, R.Y., Ago, Y., Liu, W.J., Mitani, N., Feldmann, J., McGrath, S.P., Ma, J.F., and Zhao, F.J. 2009. The rice aquaporin Lsi1 mediates uptake of methylated arsenic species. *Plant Physiol.* 150:2071–2080.

Li, X., Poon, C., and Liu, P.S. 2001. Heavy metal contamination of urban soils and street dusts in Hong Kong. *Appl. Geochem.* 16:1361–1368.

Lindqvist, O., Johansson, K., Bringmark, L., Timm, B., Aastrup, M., Andersson, A., Hovsenius, G., Hakanson, L., Iverfeldt, A., and Meili, M. 1991. Mercury in the Swedish environment—Recent research on causes, consequences and corrective methods. *Water Air Soil Pollut.* 55:xi–261.

Loganathan, P., Hedley, M.J., and Grace, N.D. 2008. Pasture soils contaminated with fertilizer derived cadmium and fluoride: Livestock effects. *Rev. Environ. Contam. Toxicol.* 192:29–66.

Loganathan, P, Hedley, M.J., Gregg, P.E.H., and Currie, L.D. 1996. Effect of phosphate fertiliser type on the accumulation and plant availability of cadmium in grassland soils. *Nutr. Cycl. Agroecosyst.* 46:169–179.

Loganathan, P., Vigneswaran, S., Kandasamy, J., and Naidu, R. 2012. Cadmium sorption and desorption in soils: A review. *Crit. Rev. Environ. Sci. Technol.* 42:489–533.

Loseto, L.L., Siciliano, S.D., and Lean, D.R.S. 2004. Methylmercury production in high Arctic wetlands. *Environ. Toxicol. Chem.* 23:17–23.

Lou, L., Ye, Z., Lin, A., and Wong, M. 2010. Interaction of arsenic and phosphate on their uptake and accumulation in Chinese brake fern. *Int. J. Phytoremediation* 12:487–502.

Ma, L.Q., and Rao, G.N. 1997. Effects of phosphate rock on sequential chemical extraction of lead in contaminated soils. *J. Environ. Qual.* 26:788–794.

Ma, L.Q., Santos, J., Cao, X., Saha, U., and Harris, W. 2008. Field application of phosphate rock for remediation of metal-contaminated soils. Final report, prepared for Florida Institute of Phosphate Research, Publication Number: 01-148-226.

Ma, L.Q., Traina, S.J., Logan, T.J., and Ryan, J.A. 1993. In situ Pb immobilization by apatite. *Environ. Sci. Technol.* 27:1803–1810.

MacDonald, A.M., Fordyce, F.M., Shand, P., and Dochartaigh, Ó.B.E. 2005. Using geological and geochemical information to estimate the potential distribution of trace elements in Scottish groundwater. British Geological Survey, groundwater programme commissioned report CR/05/238N.

Mahimairaja, S., Bolan, N.S., Adriano, D., and Robinson, B. 2005. Arsenic contamination and its risk management in complex environmental settings. *Adv. Agron.* 86:1–82.

Mauric, A., and Lottermoser, B.G. 2011. Phosphate amendment of metalliferous waste rocks, Century Pb–Zn mine, Australia: Laboratory and field trials. *Appl. Geochem.* 26:45–56.

Mavropoulos, E., Rocha, N.C., Moreira, J.C., Rossi, A.M., and Soares, G.A. 2004. Characterization of phase evolution during lead immobilization by synthetic hydroxyapatite. *Mater. Charact.* 53:71–78.

McBride, M.B., and Spiers, G. 2001. Trace element content of selected fertilizers and dairy manures as determined by ICP–MS. *Commun. Soil Sci. Plant Anal.* 32:139–156.

McGowen, S., Basta, N., and Brown, G. 2001. Use of diammonium phosphate to reduce heavy metal solubility and transport in smelter-contaminated soil. *J. Environ. Qual.* 30:493–500.

McLaughlin, M.J., Tiller, K.G., Naidu, R., and Stevens, D.P. 1996. Review: The behaviour and environmental impact of contaminants in fertilizers. *Aust. J. Soil Res.* 34:1–54.

Meier, S., Borie, F., Bolan, N.S., and Cornejo, P. 2012a. Phytoremediation of metal-polluted soils by arbuscular mycorrhizal fungi. *Crit. Rev. Environ. Sci. Technol.* 42:741–775.

Meier, S., Borie, F., Curaqueo, G., Bolan, N.S., and Cornejo, P. 2012b. Effects of arbuscular mycorrhizal inoculation on metallophyte and agricultural plants growing at increasing copper levels, *Appl. Soil Ecol.* 61:280–287.

Mendes, A.M.S., Duda, G.P., do Nascimento, C.W.A., and Silva, M.O. 2006. Bioavailability of cadmium and lead in a soil amended with phosphorus fertilizers. *Scientia Agricola* 63:328–332.

Middelburg, J.J., and Comans, M.J. 1991. Sorption of cadmium on hydroxyapatite. *Chem. Geol.* 90:45–53.

Mignardi, S., Corami, A., and Ferrini, V. 2012. Evaluation of the effectiveness of phosphate treatment for the remediation of mine waste soils contaminated with Cd, Cu, Pb, and Zn. *Chemosphere* 86:354–360.

Miretzky, P., and Fernandez-Cirelli, A. 2008. Phosphates for Pb immobilization in soils: A review. *Environ. Chem. Lett.* 6:121–133.

Mohammad, M.J., Malkawi, H.I., and Shibli, R. 2003. Effects of arbuscular mycorrhizal fungi and phosphorus fertilization on growth and nutrient uptake of barley grown on soils with different levels of salts. *J. Plant Nutr.* 26:125–137.

Molina, M., Aburto, F., Calderon, R., Cazanga, M., and Escudey, M. 2009. Trace element composition of selected fertilizers used in Chile: Phosphorus fertilizers as a source of long-term soil contamination. *Soil Sediment Contam.* 18:497–511.

Mora, M.L., Pinilla, L., Rosas, A., and Cartes, P. 2008. Selenium uptake and its influence on the antioxidative system of white clover as affected by lime and phosphorus fertilization. *Plant Soil* 303:139–149.

Mouta, E.R., Soares, M.R., and Casagrande, J.C. 2008. Copper adsorption as a function of solution parameters of variable charge soils. *J. Braz. Chem. Soc.* 19:996–1009.

Munksgaard, N., and Lottermoser, B. 2013. Phosphate amendment of metalliferous tailings, Cannington Ag–Pb–Zn mine, Australia: Implications for the capping of tailings storage facilities. *Environ. Earth Sci.* 68:33–44.

Naidu, R., and Bolan, N.S. 2008. Contaminant chemistry in soils: Key concepts and bioavailability. In *Chemical Bioavailability in Terrestrial Environment*, R. Naidu (ed.). Amsterdam: Elsevier, 9–38.

Naidu, R., Bolan, N.S., Kookana, R.S., and Tiller, K.G. 1994. Ionic-strength and pH effects on the adsorption of cadmium and the surface charge of soils. *Eur. J. Soil Sci.* 45:419–429.

Naidu, R., Kookana, R.S., Oliver, D.P., Rogers, S., and McLaughlin, M.J. 1996. *Contaminants and the Soil Environment in the Australasia–Pacific Region*. London: Kluwer Academic Publishers.

Naidu, R., Kookana, R.S., Sumner, M.E., Harter, R.D., and Tiller, K.G. 1997. Cadmium sorption and transport in variable charge soils: A review. *J. Environ. Qual.* 26:602–617.

Naja, G.M., and Volesky, B. 2009. Toxicity and sources of Pb, Cd, Hg, Cr, As, and radionuclides in the environment. In *Heavy Metals in the Environment*, L.K. Wang, J.P. Chen, Y.T. Hung, and N.K. Shammas (eds.). Boca Raton, FL: Taylor & Francis and CRC Press, 13–62.

Nakamaru, Y.M., and Sekine, K. 2008. Sorption behavior of selenium and antimony in soils as a function of phosphate ion concentration. *Soil Sci. Plant Nutr.* 54:332–341.

Nakamaru, Y., Tagami, K., and Uchida, S. 2006. Effect of phosphate addition on the sorption–desorption reaction of selenium in Japanese agricultural soils. *Chemosphere* 63:109–115.

Neunhäuserer, C., Berreck, M., and Insam, H. 2001. Remediation of soils contaminated with molybdenum using soil amendments and phytoremediation. *Water Air Soil Pollut.* 128:85–96.

Neupane, G., and Donahoe, R.J. 2013. Calcium–phosphate treatment of contaminated soil for arsenic immobilization. *Appl. Geochem.* 28:145–154.

Ok, Y.S., Oh, S.E., Ahmad, M., Hyun, S., Kim, K.R., Moon, D.H., Lee, S.S., Lim, K.J., Jeon, W.T., and Yang, J.E. 2010. Effects of natural and calcined oyster shells on Cd and Pb immobilization in contaminated soils. *Environ. Earth Sci.* 61:1301–1308.

Oliva, J., Cama, J., Cortina, J.L., Ayora, C., and De Pablo, J. 2012. Biogenic hydroxyapatite (Apatite II^{TM}) dissolution kinetics and metal removal from acid mine drainage. *J. Hazard. Mater.* 213–214: 7–18.

Park, J.H., Lamb, D., Paneerselvam, P., Choppala, G., Bolan, N., and Chung, J.W. 2011a. Role of organic amendments on enhanced bioremediation of heavy metal(loid) contaminated soils. *J. Hazard. Mater.* 185:549–574.

Park, J.H., Bolan, N., Megharaj, M., and Naidu, R. 2011b. Isolation of phosphate solubilizing bacteria and their potential for lead immobilization in soil. *J. Hazard. Mater.* 185:829–836.

Park, J.H., Bolan, N., Chung, J.W., Naidu, R., and Megharaj, M. 2011c. Environmental monitoring of the role of phosphate compounds in enhancing immobilization and reducing bioavailability of lead in contaminated soils. *J. Environ. Monit.* 13:2234–2242.

Park, J.H., Bolan, N., Megharaj, M., and Naidu, R. 2012. Relative value of phosphate compounds in reducing the bioavailability and toxicity of lead in contaminated soils. *Water Air Soil Pollut.* 223:599–608.

Pérez-Novo, C., Bermúdez-Couso, A., López-Periago, E., Fernández-Calviño, D., and Arias-Estévez, M. 2011. Zinc adsorption in acid soils: Influence of phosphate. *Geoderma* 162:358–364.

Peryea, F.J. 1991. Phosphate-induced release of arsenic from soils contaminated with lead arsenate. *Soil Sci. Soc. Am. J.* 55:1301–1306.

Peryea, F., and Kammereck, R. 1997. Phosphate-enhanced movement of arsenic out of lead arsenate-contaminated topsoil and through uncontaminated subsoil. *Water Air Soil Pollut.* 93:243–254.

Rabie, G.H. 2005. Contribution of arbuscular mycorrhizal fungus to red kidney and wheat plants tolerance grown in heavy metal-polluted soil. *Afr. J. Biotechnol.* 4:332–345.

Randall, P.M., and Chattopadhyay, S. 2013. Mercury contaminated sediment sites—An evaluation of remedial options. *Environ. Res.* 125:131–149.

Rakib, M.A., Ali, M., Akter, M.S., and Bhuiyan, M.A.H. 2014. Assessment of heavy metal (Pb, Zn, Cr, and Cu) content in roadside dust of Dhaka metropolitan city, Bangladesh. *Int. Res. J. Environ. Sci.* 3:1–5.

Ravichandran, M. 2004. Interactions between mercury and dissolved organic matter—A review. *Chemosphere* 55:319–331.

Ren, X., Yang, S., Tan, X., Chen, C., Sheng, G., and Wang, X. 2012. Mutual effects of copper and phosphate on their interaction with g-Al2O3: Combined batch macroscopic experiments with DFT calculations. *J. Hazard. Mater.* 237–238: 199–208.

Rietra, R.P., Hiemstra, T., and van Riemsdijk, W.H. 2001. Interaction between calcium and phosphate adsorption on goethite. *Environ. Sci. Technol.* 35:3369–3374.

Roberts, A.H.C., Longhurst, R.D., and Brown, M.W. 1994. Cadmium status of soils, plant and grazing animals in New Zealand. *N. Z. J. Agric. Res.* 37:119–129.

Robinson, B., Greven, M., Green, S., Sivakumaran, S., Davidson, P., and Clothier, B. 2006. Leaching of copper, chromium and arsenic from treated vineyard posts in Marlborough, New Zealand. *Sci. Total Environ.* 364:113–123.

Rodríguez Martín-Doimeadios, R., Tessier, E., Amouroux, D., Guyoneaud, R., Duran, R., Caumette, P., and Donard, O. 2004. Mercury methylation/demethylation and volatilization pathways in estuarine sediment slurries using species-specific enriched stable isotopes. *Mar. Chem.* 90:107–123.

Rosskopfová, O., Galamboš, M., Pivarčiová, L., Čaplovičová, M., and Rajec, P. 2013. Adsorption of nickel on synthetic hydroxyapatite from aqueous solutions. *J. Radioanal. Nucl.* 295:459–465.

Rosskopfová, O., Galamboš, M., and Rajec, P. 2011. Study of sorption processes of strontium on the synthetic hydroxyapatite. *J. Radioanal. Nucl.* 287:715–722.

Rothbaum, H.P., Goguel, R.L., Johnson, A.E., and Mattingly, G.E.G. 1986. Cadmium accumulation in soils from long continued applications of superphosphate. *J. Soil Sci.* 37:99–107.

Rupa, T.R., Rao, C.S., Rao, A.S., and Singh, M. 2003. Effects of farmyard manure and phosphorus on zinc transformation and phyto-availability in two alfisols of India. *Biores. Technol.* 87:279–288.

Sanderson, P., Naidu, R., and Bolan, N. 2014. Ecotoxicity of chemically stabilised metal(loid)s in shooting range soils. *Ecotoxicol. Environ. Saf.* 100:201–208.

Santillan-Medrano, J., and Jurinak, J.J. 1975. The chemistry of lead and cadmium in soil: Solid phase formation. *Soil Sci. Soc. Am. Proc.* 39:851–856.

Schwer III, R., and McNear, D.H. 2011. Chromated copper arsenate-treated fence posts in the agronomic landscape: Soil properties controlling arsenic speciation and spatial distribution, *J. Environ. Qual.* 40:1172–1181.

Schipper, L.A., Sparling, G.P., Fisk, L., Dodd, M., Power, I., and Littler, R.A. 2011. Rates of accumulation of cadmium and uranium in a New Zealand hill farm soil as a result of long-term use of phosphate fertilizer. *Agric. Ecosyst. Environ.* 144:95–101.

Seaman, J.C., Arey, J.S., and Bertsch, P.M. 2001. Immobilization of nickel and other metals in contaminated sediments by hydroxyapatite addition. *J. Environ. Qual.* 30:460–469.

Selin, N.E. 2009. Global biogeochemical cycling of mercury: A review. *Annu. Rev. Environ. Resour.* 34:43–63.

Siebielec, G., and Chaney, R.L. 2012. Testing amendments for remediation of military range contaminated soil. *J. Environ. Manage.* 108:8–13.

Sharma, K.R., Agrawal, M., and Marshall, F. 2007. Heavy metal contamination of soil and vegetables in suburban areas of Varanasi, India. *Ecotoxicol. Environ. Saf.* 66:258–266.

Smith, E., Naidu, R., and Alston, A.M. 2002. Chemistry of arsenic in soils: II. Effect of pH and ionic strength. *J. Environ. Qual.* 31:137–143.

Sneddon, I., Orueetxebarria, M., Hodson, M., Schofield, P., and Valsami-Jones, E. 2006. Use of bone meal amendments to immobilise Pb, Zn, and Cd in soil: A leaching column study. *Environ. Pollut.* 144:816–825.

Soon, Y.K. 1981. Solubility and sorption of cadmium in soils amended with sewage sludge. *J. Soil Sci.* 32:85–95.

Stahl, R.S., and James, B.R. 1991. Zinc sorption by B horizon soils as a function of pH. *Soil Sci. Soc. Am. J.* 55:1592–1597.

Stietiya, M.H., and Wang, J.J. 2014. Zinc and cadmium adsorption to aluminum oxide nanoparticles affected by naturally occurring ligands. *J. Environ. Qual.* 43:498–506.

Street, J.J., Sabey, B.R., and Lindsay, W.L. 1978. Influence of pH, phosphorus, cadmium, sewage sludge, and incubation time on the solubility and plant uptake of cadmium. *J. Environ. Qual.* 7:286–290.

Syers, J.K., MacKay, A.D., Brown, M.W., and Currie, L.D. 1986. Chemical and physical characteristics of phosphate rock materials of ranging reactivity. *J. Sci. Food Agric.* 37:1057–1064.

Tang, X., Yang, J., Goyne, K.W., and Deng, B. 2009. Long-term risk reduction of lead-contaminated urban soil by phosphate treatment. *Environ. Eng. Sci.* 26:1747–1754.

Tiberg, C., Sjöstedt, C., Persson, I., and Gustafsson, J.P. 2013. Phosphate effects on copper (II) and lead (II) sorption to ferrihydrite. *Geochim. Cosmochim. Acta* 120:140–157.

Ugbaje, E.M., and Abdu, N. 2013. Phosphate-induced cadmium adsorption in a tropical savannah soil. *Afr. J. Biotechnol.* 12:3822–3827.

Ultra Jr, V., Tanaka, S., Sakurai, K., and Iwasaki, K. 2007. Effects of arbuscular mycorrhiza and phosphorus application on arsenic toxicity in sunflower (*Helianthus Annuus* L.) and on the transformation of arsenic in the rhizosphere. *Plant Soil* 290:29–41.

van Kauwenbergh, S.J. 2002. Cadmium content of phosphate rocks and fertilizers. International Fertilizer Industry Association (IFA) Technical Conference, 31, India.

Violante, A., del Gaudio, S., Pigna, M., Ricciardella, M., and Banerjee, D. 2007. Coprecipitation of arsenate with metal oxides. 2. Nature mineralogy, and reactivity of iron (III) precipitates. *Environ. Sci. Technol.* 41:8275–8280.

Vistoso, E., Theng, B., Bolan, N., Parfitt, R., and Mora, M. 2012. Competitive sorption of molybdate and phosphate in andisols. *J. Soil Sci. Plant Nutri.* 12:59–72.

Vogel-Mikuš, K., Pongrac, P., Kump, P., Nečemer, M., and Regvar, M. 2006. Colonisation of a Zn, Cd, and Pb hyperaccumulator *Thlaspi Praecox* Wulfen with indigenous arbuscular mycorrhizal fungal mixture induces changes in heavy metal and nutrient uptake. *Environ. Pollut.* 139:362–371.

Wang, B., Xie, Z., Chen, J., Jiang, J., and Su, Q. 2008. Effects of field application of phosphate fertilizers on the availability and uptake of lead, zinc and cadmium by cabbage (*Brassica Chinensis* L.) in a mining tailing contaminated soil. *J. Environ. Sci.* 20:1109–1117.

Wang, K., and Xing, B. 2004. Mutual effects of cadmium and phosphate on their adsorption and desorption by goethite. *Environ. Pollut.* 127:13–20.

Wang, L., and Duan, G. 2009. Effect of external and internal phosphate status on arsenic toxicity and accumulation in rice seedlings. *J. Environ. Sci.* 21:346–351.

Wang, X., and Zhang, Y. 2014. Enhanced adsorption and transformation of Cu and Zn in sediments by hydroxyapatite. *Asian J Chem.* 26:186–190.

Waterlot, C., Pruvot, C., Ciesielski, H., and Douay, F. 2011. Effects of a phosphorus amendment and the pH of water used for watering on the mobility and phytoavailability of Cd, Pb, and Zn in highly contaminated kitchen garden soils. *Ecol. Eng.* 37:1081–1093.

Wiatrowski, H.A., Ward, P.M., and Barkay, T. 2006. Novel reduction of mercury (II) by mercury-sensitive dissimilatory metal reducing bacteria. *Environ. Sci. Technol.* 40:6690–6696.

Wightwick, A.M., Salzman, S.A., Reichman, S.M., Allinson, G., and Menzies, N.W. 2013. Effects of copper fungicide residues on the microbial function of vineyard soils. *Environ. Sci. Pollut. Res.* 20:1574–1585.

Xenidis, A., Stouraiti, C., and Papassiopi, N. 2010. Stabilization of Pb and As in soils by applying combined treatment with phosphates and ferrous iron. *J. Hazard. Mater.* 177:929–937.

Xie, R.J., Mackenzie, A.F., and Lou, Z.J. 1993. Causal-modeling pH and phosphate effects on molybdate sorption in 3 temperate soils. *Soil Sci.* 155:385–397.

Xiong, L.M. 1995. Influence of phosphate on cadmium adsorption by soils. *Fert. Res.* 40:31–40.

Xu,Y., Schwartz, F.W., and Traina, S.J. 1994. Sorption of Zn^{2+} and Cd^{2+} on hydroxyapatite surfaces. *Environ. Sci. Technol.* 28:1472–1480.

Yan-Chu, H. 1994. Arsenic distribution in soils. In *Arsenic in the Environment: Part I. Cycling and Characterization*, J.O. Nriagu (ed.). New York: John Wiley & Sons Inc., 17–49.

Yan, X., Zhang, M., Liao, X., and Tu, S. 2012. Influence of amendments on soil arsenic fractionation and phytoavailability by *Pteris Vittata* L. *Chemosphere* 88:240–244.

Yee, N., Ma, J., Dalia, A., Boonfueng, T., and Kobayashi, D.Y. 2007. Se(VI) reduction and the precipitation of Se(0) by the facultative bacterium *Enterobacter Cloacae* SLD1a-1 are regulated by FNR. *Appl. Environ. Microbiol.* 73:1914–1920.

Yin, X.X., Chen, J., Qin, J., Sun, G.X., Rosen, B.P., and Zhu, Y.G. 2011. Biotransformation and volatilization of arsenic by three photosynthetic cyanobacteria. *Plant Physiol.* 156:1631–1638.

Yu, G., Sun, D., and Zheng, Y. 2007. Health effects of exposure to natural arsenic in groundwater and coal in China: An overview of occurrence. *Environ. Health Perspect.* 115:636–642.

Zaefarian, F., Rezvani, M., Ardakani, M.R., Rejali, F., and Miransari, M. 2013. Impact of mycorrhizae formation on the phosphorus and heavy-metal uptake of Alfalfa. *Commun. Soil Sci. Plant Anal.* 44:1340–1352.

Zahedifar, M., Karimian, N., and Yasrebi, J. 2010. Zinc desorption of calcareous soils as influenced by applied zinc and phosphorus and described by eight kinetic models. *Commun. Soil Sci. Plant Anal.* 41:897–907.

Zeng, X., Wu, P., Su, S., Bai, L., and Feng, Q. 2012. Phosphate has a differential influence on arsenate adsorption by soils with different properties. *Plant Soil Environ.* 58:405–411.

Zhang, M., and Pu, J. 2011. Mineral materials as feasible amendments to stabilize heavy metals in polluted urban soils. *J. Environ. Sci.* 23:607–615.

Zhang, Y., and Frankenberger, W.T. 2003. Factors affecting removal of selenate in agricultural drainage water utilizing rice straw. *Sci. Total Environ.* 305:207–216.

Zhou, Y.F., and Haynes, R.J. 2010. Sorption of heavy metals by inorganic and organic components of solid wastes: Significance to use of wastes as low-cost adsorbents and immobilizing agents. *Crit. Rev. Env. Sci. Tec.* 40:909–977.

Zhu, K., Qiu, J., Ji, H., Yanagisawa, K., Shimanouchi, R., Onda, A., and Kajiyoshi, K. 2010. Crystallographic study of lead-substituted hydroxyapatite synthesized by high-temperature mixing method under hydrothermal conditions. *Inorg. Chim. Acta* 363:1785–1790.

Zupančič, M., Lavrič, S., and Bukovec, P. 2012. Metal immobilization and phosphorus leaching after stabilization of pyrite ash contaminated soil by phosphate amendments. *J. Environ. Monitor.* 14:704–710.

2

Influence of Phosphates on Copper and Zinc Retention Processes in Acid Soils

David Fernández-Calviño, Cristina Pérez-Novo, and Manuel Arias-Estévez

CONTENTS

Copper (Cu) and zinc (Zn) are microelements essential to plants, and therefore plants may be affected by deficiencies when Cu and Zn concentrations are low in the soil solution. However, both elements may cause phytotoxicity when their concentrations increase. Cu and Zn concentrations in soil are generally dependent on parent material, showing higher levels in basic rocks and lower levels in granite and limestone (Alloway 1990). In addition to being natural constituents, Cu and Zn enter soil via beneficial agricultural additives such as lime, fertilizers, manure, herbicides, fungicides, sewage sludge, and compost. Other sources, such as mine wastes, fly ash, and atmospheric deposits, may increase the heavy metal concentration in soils (Arias et al. 2001; Kabata-Pendias and Pendias 2001) and may cause damage to plants, diminish agricultural yields, and decrease the quality of natural waters in the environment. Cu and Zn can cause toxicity to a broad range of organisms and may accumulate in the food chain, causing deleterious effects in terrestrial and aquatic ecosystems. Cu toxicity affects annelids, algae, and fungi in a very relevant manner, while Zn toxicity is generally low (unless that affecting to annelids) as compared to that of other heavy metals and metalloids (Ross and Kaye 1994).

The severity of environmental problems resulting from the accumulation of metals depends on their mobility and transport in soil, which is in turn dependent on their particular adsorption–desorption reactions with soil colloids. The adsorption behavior of Cu and Zn varies among soil types and is a function of one or more soil properties. In the absence of other metals, Cu is generally considered to be adsorbed in greater quantities than Zn (Elliott et al. 1986), although they behave more similarly in soils with low organic matter content (Karathanasis 1999). Selective adsorption of Cu and Zn has been related to their susceptibility to hydration, their charge/radius ratio, their electronegativity, and their softness (McBride 1989). Competition between the two metals for binding sites in soil colloids is especially influential on their mobility. Thus, Gomes et al. (2001) found Cu to be preferentially adsorbed in a competitive manner over Zn in ultisols, oxisols, and alfisols. Agbenin and Olojo (2004) and Arias et al. (2006) found similar results, also indicating that Cu adsorbs in higher proportion than Zn in competitive experiments. This behavior is especially relevant in acid soils, where the mobility of these two metals is clearly enhanced; in fact, some regulations regarding heavy metals in soil take into consideration pH values when establishing threshold levels. However, some other chemical and physicochemical characteristics of soils influence the mobility of metals in the soil matrix, which should be taken into account when establishing reference levels to differentiate between polluted and nonpolluted soils. One of these soil characteristics is phosphate content, the subject of this chapter.

Acid soils can be found all over the world, with broad extensions in the tropical and subtropical regions of Africa, Asia, and Central and South America. Usually, these soils are edaphically evolved and have low natural fertility, with nitrogen (N) and phosphate (P) contents that suppose a fundamental constraint. N limitation can be surpassed by means of biologic fixation, harvest wastes, and other organic sources. However, these soils usually have very low P concentrations that are aggravated when irreversible retention processes take place. These facts make it necessary to spread high amounts of phosphorus fertilizers in order to achieve optimum yields with regard to food and fibers (Date et al. 1995; Sánchez and Buol 1975). Water-soluble phosphorus fertilizers, such as superphosphate, are a frequent choice to correct P deficiencies. However, most developing countries have to import these fertilizers, sometimes suffering from limited supply and difficult access for low-income farmers. Additionally, intensive agriculture requires phosphorus fertilizers to increase production but also to improve P content in soils, thus preventing further degradation. As a consequence, the search for alternatives is imperative, and in this context, in certain soil and climate conditions, the direct application of phosphoric rocks to soil could be an agronomic and economic option compared to the more expensive use of superphosphate (Chien et al. 1990).

Another aspect to be considered is that Cu, Zn, and P may be spread simultaneously to increase crop yields. When this occurs, phosphates can affect Cu and Zn dynamics through pH changes or through modifications operating on the electric charge of the colloids, increasing negative charge and then conditioning sorption of new adsorbates.

Some experiments focusing on Cu and Zn retention on acid soils are presented, paying special attention to the effects on Cu and Zn dynamics due to the simultaneous presence of P.

Influence of P on Cu and Zn Retention

Phosphorus is an essential element for plants and is classified as a major nutrient, meaning that plants need it in high amounts ranging from 0.1% to 0.5%. This element is absorbed by plants from the soil solution as monovalent orthophosphate $\left(H_2PO_4^-\right)$ and/or divalent orthophosphate $\left(HPO_4^{-2}\right)$ as a function of soil pH. At pH values between 4 and 6, the anion $H_2PO_4^-$ represents almost 100% of the total phosphorus in the soil solution, whereas at pH 8, HPO_4^{-2} represents around 80% (Black 1968). In order to achieve high production yields, plants should absorb adequate amounts of phosphorus from the soil solution during the plant growth cycles. For that, the amount of phosphorus in the soil solution needs to be constantly renewed during the plant growth period.

The chemistry and physicochemistry of P in soils are a complex matter due to the simultaneous occurrence of reactions such as dissolution/precipitation, adsorption/desorption, and reduction/oxidation, some of which take place in a very rapid or even instantaneous manner. The soluble P compounds show very high reactivity and low solubility and mobility. In acid soils, P tends to adsorb on iron (Fe) and aluminum (Al) oxides irreversibly or to precipitate as Fe and Al phosphates having low solubility. These types of reactions, as well as the high P concentrations needed by plants, mean that P deficiencies are relatively frequent, which in turn necessitates frequent spreading of P fertilizers to reach optimum yields.

Interactions between P and Cu or Zn are especially relevant due to the coexistence of two fundamental aspects:

1. P deficiencies are frequent in acid soils, and they are partially corrected by the spreading of fertilizers, which makes P inputs into these types of soils very relevant.
2. Cu and Zn solubility increases notably in acid soils, with enhanced risks of causing environmental pollution that can even affect the food chain. P interacting with Cu and Zn can affect the mobility and transport of both metals.

Three main types of interactions between P and Cu or Zn may take place in soils:

1. Cu or Zn interacting with minerals containing P
2. Interactions through soluble forms present in the soil solution
3. Cu or Zn interacting with P adsorbed on soil colloids

Cu and Zn Interactions with P Minerals

P content in rocks, expressed as P_2O_5, ranges from 0.2% (granites) to 30% (phosphoric rocks). The most important P mineral is apatite, which can be in rather high content in some igneous rocks but is scarce in acid soils. Acid soils may contain Al phosphates such as variscite or Fe phosphates as strengite. P minerals are more relevant in basic soils. In this way, Larsen (1967) indicates that apatite is the only calcium phosphate that can exist permanently in basic, neutral, or slightly acid soils. The formation of secondary apatite is favored by alkaline pH and the presence of high calcium (Ca) concentrations, while its dissolution is favored by acid pH and low Ca concentrations in the soil solution (Mengel 1985).

Although P minerals are not very abundant in most soils, in the last decades, they have become important due to the following causes:

a. Massive fertilization, including P products, one of them being direct spreading of phosphoric rocks
b. Utilization of various P minerals as low-cost biosorbents for heavy metals or other pollutants

Item a relates to soil fertility and the presence of P in the soil solution, mainly affecting its bioavailability for plants, although certain phosphoric rocks may contain high amounts of heavy metals, thus contributing to an increase in heavy metal concentrations in agricultural soils.

Item b relates to heavy metal retention capacity, thus affecting Cu and Zn, considering the potential of P minerals for soil and water remediation, which is more relevant in this chapter and is detailed in the following.

Various studies have shown evidence indicating that P compounds may be very important in the retention of heavy metals such as Cu and Zn. Some studies have focused on phosphoric rocks, while other studies have focused on minerals such as apatite. Batch-type experiments have been used to describe adsorption kinetics on hydroxyapatite (Kadouche et al. 2012) and on phosphoric rocks (Sarioglu et al. 2005), which were moderately rapid (around 40 min).

Regarding individual adsorption of heavy metals, Sarioglu et al. (2005) indicate that phosphate rocks can be used as a sorbent for the removal of copper from aqueous solutions, especially for low Cu concentrations. Saxena and D'Souza (2006) also found high heavy metal adsorption on phosphoric rocks, showing the following affinity sequence: $Pb^{+2} > Cu^{+2} > Zn^{+2} > Co^{+2}$, confirmed by Cao et al. (2004). Kadouche et al. (2012) studied Cu and Zn adsorption on hydroxyapatite, finding 88%–95% efficacy in the extraction from the soil solution, with the highest adsorption corresponding to Cu. Corami et al. (2007) investigated Cu and Zn adsorption on synthetic hydroxyapatite, finding 97.3%–98.6% efficacy for Cu removal from the solution, while it was 94.2%–97.7% for Zn.

Soil column experiments using hydroxyapatite as an adsorbent also indicate that some P minerals are very efficient in retaining Cu (Fernane et al. 2008; Oliva et al. 2011) and Zn (Oliva et al. 2010).

The retention capacity is dependent on many variables, such as pH, metal/adsorbent ratio, and characteristics of the minerals, and thus results may differ for different types of experiments. Therefore, Cu retention on natural and synthetic apatite is variable, ranging from 0.1 to 1.6 mmol g^{-1} and ranging from 0.1 to 0.9 mmol g^{-1} in the case of Zn (Oliva et al. 2010). In many cases, the maximum retention values are obtained from the Langmuir equation applied to the experimental data, and usually they are higher than the real ones. Regarding the effect of pH, Chen et al. (1997a) indicate that Zn retention on apatite increases as a function of increasing pH, and other authors found the same effect for Cu (Kandah 2002; Sarioglu et al. 2005). However, Peld et al. (2004) signal that the pH value has low influence on heavy metal retention when interacting with apatite, giving more relevance to the surface area.

Other materials containing P have been used to adsorb heavy metals. This is the case for bone char, which is composed of calcium hydroxyapatite, carbon, and calcium carbonate, and is considered useful to retain Cu and Zn (Cheung et al. 2001, 2002).

Some of these P-rich materials have been added to soils to improve their heavy metal retention potential. Chen et al. (1997b) concluded that mineral apatite was effective in removing dissolved Pb, Cd, and Zn leached from the contaminated soil they studied, and it was more effective at high pH values. Cao et al. (2003) found that phosphate was more effective in transforming soil Pb (53%) from the nonresidual to the residual phase than soil Zn (15%) and soil Cu (13%). Kumpiene et al. (2008) signal that Cu and Zn can be retained in higher proportion in soils treated with phosphate and other amendments. Basta et al. (2001) found that the addition of phosphoric rock decreased the bioavailability of Zn and other metals.

Mechanisms for Cu and Zn Retention on P Minerals

Although the mechanisms operating on heavy metal retention on phosphoric rocks and apatite are not very well known, the following processes may be implicated: formation of surface complexes, ionic exchange, diffusion through the inner pores of the phosphoric rock, and precipitation of new mineral phases.

In acid medium, phosphoric rocks suffer dissolution and then increase both Ca and phosphate in solution. This dissolution process can be expressed as follows (Saxena and D'Souza 2006):

$$Ca_{10}(PO_4)_6F_2 + 12H^+ \rightarrow 10Ca^{2+} + 6H_2PO_4^- + F^- \tag{2.1}$$

The dissolved phosphate could react with Cu or Zn giving a precipitate (more or less insoluble) by means of the following chemical reaction:

$$10M^{2+} + 6H_2PO_4^- + F^- \rightarrow M_{10}(PO_4)_6F_{2+}12H^+ \tag{2.2}$$

where M means metal (Cu or Zn in this case).

Taking into account this process, the higher the solubility of the phosphoric rock, the greater its efficacy to retain Cu and Zn, as the reaction displaces to the right. Also, more Ca dissolved facilitates further precipitation of other phases, and some ionic exchange reactions could take place, affecting metals, as proposed by Kadouche et al. (2012). This ionic exchange process is a function of the molar ratio (Qs) between the metal bound to the phosphoric rock or mineral and the Ca dissolved from the rock or mineral. When $Qs = 1$, the amounts of sorbed and released cation are coincident, indicating the possibility of ionic exchange between the phosphoric rock or mineral and the solution; when $Qs > 1$, the amount of metal adsorbed is higher than that released, suggesting that the formation of surface complexes is the main sorption mechanism; and when $Qs < 1$, the precipitation of a new mineral phase may occur, as previously indicated (Corami et al. 2007).

When Cu or Zn contact with a phosphoric mineral, surface complex formation can take place, releasing H^+ from the mineral structure, causing lower pH in the medium and also increased Ca release. Figure 2.1 shows the relationship between Ca released to the solution and Cu or Zn disappearing from the same solution (Corami et al. 2007). This surface complexation mechanism has also been signaled by Cao et al. (2004), indicating that 74.5% and 95.7% of Cu^{2+} and Zn^{2+}, respectively, can be retained this way. The above-mentioned H^+ release causing subsequent acidification can be described as follows:

$$\equiv POH + Cu^{2+} \rightarrow \equiv POCu^+ + H^+ \tag{2.3}$$

$$\equiv POH + Zn^{2+} \rightarrow \equiv POZn^+ + H^+ \tag{2.4}$$

The decrease in pH would cause a displacement to the right in Equation 2.2, thus favoring the precipitation of new mineral phases.

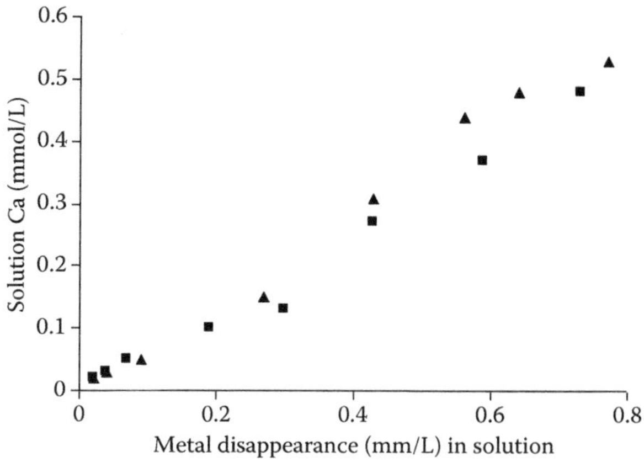

FIGURE 2.1
Relationship between metal disappearance and Ca concentration in the final solution. ▲ = copper, ■ = zinc. (From Corami, A. et al., *Journal of Hazardous Materials* 146, 164–170, 2007. With permission.)

Interaction through Soluble Forms Present in Solution

In acid soils, the presence of P in solution is affected by soil organic matter dynamics, P being a biofilic element. Low P concentrations in the soil solution are usually enough to allow normal development of plants. Fox and Kamprath (1970), as well as Barber (1995), have suggested that P concentrations around 0.2 mg L^{-1} are sufficient to reach optimum development. However, to achieve enough P absorption by plants, P concentration in the soil solution contouring roots needs continuous renovation all along the growth cycle, thus making the spreading of P fertilizers common. This P fertilization can increase notably P concentration in the soil solution, and dissolved P may interact with Cu and Zn causing modifications in the Cu and Zn species in solution. The formation of Cu–P and Zn–P species in solution could affect the affinity of both metals for the organic and inorganic soil colloids, thus impacting on metal retention/release.

This possible effect can be simulated by means of a speciation program such as MINTETQA2 (Allison et al. 1991). Table 2.1 shows the results of using this program considering the pH range 3–7. At pH 3–4, the species dominating in solution are Cu^{2+} and Zn^{2+} (>99%) both in absence and presence of P; at pH 5, Zn behaves in the same way, with Zn^{2+} dominating, and in the case of Cu, the species $CuHPO_4$ appears in percentages of 1% and 2% for P concentrations of 0.155 and 0.310 mM, respectively. At pH 6, $CuHPO_4$ appears in clearly higher proportions (9% and 16% for P concentrations 0.155 and 0.310 mM, respectively), whereas $ZnHPO_4$ appears in lower levels (2% and 3% for P concentrations 0.155 and 0.310 mM, respectively). Finally, at pH 7, the distribution shows higher changes, with 23% and 39% of $CuHPO_4$, and with 9% and 16% of $ZnHPO_4$, for P concentrations 0.155 and 0.310 mM, respectively (Table 2.1). In

TABLE 2.1

Cu and Zn Speciation (%) in the Absence and Presence of Phosphate Resulting from Simulations with the MinteqA2 Program

pH		Majority Species (Cu)[a]	%	Majority Species (Zn)	%
3	A	Cu^{2+}	99.96	Zn^{2+}	99.96
	B	Cu^{2+}	99.95	Zn^{2+}	99.96
	C	Cu^{2+}	99.94	Zn^{2+}	99.96
4	A	Cu^{2+}	99.96	Zn^{2+}	99.96
	B	Cu^{2+}	99.81	Zn^{2+}	99.95
	C	Cu^{2+}	99.70	Zn^{2+}	99.93
5	A	Cu^{2+}	99.65	Zn^{2+}	99.96
	B	Cu^{2+}	98.54	Zn^{2+}	99.78
		$CuHPO_4$	1.12		
	C	Cu^{2+}	99.47	Zn^{2+}	99.61
		$CuHPO_4$	2.19	Zn^{2+}	
6	A	Cu^{2+}	96.23	Zn^{2+}	99.87
		$CuOH^+$	2.87		
	B	Cu^{2+}	87.91	Zn^{2+}	98.25
		$CuHPO_4$	8.73	$ZnHPO_4$	1.62
		$CuOH^+$	2.61		
	C	Cu^{2+}	80.92	Zn^{2+}	96.72
		$CuHPO_4$	16.06	$ZnHPO_4$	3.16
		$CuOH^+$	2.39		
7	A	Cu^{2+}	49.72	Zn^{2+}	98.92
		$CuOH^+$	14.93		
		$Cu_2(OH)_2^{+2}$	22.71		
		$Cu_3(OH)_4^{+2}$	12.33		
	B	Cu^{2+}	41.76	Zn^{2+}	90.31
		$CuHPO_4$	22.57	$ZnHPO_4$	8.72
		$CuOH^+$	12.43		
		$Cu_2(OH)_2^{+2}$	15.84		
		$Cu_3(OH)_4^{+2}$	7.15		
	C	Cu^{2+}	35.21	Zn^{2+}	83.22
		$CuHPO_4$	38.87	$ZnHPO_4$	15.89
		$CuOH^+$	10.39		
		$Cu_2(OH)_2^{+2}$	11.13		
		$Cu_3(OH)_4^{+2}$	4.18		

Note: A = 0.155 mM Cu or Zn in absence of P, B = 0.155 mM Cu or Zn and 0.155 mM P, C = 0.155 mM Cu or Zn and 0.31 mM P.

[a] Dissolution is saturated with P for Cu and for Zn from pH 6.

view of these results, the presence of P would not cause significant changes in the speciation, mostly in the acid range, thus suggesting that the influence of P is not a key factor in Cu and Zn retention (mainly when pH is acidic). This is in accordance with the electron spin resonance results from Clark and McBride (1984), who found no evidence of interaction between Cu and P. However, the presence of natural or added P in solution may affect pH, and any increase in the pH value can increase the retention of cationic metals due to the increase in the negative charge of substrates where variable charge colloids dominate (Spark et al. 1995a,b, 1997) and also to some intrinsic characteristics of metals, such as their hydroxylation capacity. This increased pH may also cause an increase in Cu–P species. In this way, Pérez-Novo et al. (2009) found that adding P in molar ratios 1:1 and 1:2 (P:Cu) causes an increase in pH after a 24-h contact period in acid soils over granite and amphibolites (Figure 2.2). In that study, the change in the pH value caused an increase in the presence of the $CuHPO_4$ species, which was 6.5% and 12% for molar rations 1:1 and 1:2 (P:Cu), respectively. However, using the same conditions for Zn as for Cu, the addition of P did not have an effect on pH (Figure 2.3), showing that in this case there was no effect on Zn retention due to pH change.

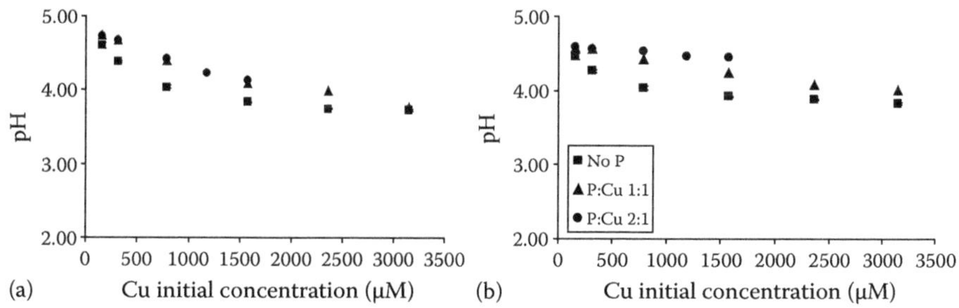

FIGURE 2.2
Effect of phosphate on the pH of two soils in the presence of various concentrations of Cu. (a) Granite soil and (b) amphibolite soil. (From Pérez-Novo, C. et al., *Geoderma* 150, 166–170, 2009. With permission.)

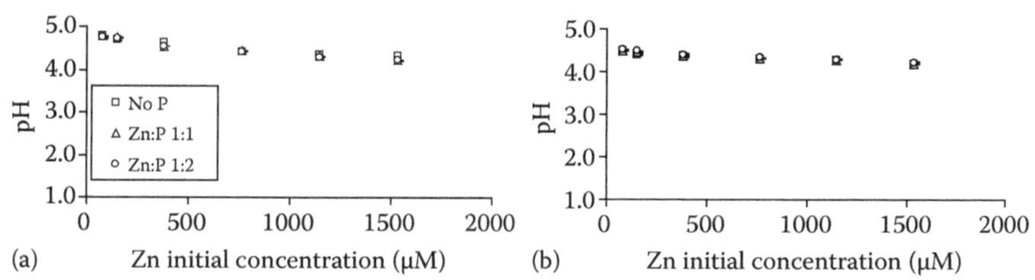

FIGURE 2.3
Effect of phosphate on the pH of two soils in the presence of various concentrations of Zn. (a) Granite soil and (b) amphibolite soil. (From Pérez-Novo, C. et al., *Geoderma* 162, 358–364, 2011. With permission.)

Cu and Zn Interactions with P Adsorbed on Soil Colloids

Once in the soil solution, phosphates are strongly adsorbed on the soil colloids. This P sorption can increase the negative charge of the surface layer, and hence, it can also increase the Cu and Zn adsorption in the soil. This effect was observed in soils with high amounts of Fe and Al oxyhydroxides (Del Nero et al. 2010) in a pH range from 3 to 10, even reversing the potential from positive to negative. However, in acid soils with high organic matter, this effect was not detected. As well, data from Pérez-Novo et al. (2009) showed no changes in the Z-potential after adding phosphates to soils developed over granite and amphibolite (Figure 2.4).

The addition of phosphates to the soil may have contradictory effects on Cu and Zn adsorption. In some types of soils and soil minerals, the mobility of Cu and Zn increased (McBride 1985), while it decreased in other soils (Pérez-Novo et al. 2009). This fact can be attributed to the effect of phosphates on soil pH. The presence of phosphates can modify the pH and hence the Cu and Zn

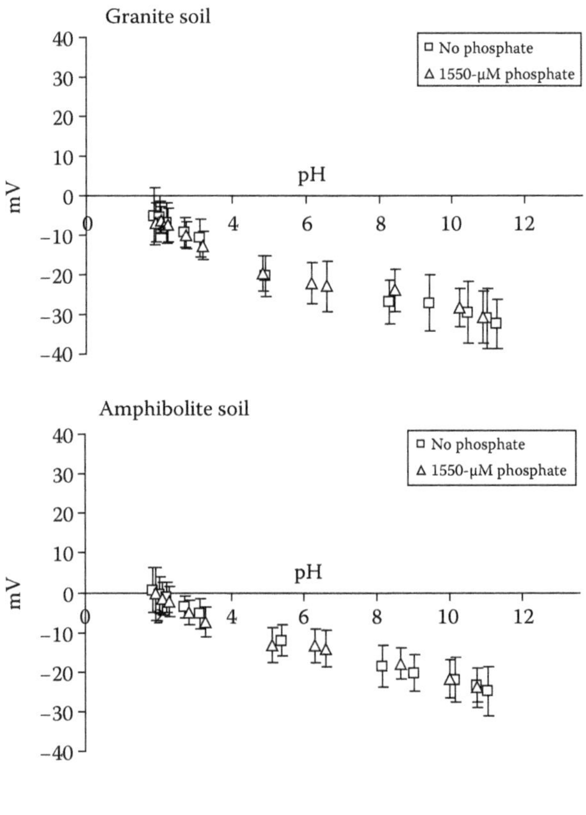

FIGURE 2.4
Z-potentials corresponding to the clay fractions of acid soils at various pH values in the presence and absence of 1.5-μM phosphate. (From Pérez-Novo, C. et al., *Geoderma* 150, 166–170, 2009. With permission.)

mobility in the soil. However, the effect of phosphates on pH can be different as a function of soil type; in alkaline soils, the presence of P can diminish the pH value, and hence the adsorption of Cu and Zn in the soil may decrease; but in acid soils, the presence of phosphates can have the opposite effect, increasing the pH and, as a consequence, the Cu and Zn retention in the soil (Barrow 1987).

In order to study the possible mechanisms in relation to phosphate effects on Cu and Zn adsorption, three different approximations can be used:

1. Via batch studies
2. Via stirred flow chamber experiments
3. Via column studies

Batch Experiments

Although data on the influence of P on Cu and Zn adsorption is scarce, batch experiments performed in acid soils have shown that the presence of P increases Cu and Zn adsorption. Figure 2.5 (Pérez-Novo et al. 2009, 2011a) shows this for soils developed over granite (low Fe and Al oxyhidroxide content) and amphibolite (high Fe and Al oxyhidroxide content). Similar Zn adsorption increases were shown by Selim et al. (2013) in soils close to neutrality. Also, in a study focusing on the effect of different anions on Zn adsorption in acid soils, Wang and Harrell (2005) showed that phosphates have higher effects on Zn adsorption increases than other anions such as nitrate or chloride.

The adsorption increases in the presence of phosphates are higher for Cu than for Zn. These discrepancies may be in relation to different mechanisms causing sorption increases. One of four mechanisms can occur: pH increases when P is in the system, the formation of P–Cu species with higher affinity to bind with soil colloids, changes in the surface potential in soils with high amounts of Fe and Al oxyhydroxides, or the existence of new specific sites as a result of P complexes formed on adsorption surfaces. In acid soils, Pérez-Novo et al. (2009) discarded the formation of P–Cu species in solution as well as changes in the surface potential in soils. However, they showed evidence of pH increases when phosphorus was added, and also that other mechanisms (such as the existence of new specific sites as a result of P complexes formed on adsorption surfaces) cannot be discarded. For Zn, adsorption causes an increase in the formation of P–Zn species in solution; furthermore, changes in the Z-potential were discarded, and no evidence of pH changes was found (Pérez-Novo et al. 2011a). In this case, the formation of P–Zn complexes in the colloid surfaces may be the main mechanism. In fact, a linear relationship was found between the phosphate adsorption and the increase in Zn adsorption ($R^2 = 0.637$) (Pérez-Novo et al. 2011a).

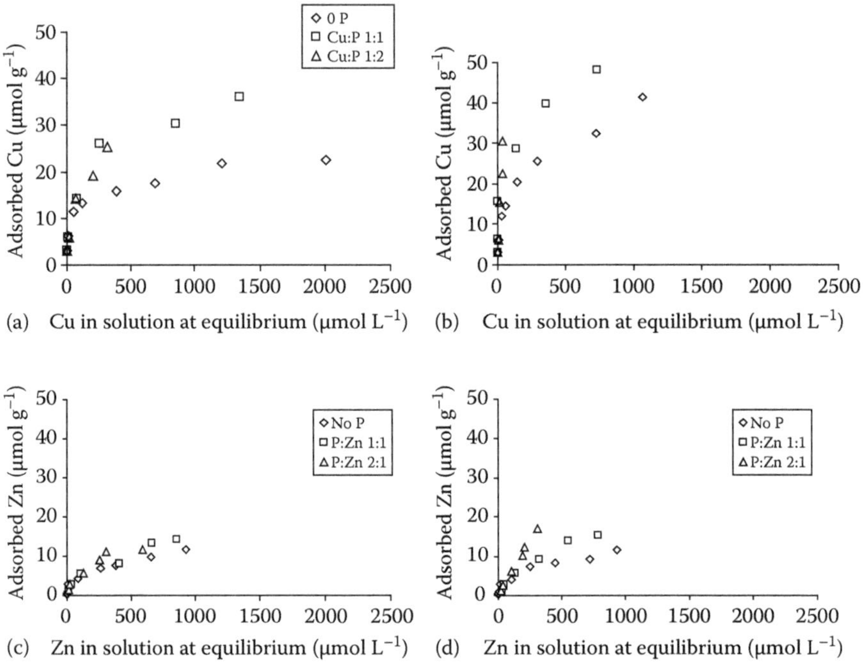

FIGURE 2.5
Sorption of Cu ([a] granite soil and [b] amphibolite soil) (From Pérez-Novo, C. et al., *Geoderma* 150, 166–170, 2009. With permission.) and Zn ([c] granite soil and [d] amphibolite soil) (From Pérez-Novo, C. et al., *Geoderma* 162, 358–364, 2011. With permission.) in presence and absence of phosphate.

Another possible explanation for Cu and Zn adsorption increases in the presence of phosphates is that Cu and Zn can act as bridging elements between phosphorus and soil organic matter. In this respect, when Cu and P are in the soil solution, the amount of P adsorbed also increases (Table 2.2). Ahumada et al. (1997) showed that for Zn, the presence of P increased the fraction of Zn bound to organic matter in the soil.

TABLE 2.2
Percentage of Adsorbed P for an Initial P Concentration of 1615 mM as a Function of the Concentration of Cu in the Soil Solution in Two Soils

[Cu] mM	Granite Soil	Amphibolite Soil
0.000	29.5 ± 3.5	86.1 ± 0.7
1.570	63.4 ± 0.4	95.4 ± 0.4
3.100	71.1 ± 1.4	98.1 ± 0.5

Source: Rodriguez, O. Interaccion na adsorción de fósforo e cobre. Estudio sobre solos desenvolvidos sobre granito e anfibolita. Facultad de Ciencias. Universidad de Vigo, Spain, 2002.

Stirred Flow Chamber Experiments

The Cu and Zn adsorption processes in acid soils are kinetically dependent (i.e., they are not instantaneous processes) (López-Periago et al. 2008; Zhao and Selim 2010). Therefore, in order to achieve better knowledge about phosphate–metal interactions, sorption kinetic experiments are important. Kinetic experiments allow us to discriminate between fast and slow reactions associated with different adsorption processes. However, only a few works studied the effect of phosphates on Cu or Zn adsorption–desorption kinetics in the soil (Pérez-Novo et al. 2011b,c; Selim et al. 2013; Zhao and Selim 2010). Generally, fast reactions are most likely adsorption via electrostatic attraction and/or inner-sphere complexation with soil components (Strawn and Sparks 2000). The slow reactions are of high importance on the extent and reversibility of heavy metal sorption on soils (Strawn and Sparks 1999). The sorption mechanisms for the slow reactions are poorly understood (Glover et al. 2002) and can include intraparticle diffusion, rate-limited adsorption processes, and precipitation reactions on surfaces (Strawn and Sparks 2000). Different methods can be employed to study heavy metal adsorption–desorption kinetics in soils, such as batch and flow methods. The main difference between the flow and batch methods is that flow methods are inherently open systems. In the stirred flow chamber method, the flow through the reactor is kept constant by a pump, and a fraction collector is used to collect the reactor effluent. In the adsorption processes, the equilibrium is achieved at inflow concentration, and during the desorption experiments, the products released to the solution are continually removed. As well, the stirring eliminates film diffusion. The main difference with other flow methods is the reactor, where a suspension of soil and solution is continuously stirred.

There are only two works evaluating phosphate effects on Cu and Zn kinetics by means of stirred flow chamber experiments (Pérez-Novo et al. 2011b,c). In these works, evidences of Cu and Zn adsorption enhancements in the presence of phosphates, via different sites, were found. Also, changes in Cu and Zn bounding energy in the presence of phosphates were found. Cu, in the absence of phosphates, was adsorbed via two types of sites, fast reaction sites and slow reaction sites (Figure 2.6), both in soils with low and with high Fe and Al oxyhydroxide content. Cu adsorption increases in the presence of phosphates, but in different sites as a function of Fe and Al oxyhydroxide content. When it is low, the increase in Cu adsorption takes place mainly via slow reaction sites, whereas when it is high, the Cu adsorption increases were produced via both fast and slow reaction sites.

In the case of Zn, and irrespectively of the Fe and Al oxyhydroxide content, the adsorption in the absence of phosphates takes place only via fast adsorption sites (Figure 2.7). However, the increase in Zn adsorption produced in

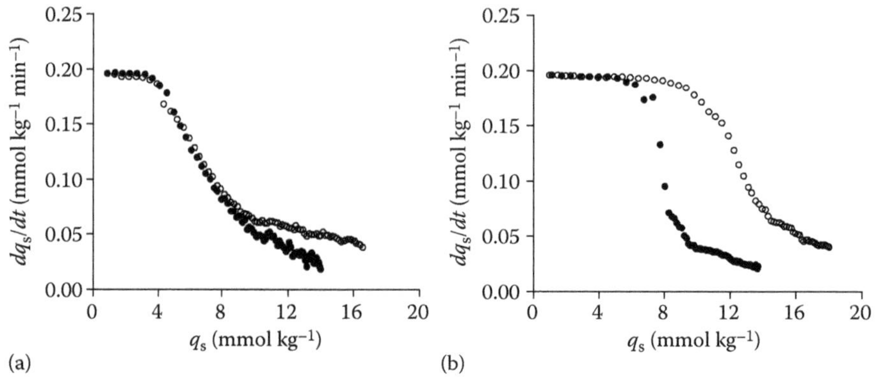

FIGURE 2.6
Copper adsorption rate as a function of Cu cumulative adsorption in the absence of P (closed circles) and in its presence (open circles). (a) Granite soil and (b) amphibolite soil. (From Pérez-Novo, C. et al., *Journal of Hazardous Materials* 185, 220–226, 2011. With permission.)

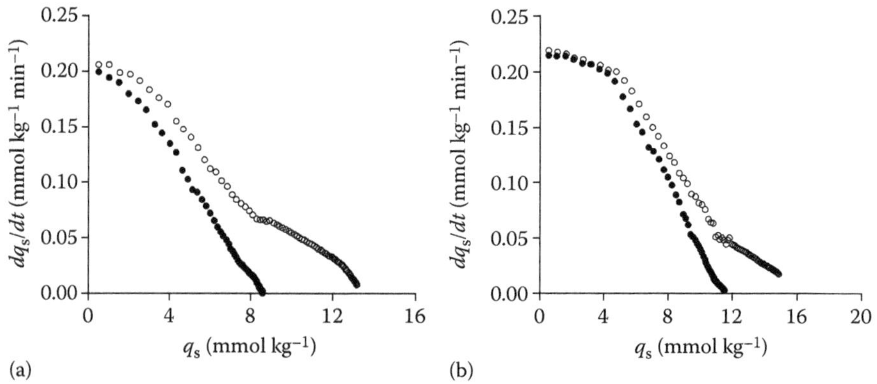

FIGURE 2.7
Zinc adsorption rate as a function of Zn cumulative adsorption in the absence of P (closed circles) and in its presence (open circles). (a) Granite soil and (b) amphibolite soil. (From Pérez-Novo, C. et al., *Chemosphere* 1028–1034, 2011. With permission.)

the presence of phosphates was mainly due to the formation of slow adsorption sites regardless of the amount of Fe and Al oxyhydroxides in the soil.

The results of these works suggest that when Cu and Zn are present in soil simultaneously with phosphates, the main mechanism responsible for the increased adsorption affecting both metals is that Cu and Zn act as bridging elements between P and soil organic matter.

Column Experiments

Column experiments are a type of flow method in which the reactor is fully filled with soil and the soil solution flows through the soil particles. In this case, the conditions have more similarity with real soil in the environment than other methods. However, film diffusion is not avoided. As in the case of the stirred flow chamber, the flow through the device is kept constant by a pump, and a fraction collector is used to collect the column effluent.

The experiments performed in soil columns allow us to extract relevant information about the retention/release of heavy metals and other adsorbates in different soils. The advantage of these types of experiments is that they suppose an approximation to soil field conditions, although most of the experiments are performed at saturation with packed columns in the laboratory.

Regarding the study of the influence of P on Cu and Zn retention, information is scarce and recent. Selim et al. (2013) modeled the influence of P on Zn retention in soils with a pH close to neutrality. Their results were similar to those obtained using the stirred flow chamber and batch experiments discussed previously. Increases in Zn retention in the presence of phosphates were observed in the breakthrough curves via the increase in the time of arrival of the Zn front in the effluent solution as well as the extent of mailing during desorption (i.e., an increase in the nonreversible adsorption during the Zn transport in the presence of P). That behavior was found in a soil with a pH value of 6.1, whereas the opposite was found in a soil with a pH value of 6.9.

Summary and Future Prospects

In this chapter we analyzed the effect of phosphorus on copper and zinc adsorption in acid soils.

Mineral P compounds have a high capacity for Cu and Zn adsorption. The main mechanisms in this regard are surface complex formation and precipitation of secondary minerals derived from the interaction between orthophosphate and Cu or Zn. This fact has caused many authors to propose these minerals as bioadsorbent materials for heavy metals. However, more studies are needed in order to verify the dissolution of these minerals, mainly in acid media, given that this dissolution process affects their adsorption efficacy in medium and long terms.

The presence of phosphates in the solution increases the retention of both Cu and Zn on acid soils, and hence the possible mechanisms involved in this enhancement were also discussed. Different processes could contribute to the enhancement of Cu and Zn retention in soils in the presence of P compounds: (a) increases in the negative charge in relation with the phosphate adsorbed on soil particles; (b) the formation, in solution, of P-metal species with greater affinity for the soil surfaces than the metal itself; (c) coprecipitation processes; (d) the influence of phosphates on the pH of the soil solution; and (e) the existence of new specific sites as a result of P complexes formed on adsorption surfaces.

We presented evidence of various enhanced Cu and Zn adsorption processes. In the absence of phosphates, Cu is adsorbed via two types of sites: fast reaction sites and slow reaction sites, whereas Zn is adsorbed only via fast adsorption sites. When phosphates are in the medium, Cu adsorption can increase for different types of sites. In soils with low Fe and Al oxyhydroxide content, the increase in Cu adsorption takes place mainly via slow reaction sites, whereas in soils containing high amounts of Fe and Al oxyhydroxides, the Cu adsorption increase is in relation to both fast and slow reaction sites. The increase in Zn adsorption takes place mainly via slow adsorption sites, regardless of the amount of Fe and Al oxyhydroxides in the soil.

Regarding the possible mechanisms involved in the increase in Cu and Zn adsorption in the presence of phosphates, the increase in negative charge due to the phosphate adsorbed on soil particles was discarded because no significant changes were detected on the Z-potential values corresponding to the clay fractions of the acid soils studied. Other processes that were also discarded were the formation of P-metal species in solution and coprecipitation, in view of the results of MINTEAQ2 simulations at the corresponding soil pHs giving increased adsorption for both Cu and Zn. The changes in the pH value of the soil solution are unequal for Zn and Cu in adsorption tests when phosphate is present. For Zn, no differences are detected in pH values when comparing the results of experiments with and without phosphates. In the case of Cu adsorption trials, the presence of phosphates can sustain higher pH values than in its absence, especially in soils with high Fe and Al oxyhydroxide concentrations. However, this higher pH is observed when just fast adsorption site increases take place, whereas the main increase in the pH value in relation to Cu adsorption in the presence of phosphates is achieved via slow adsorption sites.

As the previous mechanisms mentioned above have failed, we propose that Cu and Zn, considered as Lewis acids, could act as bridging elements between P and soil organic matter when these metals and phosphates are added simultaneously to the soil. However, the presence of other types of specific sites, as a result of P-complexes formed on adsorption surfaces, cannot be discarded, and more research is needed in this area. Also, future research should take into account the effects of adding phosphates to soil before Cu and Zn, including the effects on Cu and Zn availability, which can be of great importance in order to improve our knowledge about the fate of these metals in the environment and the subsequent appropriate management of agricultural soils to minimize the potential risks of heavy metal pollution. To perform future research, stirred flow chamber and column experiments focusing on the influence of P on Cu and Zn retention should be considered as essential because of the few results available at this point in time.

In the last decades, legal regulations on contaminated soils have included variables such as soil pH mainly due to the effect of the pH value on the mobility of heavy metals and other pollutants. As an example, regulations focusing on sewage sludge are more restrictive in the case of acid soils in relation to the increased mobility experienced by most metals in this context. Given that P can modify Cu and Zn retention, this should be taken into account in future regulations dealing with heavy metal pollution, especially regarding agricultural soils affected by massive P fertilization. Further studies about the influence of P on metal retention in different types of soils should be performed, given that it could favor the retention of heavy metal (mostly on acid soils) or, conversely, their mobility (mainly in neutral or basic soils).

References

Agbenin, J.O., Olojo, L.A. 2004. Competitive adsorption of copper and zinc by a Bt horizon of a savanna alfisol as affected by pH and selective removal of hydrous oxides and organic matter. *Geoderma* 119, 257–265.

Ahumada, I.T., Bustamante, A., Schalscha, E.B. 1997. Zinc speciation in phosphate-affected soils. *Communications in Soil Science and Plant Analysis* 28, 989–995.

Allison, J.D., Brown, D.S., Novo-Gradac, K.J. 1991. MINTEQA2/PRODEFA2: A geochemical assessment model for environmental systems (Version 3.0), U.S. Environmental Protection Agency, Athens, GA.

Alloway, B.J. 1990. The origin of heavy metals in soils. In *Heavy Metals in Soils*, B.J. Alloway (ed.). London: Blackie Academic & Professional, pp. 29–39.

Arias, M., Núnez, A., Rubinos, D., Barral, M.T., Diaz-Fierros, F. 2001. Contamination of soils near a road construction site caused by mine runoffs. *Agrochimica* 45, 134–146.

Arias, M., Pérez-Novo, C., López, E., Soto, B. 2006. Competitive adsorption and desorption of copper and zinc in acid soils. *Geoderma* 133, 151–159.

Barber, S.A. 1995. *Soil Nutrient Bioavailability. A Mechanistic Approach.* New York: John Wiley & Sons.

Barrow, N.J. 1987. The effects of phosphate on zinc sorption by a soil. *Journal of Soil Science* 38, 453–459.

Basta, N.T., Gradwohl, R., Snethen, K.L., Schroder, J.L. 2001. Chemical immobilization of lead, zinc, and cadmium in smelter contaminated soils using biosolids and rock phosphate. *Journal of Environmental Quality* 30, 1222–1230.

Black, C.A. 1968. *Soil-Plant Relationships.* New York: John Wiley & Sons.

Cao, X., Ma, L.Q., Chen, M., Singh, S.P., Harris, W.G. 2003. Phosphate-induced metal immobilization in a contaminated site. *Environmental Pollution* 122, 19–28.

Cao, X., Ma, L.Q., Rhue, D.R., Appel, C.S. 2004. Mechanism of lead, copper, and zinc retention by phosphate rock. *Environmental Pollution* 131, 435–444.

Chen, X., Wright, V., Conca, J.L., Peurrung, L.M. 1997a. Effects of pH on heavy metal sorption on mineral apatite. *Environmental Science & Technology* 31, 624–630.

Chen, X., Wright, V., Conca, J.L., Peurrung, L.M. 1997b. Evaluation of heavy metal remediation using mineral apatite. *Water, Air, and Soil Pollution* 98, 57–78.

Cheung, C.W., Chan, C.K., Porter, J.F., McKay, G. 2001. Combined diffusion model for the sorption of cadmium, copper, and zonc ions onto bone char. *Environmental Science & Technology* 35, 1511–1522.

Cheung, C.W., Porter, J.F., McKay, G. 2002. Removal of Cu(II) and Zn(II) ions by sorption onto bone char using batch agitation. *Langmuir* 18, 650–656.

Chien, S.H., Sale, P.W.G., Friesen, D.K. 1990. A discussion of the methods for comparing the relative effectiveness of phosphate fertilizers varying in solubility. *Fertilizer Research* 24, 149–157.

Clark, C.J., McBride, M.B. 1984. Adsorption of Cu(II) by allophone as affected by phosphate. *Soil Science* 38, 412–421.

Corami, A., Mignardi, S., Ferrini, V. 2007. Copper and zinc decontamination from single- and binary-metal solutions using hydroxyapatite. *Journal of Hazardous Materials* 146, 164–170.

Date, R.A., Grundon, N.J., Rayment, G.E., Probert, M.E. (eds.). 1995. *Plant-Soil Interactions at Low pH: Principles and Management. Developments in Plant and Soil Sciences*, Volume 64. Dordrecht, The Netherlands: Kluwer Academic Publishers.

Del Nero, M., Galindo, C., Barillon, R., Halter, E., Madé, B. 2010. Surface reactivity of α-Al_2O_3 and mechanisms of phosphate sorption: In situ ATR-FTIR spectroscopy and ζ potential studies. *Journal of Colloid and Interface Science* 342, 437–444.

Elliott, H.A., Liberati, M.R., Huang, C.P. 1986. Competitive adsorption of heavy metals by soils. *Journal of Environmental Quality* 15, 214–219.

Fernane, F., Mecherri, M.O., Sharrockc, P., Haidioui, M., Lounici, H., Fedoroff, M. 2008. Sorption of cadmium and copper ions on natural and synthetic hydroxylapatite particles. *Materials Characterization* 59, 554–559.

Fox, R.L., Kamprath, E.J. 1970. Phosphorus sorption isotherms for evaluating the phosphate requirements of soils. *Soil Science Society of America Proceedings* 34, 902–907.

Glover, L.J., Eick, M.J., Brady, P.V. 2002. Desorption kinetics of cadmium 2+ and lead 2+ from goethite: Influence of time and organic acids. *Soil Science Society of America Journal* 66, 797–804.

Gomes, P.C., Fontes, M.P.F., da Silva, A.G., Mendonça, E.S., Netto, A.R. 2001. Selectivity sequence and competitive adsorption of heavy metals by Brazilian soils. *Soil Science Society of America Journal* 65, 1115–1121.

Kabata-Pendias, A., Pendias, H. 2001. *Trace Elements in Soils and Plants*, Third Edition. Boca Raton, FL: CRC Press LLC.

Kadouche, S., Zemmouri, H., Benaoumeur, K., Drouiche, N., Sharrock, P., Lounici, H. 2012. Metal ion binding on hydroxyapatite (Hap) and study of the velocity of sedimentation. *Procedia Engineering* 33, 377–384.

Kandah, M.I. 2002. The potential use of low-grade phosphate rocks as adsorbent. *Chemical Engineering & Technology* 25, 921–924.

Karathanasis, A.D. 1999. Subsurface migration of copper and zinc mediated by soil colloids. *Soil Science Society of America Journal* 63, 830–838.

Kumpiene, J., Lagerkvist, A., Maurice, C. 2008. Stabilization of As, Cr, Cu, Pb, and Zn in soil using amendments—A review. *Waste Management* 28, 215–225.

Larsen, S. 1967. Soil phosphorus. *Advances in Agronomy* 19, 151–210.

López-Periago, J.E., Arias-Estévez, M., Nóvoa-Muñoz, J.C., Fernández-Calviño, D., Soto, B., Pérez-Novo, C., Simal-Gándara, J. 2008. Copper retention kinetics in acid soils. *Soil Science Society of America Journal* 72, 63–72.

McBride, M. 1989. Reactions controlling heavy metal solubility in soils. *Advances in Soil Science* 10, 1–56.

McBride, M.B. 1985. Sorption of copper (II) on aluminum hydroxide as affected by phosphate. *Soil Science Society of America Journal* 49, 843–846.

Mengel, K. 1985. Dynamics and availability of major nutrients in soils. In *Advances in Soil Science*, Volume 2, B.A. Stewart (ed.). New York: Springer-Verlag, pp. 96–106.

Oliva, J., de Pablo, J., Cortina, J.L., Cama, J., Ayora, C. 2010. The use of Apatite II™ to remove divalent metal ions zinc (II), lead (II), manganese (II) and iron (II) from water in passive treatment systems: Column experiments. *Journal of Hazardous Materials* 184, 364–374.

Oliva, J., de Pablo, J., Cortina, J.L., Cama, J., Ayora, C. 2011. Removal of cadmium, copper, nickel, cobalt and mercury from water by Apatite II™. Column experiments. *Journal of Hazardous Materials* 194, 312–323.

Peld, M., Tõnsuaadu, K., Bender, V. 2004. Sorption and desorption of Cd^{+2} and Zn^{+2} ions in apatite-aqueous system. *Environmental Science & Technology* 38, 5626–5631.

Pérez-Novo, C., Bermúdez-Couso, A., López-Periago, E., Fernández-Calviño, D., Arias-Estévez, M. 2009. The effect of phosphate on the sorption of copper by acid soils. *Geoderma* 150, 166–170.

Pérez-Novo, C., Bermúdez-Couso, A., López-Periago, E., Fernández-Calviño, D., Arias-Estévez, M. 2011a. Zinc adsorption in acid soils. Influence of phosphate. *Geoderma* 162, 358–364.

Pérez-Novo, C., Fernández-Calviño, D., Bermúdez-Couso, A., López-Periago, J.E., Arias-Estévez, M. 2011b. Influence of phosphorus on Cu sorption kinetics: Stirred flow chamber experiments. *Journal of Hazardous Materials* 185, 220–226.

Pérez-Novo, C., Fernández-Calviño, D., Bermúdez-Couso, A., López-Periago, J.E., Arias-Estévez, M. 2011c. Phosphorus effect on Zn adorption-desorption kinetics in acid soils. *Chemosphere* 1028–1034.

Ross, S.M., Kaye, K.J. 1994. The meaning of metal toxicity in soil-plant systems. In *Toxic Metals in Soil-Plant Systems*, S.M. Ross (ed.). Chicester, UK: John Wiley & Sons Ltd., pp. 27–61.

Sánchez, P.A., Buol, S.W. 1975. Soils of the tropics and the world food crisis. *Science* 188, 598–603.

Sarioglu, M., Atay, Ü.A., Cebeci, Y. 2005. Removal of copper from aqueous solutions by phosphate rock. *Desalination* 181, 303–311.

Saxena, S., D'Souza, S.F. 2006. Heavy metal pollution abatement using rock phosphate mineral. *Environmental International* 32, 199–202.

Selim, H.M., Elban, T.A., Zhao, K., Xu, J., Fergusson, E.L. 2013. Miscible displacement of zinc in soil columns: Linear and nonlinear modeling. *Soil Science Society of America Journal* 77, 391–402.

Spark, K.M., Johnson, B.B., Wells, J.D. 1995a. Characterizing heavy-metal adsorption on oxides and oxihydroxides. *European Journal of Soil Science* 46, 621–631.

Spark, K.M., Wells, J.D., Johnson, B.B. 1995b. Characterizing trace metal adsorption on kaolinite. *European Journal of Soil Science* 46, 633–640.

Spark, K.M., Wells, J.D., Johnson, B.B. 1997. The interaction of a humic acid with heavy metals. *Australian Journal of Soil Research* 35, 89–101.

Strawn, D.G., Sparks, D.L. 1999. Sorption kinetics of trace elements in soils and soil materials. In *Fate and Transport of Heavy Metals in the Vadose Zone*, H.M. Selim and I.K. Iskandar (eds.). Boca Raton, FL: Lewis Publishers, CRC Press.

Strawn, D.G., Sparks, D.L. 2000. Effects of soil organic matter on the kinetics and mechanisms of Pb(II) sorption and desorption in soil. *Soil Science Society of America Journal* 64, 144–156.

Wang, J.J., Harrell, D.L. 2005. Effect of ammonium, potassium, and sodium cations and phosphate, nitrate, and chloride anions on zinc sorption and lability in selected acid and calcareous soils. *Soil Science Society of America Journal* 69, 1036–1046.

Zhao, K., Selim, H.M. 2010. Adsorption-desorption kinetics of Zn in soils: Influence of phosphate. *Soil Science* 175, 145–153.

3

Phosphate–Uranium Interactions in Soils

John C. Seaman, Shea W. Buettner, and Hyun-shik Chang

CONTENTS

In recent years, much of the interest in trace metal–phosphate interactions in soils has been driven by the desire to utilize apatite ($Ca_5(PO_4)_3X$ (X = halide, hydroxyl) and other phosphate (PO_4^{3-})-containing materials as remediating agents for the *in situ* immobilization of contaminant metals and trace elements, especially lead (Pb) (e.g., Ma and Rao 1997; Ma et al. 1994, 1995). Phosphate-containing materials have also been evaluated as remediating agents for the *in situ* immobilization of radiological soil contaminants, including uranium (U) and other actinides (Arey et al. 1999; Bostick et al. 2000; Chattanathan et al. 2013; Fuller et al. 2002, 2003; Jensen et al. 1996; Majs 2011; Nash et al. 1997, 1998a; Rui et al. 2013; Seaman et al. 2001a, 2003; Wellman et al. 2008). The primary objective of this chapter is to provide an updated review of the dynamic processes controlling the interaction of U and phosphorus (P) in contaminated soils in an effort to improve our ability to effectively manage such sites in a manner that is environmentally protective.

The major sources of U contamination of soils include U mining and processing for nuclear power (Abdelouas 2006; Landa 2004), nuclear weapon production activities (Crowley and Ahearne 2002; Riley et al. 1992), the use of depleted uranium (DU) munitions (Bleise et al. 2003; Lloyd and Gadd 2011), and the production and application of phosphorus fertilizers (Kratz and Schnug 2006; Schipper et al. 2011; Schnug and Lottermoser 2013). Uranium is by far the most common radionuclide soil contaminant throughout the U.S. Department of Energy's (DOE) nuclear weapons complex (Riley et al. 1992). For example, approximately 43,500 kg of U and other contaminants were released into the Tims Branch stream system on the DOE's Savannah River Site (SRS) near Aiken, SC, much of which accumulated in an abandoned farm pond that served as a depositional basin (Evans et al. 1992; Pickett et al. 1987; Sowder et al. 2003). In another example, the use of depleted U munitions at the Aberdeen Proving Ground in Maryland resulted in the deposition of >70,000 kg of U to soils and sediments (Dong et al. 2006).

Uranium is primordial in origin and exists in nature as three main isotopes: ^{238}U (99.3% abundance; $t_{1/2} = 4.5 \times 10^9$ y), ^{235}U (0.71%; $t_{1/2} = 7.13 \times 10^8$ y), and ^{234}U (0.0057%; $t_{1/2} = 2.47 \times 10^5$ y). Enriching the naturally fissile isotope ^{235}U is of primary interest for nuclear weapons and power generation. In natural ore deposits, U is generally found in the +4 oxidation state as uraninite (UO_2), coffinite ($USiO_4$), or the mixed valence $U_3O_{8(s)}$ ore material commonly known as pitchblende (Elless and Lee 2002; Gavrilescu et al. 2009). Approximately 17,000 metric tons of U-bearing ore is processed to produce the 20 metric tons of nuclear fuel required for the annual operation of a typical light-water reactor (Abdelouas 2006), resulting in the generation of significant mine wastes that may serve as a potential source for continued soil contamination.

Uraninite (UO_2) represents greater than 90% of the composition of spent nuclear fuel (SNF) material. When nuclear fuel rods and uraninite ore deposits are exposed to oxidizing conditions, they tend to oxidize to form the +6 oxidation state, uranyl oxycation (UO_2^{2+}). Uranyl oxide hydrates such as metaschoepite ($UO_3{\cdot}2H_2O$) are generally the first alteration products to form, serving as an important sorptive phase for other actinides in SNF (Finn et al. 1996; Gorman-Lewis et al. 2008). Uranyl silicates, which are less soluble than uranyl oxide hydrates, are likely to form as the final weathering product under circumneutral to alkaline pH conditions in the presence of dissolved silica. Although subject to greater discussion in the "Uranyl–Phosphate Precipitates" section, the general environmental importance of uranyl phosphates is evident from their common presence in the vicinity of uranium ore deposits (Burns and Finch 1999; Gorman-Lewis et al. 2008).

As alpha emitters, the main radiation hazard associated with U isotopes is through ingestion and inhalation. Uranium is generally considered more chemotoxic than radiotoxic because of the long half-lives for the three main isotopes (Sheppard 2001). However, radium-226 (^{226}Ra; $t_{1/2}$ = 1600 years), a member of both the ^{238}U and the thorium-232 (^{232}Th) decay series, represents a major health concern as the parent nuclide for radon-222 (^{222}Rn). Radon-222 ($t_{1/2}$ = 3.8 days), the most common of the radon (Rn) isotopes, is a readily mobile inert gas that decays through alpha particle emission, which represents a significant internal exposure hazard through inhalation or ingestion. However, migration and eventual exposure to ^{222}Rn, with its 3.8-day half-life, is controlled to a large degree by the parent nuclides because of their much longer half-lives. Aside from the obvious risk associated with mining operations, ^{222}Rn levels may build up in confined indoor spaces built on soils with elevated levels of U, representing a major contributing factor in the occurrence of lung cancer.

Uranium Geochemistry

For a more comprehensive discussion of U geochemistry, readers are directed to the extensive review edited by Burns and Finch (1999). Uranium partitioning and availability in the soil environment depend on several factors: U concentration, pH, redox potential, moisture status, soil texture and mineralogy, dissolved carbon dioxide (CO_2) concentrations, organic matter content, the presence and concentration of complexing anions (such as carbonate [CO_3^{2-}]), and microbial activity. The interrelated nature of these variables is clearly obvious (i.e., microbial activity, redox status, dissolved CO_2 levels, organic matter content), complicating the discussion of critical mechanisms controlling U behavior within a complex natural system, such as the soil environment. For instance, the soil profile is subject to significant changes in moisture status, which obviously impacts biological activity and dissolved gas levels, which in turn can drastically alter solution chemistry, aqueous phase U(VI) speciation, and even mineral surface reactivity. Although much of the current discussion will focus on abiotic reactions, microbes obviously play an important role in controlling U behavior in soils and aquifer systems.

Redox status is an important factor governing U behavior in the soil environment. Under oxidizing conditions, U(VI) hydrolyzes to form the soluble uranyl oxycation, (UO_2^{2+}) while under reducing conditions, U(IV) displays a strong tendency to hydrolize, with the primary aqueous species being $U(OH)^0_{4(aq)}$ (Sandino and Bruno 1992). The half-cell reduction equation for U(VI) is

$$UO_2^{2+} + 2e^- \rightarrow UO_{2(\text{uraninite})} \tag{3.1}$$

with a standard potential of 0.411 V (E^o) (Brooks et al. 2003). Uranyl reduction generally follows the complete reduction of nitrate in soil systems, but may be concurrent with the reduction of sulfate $\left(SO_4^{2-}\right)$ and ferric iron (Fe(III)) (Nyman et al. 2006; Salome et al. 2013; Wu et al. 2006).

The current discussion will largely focus on U(VI), given the high solubility and mobility of the oxidized species. However, much of the environmental research concerning U contamination has focused on the biologically mediated reduction of U(VI) to U(IV) as a means of limiting migration and bioavailability. The direct reduction of U(VI) by aqueous Fe(II) is thermodynamically feasible, but the reaction kinetics are apparently quite slow to be of practical utility (Liger et al. 1999). However, U(VI) is subject to abiotic reduction by Fe(III) oxides and silicates in the presence of Fe(II) and Fe(II)-bearing minerals (Fox et al. 2013; Nico et al. 2009; Renock et al. 2013). In addition to direct enzymatic reduction, the development of an Fe reducing zone and the reductive cycling of Fe(II)/Fe(III) has important implications for U(VI) reduction and stabilization. However, the bioreduction of U(VI) and persistence of the reduced species may be limited under low pH conditions, environments that contain significant nitrate, or when subsequently exposed to oxygen (O_2).

Newsome et al. (2014) provide an excellent review of recent studies concerning the bioremediation of U and other radionuclides, including americium (Am), iodine (I), neptunium (Np), plutonium (Pu), strontium (Sr), and technetium (Tc). The major mechanisms by which microbes control the fate of U in soils and aquifers can be generally divided into the following categories: bioreduction, bioaccumulation, biomineralization, and biosorption (Newsome et al. 2014). Even so, microbial processes such as bioaccumulation and biosorption may have limited relevance to remediation, except when viewed as possible mechanistic components of the more important biologically controlled processes (i.e., bioreduction and biomineralization).

Numerous studies have demonstrated the ability of a diverse range of microorganisms, primarily metal and SO_4^{2-}-reducing bacteria, to reduce U(VI) to the less soluble U(IV) with the subsequent formation of UO_2 and other U(VI) products (Bernier-Latmani et al. 2010; Chang et al. 2014; N'Guessan et al. 2008; Rui et al. 2013; Stylo et al. 2013). For example, Bernier-Latmani et al. (2010) observed that efforts to stimulate microbial reduction of U(VI) can lead to the formation of a range of alternate products besides UO_2, including U(VI) bound to the microbial biomass, known as biosorption, as well as U(IV)-orthophosphates such as ningyoite [$CaU(PO_4)_2$], $U_2O(PO_4)_2$, and $U_2(PO_4)(P_3O_{10})$. The precipitation of U(VI) associated with microbially generated inorganic phosphates, a form of biomineralization, will be discussed in greater detail in the "Phosphate $\left(PO_4^{3-}\right)$ and the Bioremediation of Uranium" section. Nonetheless, U(IV) produced through microbial action may be subject to reoxidation with the stability of the reduced phase critical in determining the long-term effectiveness of such remediation efforts. Furthermore, studies suggest that biomass-associated U(IV) and poorly crystalline biogenic UO_2 may be more susceptible to oxidation than more crystalline U(IV) products (Cerrato et al. 2013; Stylo et al. 2013).

Aqueous Phase Uranyl Speciation

In addition to redox status, aqueous phase UO_2^{2+} speciation is largely dictated by pH and the presence of oxygen-containing ligands that strongly bond with UO_2^{2+}. Uranyl also tends to form strong complexes with dissolved humic and fulvic acids, which can increase U solubility. Thus, soil parameters such as OM content and pH play a strong role in controlling U partitioning (Vandenhove et al. 2007).

A summary of important aqueous phase UO_2^{2+} speciation reactions is provided in Table 3.1 along with stability constants at standard 298 K and $I = 0$ M. Most of the aqueous-phase speciation constants are derived from Grenthe et al. (1992) and Guillaumont et al. (2003), with specific exceptions provided in the table footnotes. With some revision, this database has been used extensively in previous U(VI) modeling studies (e.g., Davis et al. 2004; Hsi and Langmuir 1985; Turner et al. 1996; Waite et al. 1994). For the current discussion, this database was used to produce the current aqueous phase speciation diagrams under rather idealized conditions to illustrate the general complexity of UO_2^{2+} speciation and identify specific U(VI) species that control reactive behavior as a function of the pH of the soil environment (Figures 3.1 and 3.2).

Figure 3.1 illustrates the complex distribution of aqueous-phase UO_2^{2+} species as a function of pH in the presence and absence of CO_3^{2-}. In rather simple solutions where dissolved CO_2 is absent, uranyl (UO_2^{2+}) hydrolizes with increasing pH, forming monomers, dimers, and trimers (Figure 3.1a), that is, $UO_2OH^+, UO_2(OH)_2^0, UO_2(OH)_3^-$, and so forth (Davis et al. 2002). With increasing U(VI) concentrations, the hydrated uranyl oxide schoepite (i.e., $UO_3 \cdot xH_2O_{(S)}$) may form with a minimum solubility under circumneutral pH conditions. Uranyl also forms strong complexes with various inorganic oxyanions besides hydroxyl (OH^-), such as carbonate (CO_3^{2-}) and phosphate $\left(PO_4^{3-}\right)$, which greatly influence its mobility and behavior in the soil environment. In equilibrium with air (Figure 3.1b), the aqueous-phase speciation of UO_2^{2+} in the neutral-to-alkaline pH range is dominated by various uranyl–carbonate complexes that may differ in relative importance at different partial pressures of CO_2, a factor that is readily subject to change in soil environments. The oxidation of uraninite and formation of stable uranyl complexes can be generalized as

$$UO_{2(S)} + \frac{1}{2}O_2 + 3CO_3^{2-} + H_2O \rightarrow UO_2(CO_3)_3^{4-} + 2OH^- \qquad (3.2)$$

with the formation of the uranyl–tricarbonate complex dominating in the pH range of 8 to 10, while the hydroxyl carbonate complex $(UO_2)_2CO_3(OH)_{3(aq)}^-$ dominates aqueous uranyl speciation in the pH range of 6 to 8 (Davis et al. 2002).

TABLE 3.1

Formation Constants for U(VI) Solution Species Derived from Grenthe et al. (1992)

Reactions	Log *K*
Uranyl Hydroxyl Complexes	
$UO_2^{2+} + H_2O \leftrightarrow UO_2OH^- + H^+$	–5.20
$UO_2^{2+} + 2H_2O \leftrightarrow UO_2(OH)_{2(aq)} + 2H^+$	–11.5
$UO_2^{2+} + 3H_2O \leftrightarrow UO_2(OH)_3^- + 3H^+$	–20.0
$UO_2^{2+} + 4H_2O \leftrightarrow UO_2(OH)_4^{2-} + 4H^+$	–33.0
$2UO_2^{2+} + H_2O \leftrightarrow (UO_2)_2OH^{3+} + H^+$	–2.70
$2UO_2^{2+} + 2H_2O \leftrightarrow (UO_2)_2(OH)_2^{2+} + 2H^+$	–5.62
$3UO_2^{2+} + 4H_2O \leftrightarrow (UO_2)_3(OH)_4^{2+} + 4H^+$	–11.9
$3UO_2^{2+} + 5H_2O \leftrightarrow (UO_2)_3(OH)_5^+ + 5H^+$	–15.55
$3UO_2^{2+} + 7H_2O \leftrightarrow (UO_2)_3(OH)_7^- + 7H^+$	–31.0
$4UO_2^{2+} + 7H_2O \leftrightarrow (UO_2)_4(OH)_7^+ + 7H^+$	–21.9
Uranyl Carbonate Complexes	
$UO_2^{2+} + CO_3^{2-} \leftrightarrow UO_2CO_{3(aq)}$	9.67
$UO_2^{2+} + 2CO_3^{2-} \leftrightarrow UO_2(CO_3)_2^{2-}$	16.94
$UO_2^{2+} + 3CO_3^{2-} \leftrightarrow UO_2(CO_3)_3^{4-}$	21.60
$3UO_2^{2+} + 6CO_3^{2-} \leftrightarrow (UO_2)_3(CO_3)_6^{6-}$	54.0
$2UO_2^{2+} + CO_3^{2-} + 3H_2O \leftrightarrow (UO_2)_2CO_3(OH)_3^- + 3H^+$	–0.86
$3UO_2^{2+} + CO_3^{2-} + 3H_2O \leftrightarrow (UO_2)_3CO_3(OH)_3^+ + 3H^+$	0.66
Uranyl/Calcium Carbonate Complexes[a]	
$Ca^{2+} + UO_2^{2+} + 3CO_3^{2-} \leftrightarrow CaUO_2(CO_3)_3^{2-}$	25.4
$2Ca^{2+} + UO_2^{2+} + 3CO_3^{2-} \leftrightarrow Ca_2UO_2(CO_3)_{3(aq)}$	30.55
Uranyl Phosphate Complexes	
$UO_2^{2+} + PO_4^{3-} = UO_2PO_4^-$	13.23
$UO_2^{2+} + PO_4^{3-} + H^+ = UO_2HPO_{4(aq)}$	19.59
$UO_2^{2+} + PO_4^{3-} + 2H^+ = UO_2H_2PO_4^+$	22.82
$UO_2^{2+} + PO_4^{3-} + 3H^+ = UO_2H_3PO_4^{2+}$	22.46
$UO_2^{2+} + 2PO_4^{3-} + 4H^+ = UO_2(H_2PO_4)_{2(aq)}$	44.04
$UO_2^{2+} + 2PO_4^{3-} + 5H^+ = UO_2(H_2PO_4)(H_3PO_4)^+$	45.05
Uranyl Nitrate Complex[b]	
$UO_2^{2+} + NO_3^- = UO_2NO_3^+$	0.3
Phosphate Acid-Base	
$PO_4^{3-} + H^+ = HPO_4^{2-}$	12.35
$PO_4^{3-} + 2H^+ = H_2PO_4^-$	19.56
$PO_4^{3-} + 3H^+ = H_3PO_{4(aq)}$	21.70

TABLE 3.1 (CONTINUED)

Formation Constants for U(VI) Solution Species Derived from Grenthe et al. (1992)

Reactions	Log *K*
Auxiliary Reaction	
$H_2O = OH^- + H^+$	–13.997

[a] Bernhard, G. et al., *Radiochem Acta* 89:511–518, 2001.

[b] Reprinted from *Update on the Chemical Thermodynamics of Uranium, Neptunium, Plutonium, Americium and Technetium*, OECD Nuclear Energy Agency, Guillaumont, R., T. Fanghanel, J. Fuger, I. Grenthe, V. Neck, D.A. Palmer, and M.H. Rand, Copyright 2003, with permission from Elsevier.

In addition to dominating aqueous phase speciation, the uranyl–carbonate complexes greatly reduce adsorption to soil mineral phases. In fact, alkaline leaching solutions have been effectively used as selective extractants for carbonate-bearing U ore materials and calcareous soils contaminated with U (Duff et al. 1998; Mason et al. 1997).

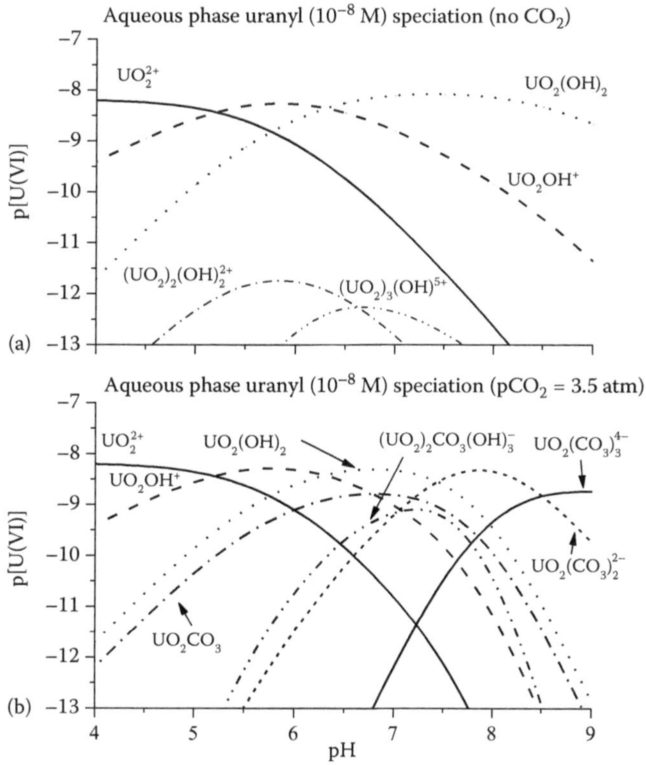

FIGURE 3.1

Aqueous phase uranyl speciation in the absence (a) and presence (b) of carbonate based on reaction constants provided in Table 3.1.

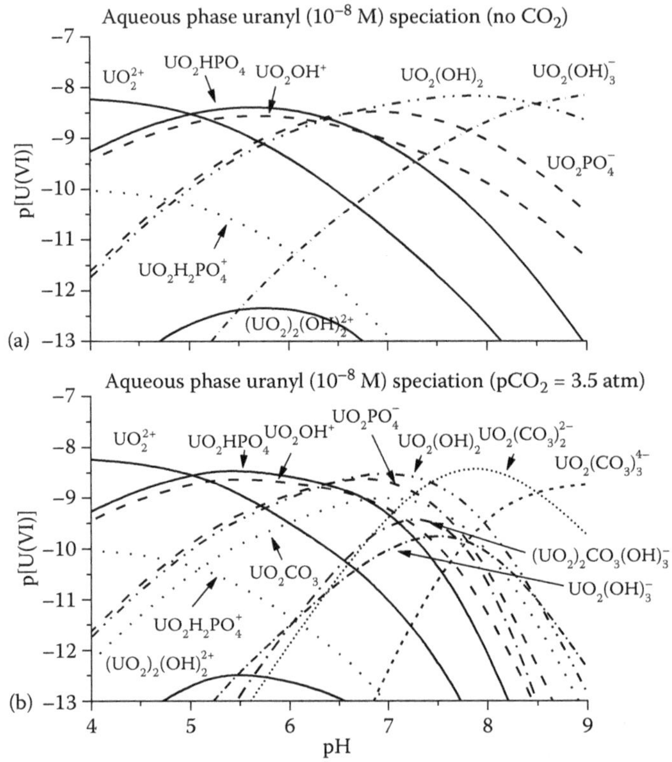

FIGURE 3.2
Impact of phosphate on the aqueous phase uranyl speciation in the absence (a) and presence (b) of carbonate based on reaction constants provided in Table 3.1.

For calcium (Ca^{2+})-dominated systems, the $CaUO_2(CO_3)_3^{2-}$ and $Ca_2UO_2(CO_3)^0_{3(aq)}$ species must be included for accurate speciation estimates as they represent the dominant dissolved species for U(VI) and may not be included in many commonly used thermodynamic databases (Bernhard et al. 2001; Davis et al. 2004). In fact, formation of the Ca–UO_2–CO_3 complex can even inhibit bacterial reduction of U(VI) (Brooks et al. 2003). However, for simplicity, Ca^{2+} was not included in the current aqueous-phase U(VI) speciation graphs (Figures 3.1 and 3.2).

Figure 3.2 illustrates UO_2^{2+} speciation as a function of pH at a fixed concentration of PO_4^{3-} in systems that are both closed (Figure 3.2a) and open (Figure 3.2b) to the atmosphere. Clearly, the presence of PO_4^{3-} further complicates aqueous-phase UO_2^{2+} speciation, especially in systems where carbonate is present. Uranyl forms several soluble uranyl–phosphate complexes that are subject to precipitation if UO_2^{2+} and PO_4^{3-} concentrations are too high and impact adsorption to various soil minerals under circumneutral-to-acidic pH conditions. At higher pH, the carbonate species still play a major role in controlling UO_2^{2+} aqueous-phase speciation and solid-phase partitioning under most environmental conditions.

Uranyl Adsorption Behavior

The complex aqueous chemistry of U(VI) can make predicting solid-phase partitioning in soil and aquifer systems quite difficult. Figure 3.3, adapted initially from Singh et al. (2012), is provided to highlight a number of important reactions controlling U(VI) partitioning in the presence of soil clays and Fe(III) oxides/oxyhydroxides serving as idealized sorbents. It is important to note that the surface complexation reactions depicted in Figure 3.3 are provided to illustrate the overall complexity of U(VI) behavior in soils and facilitate additional discussion. However, the exact reactions, especially the surface complexation reactions depicted in Figure 3.3, may not accurately represent the chemical speciation of the surface complex at the molecular scale, as this is a subject of continued investigation. Because various U(VI) partitioning reactions may dominate under specific chemical conditions, Figure 3.3 is not meant to suggest that all reactions are simultaneously active within a given system, even one as complex as most soils. However, the following discussion will highlight conditions where multiple reactions likely

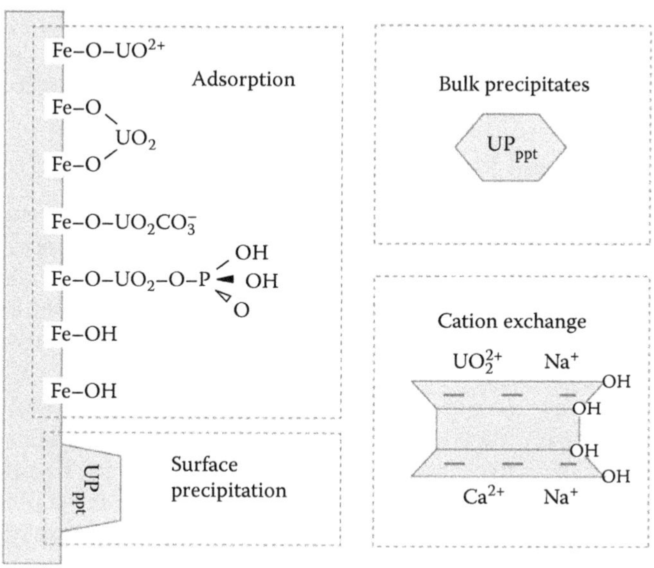

FIGURE 3.3
Idealized depiction of various U(VI) sorption processes in soils. (Revised from Singh, A. et al., *Environ Sci Technol* 46:6594–6603, 2012.)

occur or where distinguishing one mechanism from another is tenuous at best, especially when limited to bulk solution chemistry data.

Cation Exchange

Uranium(VI) may be bound to cation exchange sites associated with isomorphic substitution in soil clays, especially under low ionic strength conditions where competition for exchange sites is limited, and at low pH where the UO_2^{2+} species dominates (Boult 1998; Turner et al. 1996). Thus, U(VI) sorption by soil clays can be described by cation exchange models containing a fixed number of sorption sites determined by isomorphous substitution. For example, U(VI) may complete with alkaline and alkaline earth metals, in this case sodium (Na^+), for exchange sites:

$$X_2UO_2 + 2Na^+ \leftrightarrow 2XNa + UO_2^{2+} \quad (3.3)$$

with X representing fixed cation exchange sites.

Since Ra^{2+} is an environmentally important daughter product in the ^{238}U decay series, it is important to note some general trends with respect to its fate and transport in the soil environment. Radium, an alkali earth metal that is insensitive to redox conditions, behaves similarly to Ca^{2+} and barium (Ba^{2+}). Radium may coprecipitate with various sulfate and CO_3^{2-} minerals and readily sorbs to phyllosilicate clays and Fe and aluminum (Al) oxides in a fairly reversible manner that is not generally associated with strong surface complexation reactions. Because of its failure to form strong surface complexes with soil minerals, Ra^{2+} is considered to be more mobile in the soil environment than U and more subject to plant uptake (Zhang et al. 2002).

Surface Complexation Reactions

Despite the potential for cation exchange, UO_2^{2+} adsorption is generally dominated by surface complexation reactions at amphoteric surfaces (inorganic hydroxyl (OH) moiety) on soil minerals and ionizable functionalities on soil organic matter that are sensitive to the conditions in the solution environment (pH, ionic strength, counter ion valence, etc.), and therefore termed variable charge sites/minerals. The oxides and oxyhydroxides of Al, Fe, and other metals display this type of sorption mechanism. The silanol and aluminol functionalities on the edges of phyllosilicate clays also display this type of charge behavior, with the cation exchange capacity for kaolinite largely derived from amphoteric edge sites because of the limited isomorphic substitution within the clay structure. Thus, UO_2^{2+} sorption by soil clays can be described by a mechanistic model containing both fixed and variable charge sorption sites (Marques Fernandes et al. 2012; Turner et al. 1996).

As noted above, the tendency of U(VI) to form strong aqueous phase complexes with many common soil ligands further complicates soil partitioning behavior. In general, surface complexation modeling (SCM) has developed as a method for describing solute partitioning (i.e., adsorption) in response to changes in chemical conditions that alter both solution-phase speciation and mineral surface reactivity, making it seemingly ideal for describing U(VI) partitioning. In SCM, sorption behavior is described in terms of surface functional groups on reactive minerals, with formation constants, reaction stoichiometries, and in some cases sorption site densities determined by fitting batch sorption data, or exhaustive physicochemical characterization of the sorptive material. For example, the protonation and deprotonation of an amphoteric surface functional group on a mineral surface can be described as

$$\equiv M-OH_2^+ \leftrightarrow \equiv M-OH^0 + H^+ \quad K_{a1} \tag{3.4}$$

$$\equiv M - OH^0 \leftrightarrow \equiv M - O^- + H^+ \quad K_{a2} \tag{3.5}$$

where M represents the primary metal (Fe, Al, etc.), OH is a surface hydroxyl filling the coordination sphere of the metal, and K_{a1} and K_{a2} are the apparent acidity constants, with the mineral surface charge becoming more positive as the pH decreases and more negative as the pH increases. Equilibrium constants for the protonation/deprotonation and other surface complexation reactions can be written as

$$K = \frac{[MO^-]\gamma_{MO}(H^+)}{[MOH]\gamma_{MOH}} \tag{3.6}$$

where the brackets refer to concentration and parentheses refer to activity, and γ is the activity coefficient for the respective species. Equilibrium constants derived for surface complexation reactions differ from aqueous reactions in that the surface activity terms represent corrections for the electrostatic energy at the mineral solution interface; that is, the electrical double layer (*edl*):

$$K = K_C \times edl \tag{3.7}$$

where K_C is a conditional equilibrium constant written in terms of the activities for aqueous species and concentrations for surface sites, and the *edl* term is a model-dependent means of accounting for the surface potential or potential within the adsorptive plane (Zachara and Westall 1998). One limitation in applying SCM to natural systems is the difficulty of characterizing the electrical double-layer properties of heterogeneous soils and aquifer sediments where sorbed ions, organic matter, and oxide coatings alter mineral surface charge properties in a manner that is difficult to predict and evaluate experimentally. Several approaches to SCM have been developed (e.g., constant capacitance model, double layer model, triple layer model), generally differing in the manner in which they describe the mineral/water interface and account for columbic interactions that impact the mass action equations (Goldberg 2005).

As a consequence of the amphoteric surface behavior, both anions and cations may be electrostatically sorbed to the charged oxide surface represented by Equations 3.4 and 3.5, with the sorbed ion remaining hydrated and termed *outer sphere* complexes. Using uranyl (UO_2^{2+}) as an example, hydrous oxides can also form stronger inner sphere surface complexes associated with at least partial dehydration of the sorbed species and the displacement of protons on the mineral surface:

$$\equiv \mathrm{Fe-OH} + \mathrm{UO_2^{2+}} \leftrightarrow \equiv \mathrm{Fe-OUO_2^{+}} + \mathrm{H^{+}} \quad K \tag{3.8}$$

which can be described in a manner similar to aqueous-phase complexes through the use of a formation constant, *K*. The sorption behavior of anionic, cationic, and even uncharged aqueous phase species can be described using this convention.

Surface complexation modeling has been widely used to describe U(VI) sorption to various ideal soil mineral analogs and even natural soil/sediment materials (e.g., Cheng et al. 2004; Davis and Randall 2003; Davis et al. 2002; Hsi and Langmuir 1985; Lenhart and Honeyman 1999; Marques Fernandes et al. 2012; Payne et al. 1996, 2004; Singh et al. 2010; Waite et al. 1994, 2000; Wazne et al. 2003; Zeng et al. 2009). While numerous studies have demonstrated the ability to describe U(VI) sorption to ideal mineral phases under a broad range of chemical conditions, the application of SCMs to complex soil systems is somewhat less common (Barnett et al. 2002; Davis et al. 2004; Dong and Wan 2014; Dong et al. 2012; Logue et al. 2004; Stoliker et al. 2011), in part because of the difficulty in identifying and describing the active sorbent phase where aggregation structure, specifically sorbed ions, organic matter, and oxide coatings alter mineral surface properties when compared to ideal analogs (Davis et al. 1998; Wang et al. 2011; Zachara et al. 1992; Zhang et al. 2011). Furthermore, the relative importance of a specific mineral phase to control solid-phase UO_2^{2+} partitioning can vary dramatically with fluctuating chemical conditions in natural systems (Dong et al. 2012). Although most surface complexation studies assume equilibrium, recent studies indicate that UO_2^{2+} sorption reactions in natural systems have a significant kinetic component with rates being limited by intra- and intergranular mass transfers that control the supply of reactants and removal of reaction products from complexation sites (Liu et al. 2009, 2013).

In the absence of spectroscopic evidence, surface complexation models are often based on the assumption that the dominant solution species is the one being adsorbed, which emphasizes the importance of understanding UO_2^{2+} aqueous-phase speciation. Debate continues on the structure of the dominant sorbed uranyl species formed in even idealized oxide-solution systems, let alone complex soil and groundwater systems. Therefore, caution must be used in selecting an appropriate surface complex for describing sorption data if the modeling effort is meant to have mechanistic validity (Jove-Colon et al. 2006). Goldberg et al. (2007) provide an excellent review of the experimental techniques used in validating a mechanistic understanding of contaminant sorption processes.

Uranyl Adsorption in the Presence of CO_3^{2-}

Before discussing the impact of PO_4^{3-} on uranyl sorption, it is important to illustrate how other common aqueous-phase complexation reactions impact partitioning. Figure 3.4a, taken from Waite et al. (1994), provides an example of UO_2^{2+} sorption to ferrihydrite ($Fe(OH)_{3(s)}$) as a function of pH under atmospheric CO_2 levels. Based on extended x-ray absorption fine structure

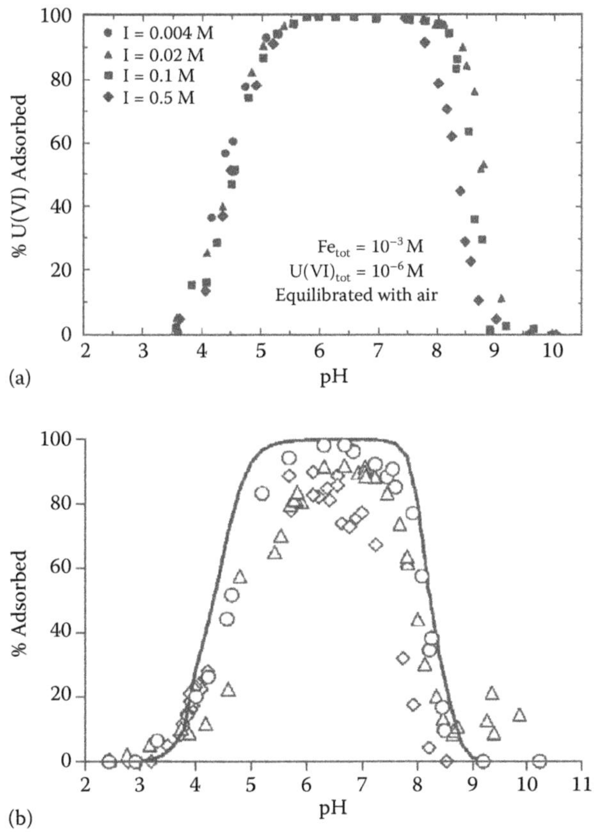

FIGURE 3.4
Impact of carbonate on uranyl sorption to ferrihydrite (a) and sediments from three DOE facilities (b). (a) Uranyl (0.24 mg U L^{-1}) sorption to ferrihydrite (1×10^{-3} M Fe) as a function of ionic strength in equilibrium with atmospheric CO_2. (From Waite, T.D. et al., *Geochim Cosmochim Acta* 58:5465–5478, 1994.) (b) Uranyl (1 mg U L^{-1}) sorption to sediments from Oak Ridge (circles), Hanford (diamonds), and Savannah River (triangles) at an ionic strength of 0.1 M in equilibrium with atmospheric CO_2. (From Barnett, M.O. et al., *Environ Sci Technol* 36:937–942, 2002.)

(EXAFS) analysis, Waite et al. (1994) suggested that the surface complexation chemistry for UO_2^{2+} is actually less complex than the aqueous-phase speciation, with adsorption to ferrihydrite under acidic conditions largely due to inner sphere mononuclear bidentate complexes (i.e., $\equiv FeO_2UO_2$) described by the following reaction:

$$\equiv Fe(OH)_2 + UO_2^{2+} \leftrightarrow \equiv FeO_2UO_2^0 + 2H^+ \tag{3.9}$$

with $\equiv Fe(OH)_2$ representing two surface hydroxyl functional groups that form an edge-sharing bidentate complex with uranyl. This reaction is responsible for the sharp increase in U(VI) sorption with increasing pH from 3 to 5 (i.e., the adsorption edge for variable charge minerals) and then a plateau with respect to U(VI) sorption from pH ≈ 5 to 8. However, UO_2^{2+} sorption within the plateau pH range may decrease with increasing carbonate concentration (and other strong complexing agents) above atmospheric levels due to the low surface affinity of the predominant uranyl–carbonate complexes and the competitive adsorption of carbonate on the mineral surface (Wazne et al. 2003). Based on zeta potential measurements, Wazne et al. (2003) proposed that UO_2^{2+} sorbed to ferrihydrite as a monodentate complex in the absence of carbonate:

$$\equiv FeOH + UO_2^{2+} \leftrightarrow \equiv FeOUO_2^+ + H^+ \tag{3.10}$$

With increasing pH in Figure 3.4a, Waite et al. (1994) attributed uranyl adsorption under neutral to alkaline pH conditions to a 1:1 uranyl carbonate ternary complex according to the following stoichiometry:

$$\equiv Fe(OH)_2 + UO_2^{2+} + CO_3^{2-} \leftrightarrow \equiv FeO_2UO_2CO_3^{2-} + 2H^+ \tag{3.11}$$

with uranyl sorption decreasing dramatically over the pH range of 8 to 9 for systems open to the environment. In the absence of carbonate, UO_2^{2+} desorption does not occur until a much higher pH (i.e., pH ≈ 10 to 11) associated with formation of the negatively charged uranyl–hydroxide complexes, a condition generally more relevant to certain caustic waste disposal scenarios than typical soil conditions.

Laboratory studies have confirmed that similar surface complexation reactions control the disposition of U(VI) in natural soil systems. For example, Barnett et al. (2002) used the model developed by Waite et al. (1994) for ferrihydite to describe UO_2^{2+} sorption to subsurface materials collected from three DOE facilities: Oak Ridge, TN, Hanford, WA, and Savannah River, SC (see Figure 3.4b). Remarkably, the Waite et al. (1994) model was able to roughly describe U(VI) sorption behavior as a function of pH for all three subsurface materials. The similar degree of sorption observed for the three materials despite their different clay mineralogy was attributed to their similar Fe oxide contents, presumably the predominate sorptive phase (Barnett et al. 2000; Bostick et al. 2002). In a related spectroscopic study, Bostick et al. (2002) observed that the formation of ternary uranyl complexes dominated sorption behavior under all soil conditions tested. Of specific note, uranyl–PO_4 complexes were found to control partitioning in subsurface materials from more acidic environments, while uranyl–carbonate complexes were associated with sorption under more neutral conditions. Logue et al. (2004) also found that the SCM model of Waite et al. (1994) was able to reasonably describe U(VI) sorption to six subsurface sediments from Oyster, VA, when using citrate–dithionite–bicarbonate (CDB) extractable Fe to estimate sorption sites (Logue et al. 2004).

In a recent spectroscopic study evaluating U(VI) sorption to goethite (i.e., α-FeOOH), Sherman et al. (2008) proposed that the bidentate $\equiv Fe(OH)_2UO_2(H_2O)_n$ complex was largely responsible for sorption in CO_2-free systems, while enhanced adsorption at low pH when open to the atmosphere can be attributed to a $\equiv FeOCO_3UO_2$ ternary complex. As noted previously, desorption occurs with increasing pH due to the formation of strong U(VI)–CO_3 and U(VI)–OH aqueous-phase complexes, which is preceded by the formation of the $\equiv Fe(OH)_2UO_2CO_3$ ternary surface complex (Sherman et al. 2008).

As demonstrated by the previous examples, Fe(III) oxyhydroxides are often the dominant U(VI) sorptive mineral in soil systems. Uranyl may coprecipitate with Fe(III) oxyhydroxides, a portion which is subject to release during the transformation to more crystalline Fe(III) oxides (Bruno et al. 1995; Ohnuki et al. 1997). Additionally, Fe(III) oxyhydroxides may serve as a nucleation site for the eventual formation of uranyl minerals such as schoepite ($(UO_2)_8O_2(OH)_{12}\cdot 12H_2O_{(S)}$) (Bruno et al. 1995), illustrating the important role Fe oxides and other soil minerals can play in catalyzing the precipitation of U(VI)-bearing solids, a phenomenon that will be discussed in greater detail with respect to uranyl–phosphate interactions. Although the presence of Fe(III) oxides is critical in facilitating the abiotic reduction of U(VI), UO_2^{2+} sorption to Fe(III) oxides may be strong enough to actually hinder microbial reduction (Jeon et al. 2005).

Uranyl Sorption Behavior in the Presence of Phosphate (PO_4^{3-})

While there are numerous studies evaluating the speciation of sorbed uranyl associated with various soil oxides and oxyhydroxides, spectroscopic studies evaluating the molecular scale identity of uranyl–phosphate sorption species are far more limited (Del Nero et al. 2011; Galindo et al. 2010; Singh et al. 2010, 2012). Studies by Cheng et al. (2004, 2007), Singh et al. (2010, 2012), and Payne et al. (1996) evaluating the impact of PO_4^{3-} on UO_2^{2+} sorption by Fe(III) oxyhydroxides clearly demonstrate the complexity of uranyl partitioning behavior even in rather well-controlled laboratory batch systems. Multiple mechanisms have been invoked to account for the decreased aqueous phase $U(VI)_{(aq)}$ and PO_4^{3-} in the presence of Fe(III) oxides, including adsorption, precipitation of a $U(VI)–PO_{4(s)}$ phase by homogeneous nucleation (i.e., in bulk solution), and precipitation of a $U(VI)–PO_{4(s)}$ phase by heterogeneous nucleation (i.e., at a mineral surface), all of which are schematically depicted in Figure 3.3.

Precipitation of a secondary phase and even precipitation induced by the presence of a sorbing mineral surface that facilitates heterogeneous nucleation may confound *apparent* macroscopic sorption behavior even in idealized laboratory systems, but more so in complex natural systems with multiple competing sorptive surfaces. Thus, the relative sensitivity of a given spectroscopic technique must be considered as the degree of reactive surface loading for both UO_2^{2+} and PO_4^{3-} is an important factor in transitioning from ternary surface complexes to surface precipitation, with surface speciation presumably differing as a function of surface loading levels of either species (Del Nero et al. 2011; Galindo et al. 2010). Adsorption and precipitation not only confound the interpretation of geochemical data, they can also complicate the productive discussion of the subject.

Enhanced uptake of U(VI) and other metals by Fe(III) oxyhydroxides in the presence of PO_4^{3-} at lower surface coverages has been widely observed and attributed to changes in the surface charge of the Fe(III) oxide and/or sorption of a ternary metal–phosphate (i.e., $M - PO_4$) complex:

$$\equiv FeOH + H^+ + PO_4^{3-} \leftrightarrow \equiv FePO_4^{2-} + H_2O \tag{3.12}$$

that electrostatically enhance sorption and the formation of specific ternary complexes with the sorbed metal species:

$$\equiv FeOH + PO_4^{3-} + UO_2^{2+} + H^+ \leftrightarrow \equiv FePO_4UO_2^0 + H_2O \tag{3.13}$$

although the exact structure is still in question (Bostick et al. 2002; Payne et al. 1996; Singh et al. 2012).

Cheng et al. (2004) evaluated the impact of PO_4^{3-} on U(VI) sorption to a goethite-coated sand at low U(VI) concentrations that were assumed to be undersaturated with respect to uranyl–phosphate solids. In batch systems closed to the atmosphere, Cheng et al. (2004) found that PO_4^{3-} increased U sorption at low pH, which they described as a ternary complex (i.e., $\equiv FePO_4UO_2$) while a decrease in U(VI) sorption at high pH was attributed to the formation of soluble $U(VI)–PO_4$ complexes, similar to the way both carbonate and hydroxyl complexes reduce uranyl sorption with increasing pH (Figure 3.5). In a related column study using goethite-coated sand, the presence of PO_4^{3-} as a cosolute with UO_2^{2+} or presorbed to the column matrix greatly increased U(VI) adsorption, further supporting the formation of a $\equiv FePO_4UO_2$ ternary complex following the inner sphere PO_4^{3-} soprtion (Cheng et al. 2007). In a system open to the atmosphere, Payne et

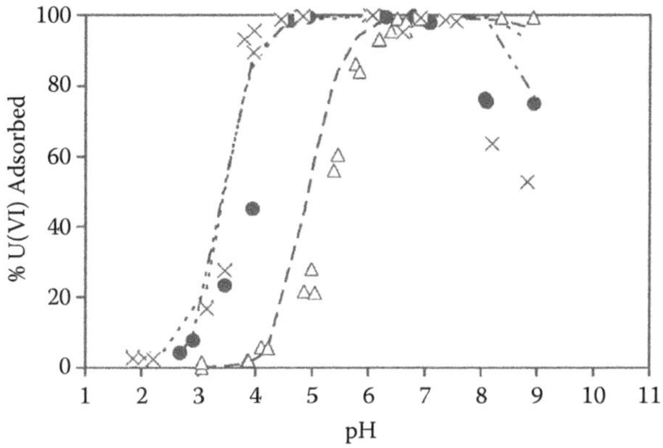

FIGURE 3.5
Uranyl (5 × 10⁻⁶ M) sorption to goethite-coated sand (total Fe 3.15 × 10⁻⁴ M) in a system closed to the atmosphere in the absence of PO_4^{3-} and presence of PO_4^{3-} at two concentrations, 5 × 10⁻⁵ and 2 × 10⁻⁴ M. Lines represent model fits. (From Cheng, T. et al., *Environ Sci Technol* 38:6059–6065, 2004.)

al. (1996) reported similar behavior for U(VI) sorption to ferrihydrite under acidic conditions, with PO_4^{3-} shifting the sorption edge to lower pH values, which they attributed to a ternary UO_2–PO_4 complex. However, there was very little effect observed with PO_4^{3-} at high pH in the presence of carbonate (Payne et al. 1996). Payne et al. also found that the presence of humic acid also increased U(VI) sorption under acidic pH conditions.

In a subsequent batch study focusing on the impact of PO_4^{3-} on UO_2^{2+} sorption to goethite, Singh et al. (2010) revised the Cheng et al. (2004) model to address a wider range of UO_2^{2+} and PO_4^{3-} concentrations by accounting for precipitation in addition to sorption through the use of batch treatments in the presence and absence of the sorptive phase (i.e., goethite). Adsorption to goethite was a major mechanism controlling the aqueous phase concentrations of both U and PO_4^{3-}, with uranyl sorption in the absence of PO_4^{3-} attributed to the $\equiv FeO_2UO_2^0$ surface complex. Phosphate had little impact on U sorption to goethite at low PO_4^{3-} concentrations (15 μM), with results suggesting that UO_2^{2+} and PO_4^{3-} sorb independently at low surface coverages. With increasing PO_4^{3-} concentrations, enhanced UO_2^{2+} sorption was attributed to a ternary surface complex; that is, $\equiv FeO_2PO_4UO_2$ possibly following initial PO_4^{3-} sorption by goethite independent of UO_2^{2+} (Cheng et al. 2006). Other studies have also suggested that preadsorption of PO_4^{3-} and even arsenate may enhance subsequent uranyl sorption (Galindo et al. 2010; Payne et al. 1996; Tang and Reeder 2009; Tang et al. 2009).

In a related study, Singh et al. (2012) evaluated the molecular-scale mechanisms of uranyl partitioning in the presence of goethite and PO_4^{3-} across a pH range of 4 to 7. Enhanced UO_2^{2+} adsorption to goethite in the presence of PO_4^{3-} was the dominant mechanism controlling partitioning at pH 4 under low to moderate U levels. However, greater precipitation in the form of the meta-autunite group, consisting of interlayer cations between sheets of coordinated phosphate tetrahedral and uranyl square bipyramids, was observed with both increasing U concentrations and pH. In the absence of PO_4^{3-}, uranyl formed bidentate edge and corner-sharing complexes. Additionally, spectroscopic results suggested the presence of a ternary UO_2–PO_4 bridging complex on the goethite at low pH and U concentrations, but could not reject the possibility of a P-bridging complex as well. In the absence of goethite, chernikovite ($UO_2HPO_4{\cdot}4H_2O_{(s)}$) precipitation was observed at all pH values (pH ≈ 4–7) over the extended equilibration period (i.e., 1 year). Less U–PO_4 precipitation in the presence of goethite was attributed to sorption by goethite limiting available $PO_{4(aq)}^{3-}$.

Surface-induced uranyl precipitation is not limited to Fe(III) oxides. Studies by Del Nero et al. (2011) and Galindo et al. (2010) looking at U(VI) sorption to alumina under acidic conditions found that PO_4^{3-} enhances surface complexation of U(VI) and the formation of U(VI)–PO_4 surface precipitates, even at trace U(VI) levels. In fact, the sorption of both U(VI) and PO_4^{3-} to secondary Fe oxide phases within weathering zones associated with U ore may facilitate the subsequent formation of U-phosphates by increasing the concentration of both U(VI) and PO_4^{3-} within a relatively localized environment.

Uranyl–Phosphate Precipitates

As noted earlier, secondary uranyl phosphates often develop in the vicinity of uranium ore deposits (Burns and Finch 1999). Uranyl phosphates such as uranyl orthophosphate are generally less soluble than uranyl silicates and oxides at pH values of less than 8 (Sandino and Bruno 1992), with metaschoepite ($UO_3{\cdot}2H_2O$) rapidly transforming to autunite hydrates (e.g., $Ca[(UO_2)(PO_4)]_2{\cdot}xH_2O$) in the presence of phosphate (Sowder et al. 1996, 1999). Over the pH range of most soil systems (pH ≈ 5–8), uranyl phosphates may control U(VI) solubility under conditions where the total concentration ratio

$$\frac{\left(PO_4^{3-}\right)_T}{\left(CO_3^{2-}\right)_T} > 10^{-1} \tag{3.14}$$

(Buck et al. 1996; Sandino and Bruno 1992). As pH increases above 7, the solubility of various apatites may inhibit the formation of uranyl phosphates because of the limited $PO_{4(aq)}^{3-}$ concentrations.

An abbreviated list of important uranyl phosphate minerals is provided in Table 3.2. Despite their obvious environmental significance, the lack of consistent thermodynamic data for many uranyl precipitates can limit the accurate prediction of mineral solubility and stability under varying environmental

TABLE 3.2

Important Uranium Minerals and Their Solubility Products at 298 K and I = 0 M

Uranium and Other Solids	Name	Log K_{sp}
$UO_3 \cdot 2H_2O_{(s)} + 2H^+ = UO_2^{2+} + 3H_2O$	Meta-schoepite	5.6
$UO_2HPO_4 \cdot 4H_2O_{(s)} = UO_2^{2+} + PO_4^{3-} + H^+ + 4H_2O$	Chernikovite	–24.20
$UO_2HPO_4 \cdot 3H_2O_{(s)} = UO_2^{2+} + PO_4^{3-} + H^+ + 3H_2O$	Uranium hydrogen phosphate	–25.52
$NaUO_2PO_4 \cdot xH_2O_{(s)} = UO_2^{2+} + Na^+ + PO_4^{3-} + xH_2O$	Sodium meta-autunite	–23.64
$UO_2(H_2PO_4)_2 \cdot 3H_2O_{(s)} = UO_2^{2+} + 2PO_4^{3-} + 4H^+ + 3H_2O$	Uranyl phosphate hydrate	–45.10
$(UO_2)_3(PO_4)_2 \cdot 4H_2O_{(s)} = 3UO_2^{2+} + 2PO_4^{3-} + 4H_2O$	Uranyl orthophosphate	–49.36

Source: Reprinted from *Update on the Chemical Thermodynamics of Uranium, Neptunium, Plutonium, Americium and Technetium,* OECD Nuclear Energy Agency, Guillaumont, R., T. Fanghanel, J. Fuger, I. Grenthe, V. Neck, D.A. Palmer, and M.H. Rand, Copyright 2003, with permission from Elsevier.

conditions (Gorman-Lewis et al. 2008). The best available thermodynamic databases related to aqueous phase uranyl speciation and mineralogy are derived from Grenthe et al. (1992), Guillaumont et al. (2003), and Sandino and Bruno (1992). A review by Gorman-Lewis et al. (2009) provides an excellent summary of available solubility data for a range of uranyl minerals, a discussion of the experimental conditions under which the requisite thermodynamic data were generated, and the inherent limitations to their application to a wide range of environmental conditions.

Uranyl phosphates represent a diverse group of minerals, with the autunite group being the most common. Autunite formation is often invoked as the mechanism controlling U(VI) solubility and mobility in soil systems (Gorman-Lewis et al. 2008). Minerals from the autunite group (i.e., hydrated 1:1 (U:P) calcium uranyl phosphate [$Ca(UO_2)_2(PO_4)_2 \cdot 5\text{-}6H_2O$] and dehydrated meta-autunite groups or the hydrogen uranyl phosphate, chernikovite [$H(UO_2)(PO_4) \cdot 4H_2O$]) are generally thought to form in soils and sediment systems with elevated uranium and phosphate (Burns and Finch 1999).

Copper (Cu) and magnesium (Mg) uranyl phosphates from the meta-autunite group have been identified in Fe-oxide rich soils overlying the U ore deposit at Koongarra, Australia (Murakami et al. 1997). In another example, uranyl–phosphates of the meta-autunite group, apparently stable for hundreds of thousands of years, formed within the soil and saprolite profile overlying the Coles Hill U deposit in Virginia, United States. Uranyl concentrations within the saturated saprolite are actually enriched when compared to that of the ore-grade deposit, while concentrations within the unsaturated vadose zone are much lower, despite U(VI) being largely associated with Al–phosphates and Fe oxyhydoxides. Weathering within the unsaturated zone, where the meta-autunite minerals are unstable, apparently resulted in the downward transport of U(VI), where it is concentrated and reprecipitated as meta-autunite at the higher PO_4^{3-} to CO_3^{2-} ratios of the saturated zone. In the unsaturated zone, U is associated with an Al–phosphate mineral in the crandallite group and Fe-oxyhydroxide coatings that are high in both U and PO_4^{3-}

(Jerden Jr. and Sinha 2006; Jerden Jr. et al. 2003). While the autunite-rich zone present at the Koongarra, Australia, site appears to be unstable, resulting in a 40 m secondary plume downgradient from the weathered profile overlying the site (Murakami et al. 1997), there is little evidence for lateral dispersion at the Coles Hill site (Jerden Jr. and Sinha 2006; Jerden Jr. et al. 2003).

The general persistence of U phosphates has led to the evaluation of apatite as a permeable reactive barrier for immobilizing U(VI) (Fuller et al. 2003). As noted for sorption reactions involving Fe(III) oxides, UO_2^{2+} partitioning in the presence of apatite is quite complex, involving adsorption as well as precipitation of U(VI)–$PO_{4(s)}$ by homogeneous nucleation (i.e., in bulk solution) and heterogeneous nucleation (i.e., at a mineral surface), similar to the depiction in Figure 3.3. In batch experiments, Fuller et al. (2002) evaluated UO_2^{2+} sorption to hydroxyapatite (HA) with the formation of chernikovite observed when sorbed U concentrations were >7000 μg g^{-1}. At lower sorption levels, uranyl formed inner sphere complexes with the HA attributed to a bidentate linkage of U(VI) with dissolved PO_4^{3-}; that is, $\equiv CaO_2UO_2PO_4$ and U(VI) bonding directly to PO_4^{3-} surface sites. No chernikovite was detected, even in treatments that were initially supersaturated with respect to the U–PO_4 mineral, and U(VI) substitution within the HA was ruled out. In addition, ternary uranyl–carbonate complexes on the HA surface were observed in the presence of carbonate. In a companion study, Fuller et al. (2003) used spectroscopic methods to evaluate UO_2^{2+} sorption to various apatite (i.e., bone char and bone meal apatite) materials in the presence and absence of dissolved CO_3^{2-}. Uranyl sorption via surface complexation was observed on the surface of the apatite for U surface loadings up to 5500 μg g^{-1}. At surface loadings ≥ 5500 μg g^{-1}, formation of the uranyl phosphate chernikovite was observed in the absence of carbonate. The presence of CO_3^{2-} suppressed the formation of chernikovite for bone meal, bone charcoal, and even HA. Similar results were observed for bone char apatite that had been used as a permeable barrier, with the presence of dissolved CO_3^{2-} inhibiting the formation of uranyl–phosphates, with much of the sorbed U(VI) remaining labile in the presence of dissolved CO_3^{2-}, readily desorbing when exposed to U(VI)-free groundwater (Fuller et al. 2003).

Using electron-based analytical techniques, Buck et al. (1996) identified a number of U-bearing solid phases in contaminated soil from the Fernald site in Ohio, a uranium ore processing facility used for nuclear weapons production. The U-bearing phases identified at the site include meta-autunite ($Ca(UO_2)_2(PO_4)_2 \cdot xH_2O$) and uraninite, plus uranium metaphosphate, $U(PO_3)_4$, which may have formed at high temperatures within incinerators (>400°C) that were part of the ore refinement process. The U metaphosphate appeared to be resistant to carbonate-based extraction tests. However, the presence of U phosphates at the Fernald site was not surprising, as tributyl phosphate was used in ore processing (Buck et al. 1996).

Ohnuki et al. (2004) studied UO_2^{2+} sorption and precipitation in the presence of single crystals of fluorapatite under acidic conditions (pH ≈ 2.3 to 4.0).

Surprisingly, UO_2^{2+} sorption increased with decreasing pH, which was attributed to the formation of autunite within the leached layer of apatite (i.e., surface mineralization), where conditions of local saturation exist despite the bulk solution being undersaturated with respect to autunite (Ohnuki et al. 2004), similar to attributing the development of uranyl–phosphates near U ore deposits to the effective microscale concentration of reactant species through sorption to mineral surfaces.

In a series of laboratory batch experiments using U-contaminated soil from the DOE's SRS facility, Arey et al. (1999), Seaman et al. (2001a,b, 2003, 2005), and Majs (2011) found that apatite addition was generally effective in reducing the solubility of UO_2^{2+} and a range of other contaminant metals (i.e., Ni and Pb). The treated soils proved to be undersaturated with respect to various autunite minerals, and enhanced UO_2^{2+} sorption was attributed to precipitation with secondary Al–phosphates that were identified by electron microscopy (Arey et al. 1999; Majs 2011; Seaman et al. 2001a,b, 2003, 2005). The observed persistence of Al–phosphates from the crandalite group in highly weathered soils and sediments from the Atlantic Coastal Plain (Figure 3.6) (Seaman 2000; Seaman et al. 1997) supports the long-term efficacy of PO_4^{3-} as a chemical agent for limiting U(VI) mobility (Seaman 2000; Seaman et al. 1997). Characterization efforts also confirmed the presence of Ba and Sr (also trace U) in association with crandallite, suggesting that these minerals represent a solid solution from the ideal structure of crandallite ($CaAl_3(PO_4)_2(OH)_5{\cdot}H_2O$) to gorceixite ($BaAl_3(PO_4)_2(OH)_5{\cdot}H_2O$). Furthermore, selective extraction results suggest that the mineral persists when exposed to both fluctuations in redox conditions and metal complexing agents, such as citrate. As discussed previously, similar Al–phosphates have been identified as the persistent resident phase for U in soils overlying the unmined Coles Hill uranium deposit in Virginia (Jerden Jr. and Sinha 2006).

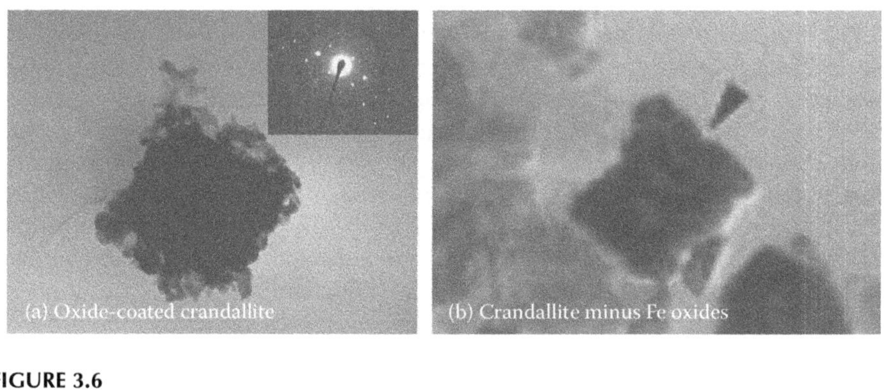

FIGURE 3.6
(a) TEM image of crandallite (inset electron diffraction) grains commonly observed in the vadose zone and water-table aquifer sediments (Irwinton Sand) on the SRS. Note the fine-textured, rod-shaped Fe oxides (goethite) that appear to coat/armor the crandallite in (a). (b) Crandallite grains in same soil subjected to CDB reductive extraction to remove Fe oxides.

Phosphate (PO_4^{3-}) and the Bioremediation of Uranium

Much of the recent work concerning the bioremediation of U has been associated with the DOE's Field Research Center (FRC) on the DOE Oak Ridge facility, a site that was contaminated with U and high levels of nitrate from more than 30 years of ^{235}U enrichment activity (Brooks 2001). The results have clearly demonstrated the importance of microbes in controlling aqueous- and solid-phase U speciation in the soil environment through either direct metabolic processes or other indirect microbial processes, such as controlling soil redox status and dissolved CO_2 levels, and the production of other complexing ligands. The general microbe–uranium interactions controlling U mobility in the subsurface environment include bioreduction, bioaccumulation, biomineralization, and biosorption. However, research evaluating the bioremediation of U has generally focused on the addition of electron donors to stimulate microbial activity and the reduction of U(VI), primarily by dissimilatory metal and SO_4^{2-}-reducing bacteria, to the less soluble U(IV) species as a means of limiting U mobility and bioavailability (e.g., Lovley and Phillips 1992; Lovley et al. 1991). The recent article by Newsome et al. (2014) provides an excellent review of the available scientific literature for bioreduction and other important microbe–uranium interaction mechanisms.

Although abiotic U(VI) reduction is possible in the presence of mineral surfaces, enzymatic reduction is generally recognized as the dominant mechanism under ambient environmental conditions. While uraninite is thought to be the primary product resulting from bioreduction, recent studies suggest that alternate products may be quite important depending on the specific chemical conditions of the system (Bernier-Latmani et al. 2010; Campbell et al. 2011; Chang et al. 2014; Rui et al. 2013). Based on spectroscopic evidence, Alessi et al. (2014) observed that the noncrystalline U(VI) products resulting from microbial reduction were generally associated with various phosphate groups, including organic phosphate groups associated with the microbial biomass and the precipitation of inorganic phosphate polymers. As always, the success of efforts to immobilize U in the reduced state is limited by the long-term stability of the final reduced product when subjected to changing environmental conditions, especially in systems where high nitrate levels hasten U(IV) oxidation.

The limited success of bioreduction has compelled researchers to evaluate other possible means of addressing U soil contamination through bioremediation. Two mechanisms of microbe–uranium interaction have only limited value with respect to environmental remediation. Microbial bioaccumulation of U is generally associated with the limited intercellular formation of uranyl–phosphates, while U uptake at the surface of dead or living cells due to various reactive chemical functionalities (i.e., carboxyl, amine, hydroxyl, and phosphate), termed biosorption, is of limited practical importance to soil remediation. However, biosorption can play a role in facilitating biomineralization through surface nucleation in a manner similar to Fe(III) oxides. It is important to note that sorption processes associated with microbial cell membranes can be described using the same surface complexation modeling techniques used to describe both UO_2^{2+} and PO_4^{3-} sorption to soils and various soil minerals.

Biomineralization is the most important microbe–uranium interaction mechanism relevant to the current discussion. Biomineralization includes the precipitation of metals associated with the release of microbially generated ligands such as sulfide (S^{2-}) or PO_4^{3-} or metal precipitation such as carbonates or hydroxides in response to localized alkaline conditions at the cell surface. Both microbes and plants secrete phosphatase enzymes as an active means of acquiring essential P from the soil organic matter and as a mechanism for reducing metal toxicity (Macaskie et al. 1992; Newsome et al. 2014). In fact, uranyl phosphate precipitates have been observed to form on microbial cell surfaces in soils that are naturally high in U (Mondani et al. 2011).

Biomineralization can be more efficient than chemical precipitation because localized concentration of sorbed reactants may serve as a nucleus for precipitation at the cell surface (i.e., surface nucleation). Furthermore, extracellular lipopolysaccharides may provide nucleation sites for the formation of tethered uranyl phosphates that do not foul the cell surface (Macaskie et al. 2000). Uranium reduced through microbial activity may associate with phosphate groups on extracellular lipopolysaccharides as well (Alessi et al. 2014).

While the current discussion focuses on the microbial aspects of biomineralization, additional logistical considerations will be discussed in the "Soluble Phosphorus Sources" section. Much of the previous work on microbial biomineralization has focused on the use of glycerol-phosphate (i.e., glycerol-2-phosphate and glyceraol-3-phosphate) as both a PO_4^{3-} and carbon source, which may not be economically feasible as an *in situ* remediating agent (Newsome et al. 2014). However, some studies have demonstrated U(VI) biomineralization using alternate sources of organic P. Patterson-Beedle et al. (2010) found that *Escherichia coli* was able to use phytate (IP6, inositol phosphate) as a phosphorus source resulting in the precipitation of hydrogen uranyl phosphate (HUO_2PO_4) (Patterson-Beedle et al. 2010). Mixed cultures have been able to use tributyl phosphate, a solvent used in processing actinides, as an organic P source (Thomas and Macaskie 1996). As represented in Figure 3.7, microbes produce phosphatase enzymes that release inorganic orthophosphate from organic compounds, represented by glycerol phosphate in Figure 3.7, that may then precipitate with UO_2^{2+} and other metals (Appukuttan et al. 2006; Beazley et al. 2007, 2009, 2011; Jroundi et al. 2007; Macaskie et al. 1992; Martinez et al. 2007; Newsome et al. 2014; Salome et al. 2013; Shelobolina et al. 2009; Thomas and Macaskie 1996). The resulting biogenic uranyl phosphate may then serve as an effective sorbent for other toxic metals that are often present in contaminated systems (Basnakova and Macaskie 2001; Patterson-Beedle et al. 2006).

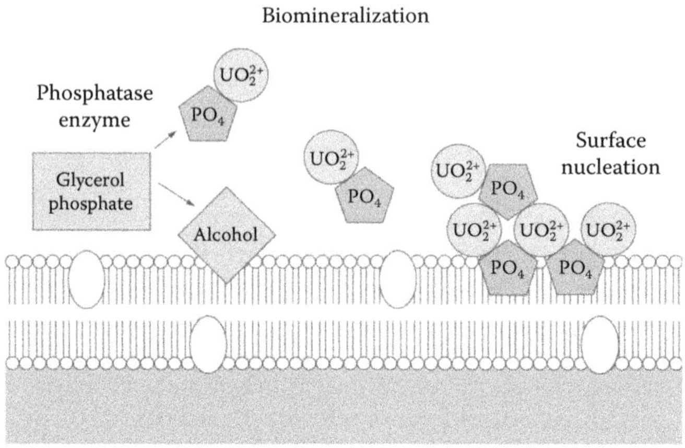

FIGURE 3.7
Schematic illustrating microbe-uranium interactions at the cell surface. (From Newsome, L. et al., *Chem Geol* 363:164–184, 2014.)

Laboratory studies have demonstrated the microbially catalyzed formation of uranyl phosphate precipitates under a broad range of chemical conditions, including both aerobic and anaerobic systems, and systems with low pH and high nitrate levels that may inhibit the stable reduction of U(IV), with several different uranyl phosphate end products having been identified (Beazley et al. 2007, 2009, 2011; Macaskie et al. 1992; Martinez et al. 2007; Salome et al. 2013; Shelobolina et al. 2009). For example, Beazley et al. (2007) found that two of three aerobic heterotrophic bacterial isolates from the DOE-FRC were able to hydrolyze sufficient glycerol-3-phosphate to precipitate autunite/meta-autunite group minerals. In anaerobic microcosms using sediments from the FRC, Salome et al. (2013) studied the impact of various terminal electron acceptors (i.e., Fe oxides, nitrate, and sulfate) on U(VI) bioreduction and biomineralization at pH 5.5 and 7.0 using glycerol-2-phosphate. Nitrate was consumed in the presence of glycerol phosphate and U remained in the oxidized state, U(VI), forming uranyl phosphates rather than reduced U products.

In a laboratory column study, Beazley et al. (2011) demonstrated the effectiveness of indigenous microbial populations present in U contaminated soils from the FRC to release PO_4^{3-} from glycerol phosphate, attributed to the production of a nonspecific acid phosphatase, with greater PO_4^{3-} production and uranyl phosphate precipitation observed at pH 5.5, while sorption to Fe(III) oxides dominated U(VI) retention at pH 7. In the absence of an organic P source, a large degree of U(VI) was retained through various sorption reactions. A facultative anaerobe isolated from the FRC, *Rahnella* sp., was able to use nitrate as a terminal electron acceptor under anaerobic conditions with glycerol-3-phophate as the sole carbon and PO_4^{3-} source to induce the precipitation U as chernikovite [$H_2(UO_2)_2(PO_4)_2$], an autunite-type mineral (Beazley et al. 2009). In addition, Martinez et al. (2007) demonstrated that *Rahnella* and *Bacillus* spp. isolates from the FRC produce sufficient phosphatase under aerobic, acidic (pH ≈ 5.0–5.5) conditions to precipitate autunite. In systems with high levels of Ca^{2+}, Shelobolina et al. (2009) observed that U(VI) coprecipitated with Ca–P through incorporation in the resulting hydroxyapatite under aerobic and nitrate reducing conditions, with the U(VI) associated with the hydroxyapatite being potentially more stable than autunite under circumneutral pH conditions. Biogenic uranyl phosphates appear to be more subject to reduction than abiotic uranyl phosphates, with dissolution as the first step in the reduction process and eventual formation of U(IV) phosphate such as ningyoite, $CaU(PO_4)_2{\cdot}H_2O$, rather than uraninite, UO_2 (Rui et al. 2013). Additionally, soil microorganisms may also enhance dissolution of phosphate-bearing minerals such as meta-autunite formed at the cell surface when nutrient P becomes limiting (Katsenovich et al. 2012; Smeaton et al. 2008).

Soluble Phosphorus Sources

The limited solubility of solid-phase PO_4^{3-} sources such as apatite and the limited mobility of orthophosphate in the soil environment constrain its practical application as a remediating agent for immobilizing U to surficial materials where it can readily be incorporated within the target zone. The development of flow-through reactive barriers to address U-contaminated groundwater, as discussed previously, has been one strategy to address such limitations (Bostick et al. 2000; Fuller et al. 2002). As discussed in the "Phosphate $\left(PO_4^{3-}\right)$ and the Bioremediation of Uranium" section, microbially produced phosphatase enzymes can stimulate the release of PO_4^{3-} from the indigenous organic P fraction, or as a means of *activating* a mobile, organic P amendment, such as glycerol phosphates and phytate, presumably resulting in the subsequent precipitation of uranyl phosphates. Thus, biomineralization is another proposed strategy for expanding the range of soluble PO_4^{3-} sources that can be used for targeted U immobilization.

While the "Phosphate $\left(PO_4^{3-}\right)$ and the Bioremediation of Uranium" section discussed the microbial aspects of biomineralization, significant research has been conducted evaluating a range of soluble P materials including trisodium trimetaphosphate (TP3; $Na_3P_3O_9$) and the organic PO_4^{3-} source, phytate (Jensen et al. 1996; Majs 2011; Nash et al. 1997, 1998a,b; Seaman et al. 2003, 2005; Wellman et al. 2006, 2007). Research evaluating the use of phytate as a remediating agent has generally focused on the remedial end products (i.e., uranyl and actinide phosphate precipitates) assuming effective enzymatic release of inorganic PO_4^{3-} (Jensen et al. 1996; Majs 2011; Nash et al.

1997, 1998a,b) or the potential environmental consequences associated with the application of the soluble phytate (Majs 2011; Seaman et al. 2003, 2005; Wellman et al. 2006).

Phytate (IP6), the hexaphosphoric acid of myoinositol, is a natural waste product of the fermentation industry and a major component of plant seeds with six orthophosphate moieties that provide 12 coordinate ligands for binding metal ions. The effectiveness of IP6 as a remediating agent is generally based on the assumption that hydrolysis and mineralization by microbially derived phosphatase enzymes will eventually release inorganic PO_4^{3-}, which results in contaminant immobilization through the formation of insoluble secondary contaminant-phosphate precipitates as discussed in the section on biomineralization ("Phosphate (PO_4^{3-}) and the Bioremediation of Uranium"). The effectiveness of TP3 is similarly based on the assumption that hydrolysis reactions yield ortho- and pyrophosphate that will react with the target contaminant.

The effectiveness of IP6 and other soluble PO_4^{3-} sources in limiting U and other metal solubility is generally controlled by the material's compatibility with the geology and groundwater chemistry of a given site, factors that can only be realistically evaluated under dynamic flowing conditions that better mimic the field (Seaman et al. 2005; Wellman et al. 2006, 2007). In batch studies, Seaman et al. (2003) and Majs (2011) both demonstrated that Na-IP6 addition—the soluble form most likely to be used as a soil amendment—actually increased U (e.g., Figure 3.8) and Ni solubility from contaminated soils, either through direct U(VI)-$IP6_{(aq)}$ complexation and/or the solubilization of natural organic matter and the dispersion of organomineral colloids.

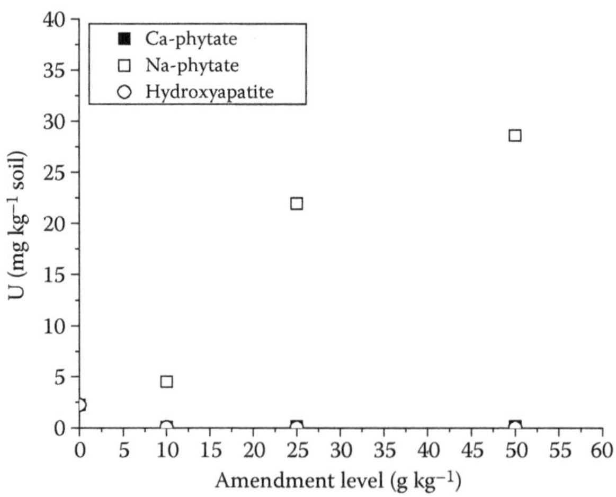

FIGURE 3.8
Residual U in solution as a function of PO_4^{3-} amendment from three sources (Ca-phytate, Na-phytate, and hydroxyapatite) to a soil from the DOE Savannah River Site. (Redrawn from Seaman, J.C. et al., *J Environ Qual* 32:153–161, 2003.)

Furthermore, Seaman et al. (2005) and Wellman et al. (2006) have both noted that IP6 decreased the hydraulic conductivity of the treated matrix material in dynamic column tests. For Wellman et al. (2006), the reduction in conductivity was attributed to the precipitation of Ca-phytate, while obvious clay dispersion was the major factor reducing hydraulic conductivity for Seaman et al. (2005), with clay dispersion also evident in batch studies as well (Majs 2011; Seaman et al. 2003). Wellman et al. (2006) observed no reduction in conductivity for tripolyphosphate, $[Na_3(PO_4)]_n$. These studies clearly demonstrate the need to evaluate the practical aspects of environmental remediation strategies under dynamic conditions that better mimic their eventual application in the field, including both saturated and unsaturated conditions when relevant.

Implications for Environmental Remediation

Up to now the discussion has largely focused on specific mechanistic processes controlling both aqueous- and solid-phase U speciation in soils as influenced by organic and inorganic phosphate, with less emphasis on their application to environmental remediation and stewardship. The primary objective of any remedial action is to reduce the risks to human health and the environment through the removal of the contaminant source (i.e., contaminated soil, sediments, groundwater, or leaky storage facility) or by blocking critical exposure pathways. The removal of contaminated soil materials and subsequent placement in a disposal facility is an obvious example of a remedial action. However, the manipulation of chemical conditions in a manner that makes the contaminant less soluble and/or bioavailable is a type of remedial action that effectively blocks potential exposure pathways.

Less expensive *in situ* remediation strategies typically rely on the introduction of chemical additives that are specifically chosen to alter target contaminant speciation in a manner that enhances solid-phase partitioning in order to reduce migration, bioavailability, and ultimately the associated exposure hazard. However, the use of *in situ* remediation techniques requires a fundamental understanding of the underlying geochemical processes and a commitment to long-term monitoring, as the contaminant remains in place with the potential to become a hazard at a later date. This is especially the case when evaluating radioactive contaminants where the hazard may exist for decades, if not much longer, and a significant dose may result from simply being in close proximity to the contaminated materials regardless of their chemical form. In contrast, the *ex situ* remediation of radioactive materials faces the additional obstacle of finding a suitable location for eventual waste disposal that is acceptable to the general public.

As noted previously, the use of PO_4^{3-} as a remediation agent for the *in situ* treatment of U-contaminated soils and sediments developed from the observation that U phosphates form in association with natural U-bearing ore deposits and U-contaminated environments (Ewing and Wang 2002; Jerden Jr. and Sinha 2003, 2006; Jerden Jr. et al. 2003; Mehta et al. 2014). However, the complex interactions of U and PO_4^{3-} in the soil environment can severely confound efforts to predict the long-term stability of reactive end products resulting from remediation efforts. A mechanistic understanding of the relevant processes is inherently limited by the complex relationship among aqueous-phase U speciation, surface complexation reactions, and precipitation reactions occurring in the general bulk solution. A wide array of microbial processes further complicates U behavior in soils. Research evaluating potential remediation strategies should focus on systems where the geographic extent of contamination makes *ex situ* remediation strategies impractical and conditions where other techniques, such as bioreduction, are of limited utility, such as low pH systems with significant nitrate levels. Laboratory and field studies that evaluate the practicality of proposed *in situ* remediation strategies involving the manipulation of PO_4^{3-} to immobilize U under the complex conditions inherent to real-world systems, including laboratory batch experiments and dynamic column studies at various treatment scales, must continue. While distinguishing the exact mechanisms of solid-phase U partitioning may be tenuous, such studies are critical in identifying pitfalls and logistical constraints that can severely hamper field-scale remediation success.

Acknowledgments

The authors acknowledge the assistance of L. Lee in figure preparation and M. Baker with manuscript preparation. This work was supported through a Cooperative Agreement (DEFC09-07-SR22506) between the Department of Energy and the University of Georgia Research Foundation.

References

Abdelouas, A. 2006. Uranium mill tailings: Geochemistry, mineralogy and environmental impact. *Elements* 2:335–341.

Alessi, D.S., J.S. Lezama-Pacheco, J.E. Stubbs, M. Janousch, J.R. Bargar, P. Persson, and R. Bernier-Latmani. 2014. The product of microbial uranium reduction includes multiple species with U(IV)–phosphate coordination. *Geochim Cosmochim Acta* 131:115–127.

Appukuttan, D., A.S. Rao, and S.K. Apte. 2006. Engineering of *Deinococcus radiodurans* R1 for bioprecipitation of uranium from dilute nuclear waste. *Appl Environ Microbiol* 72:7873–7878.

Arey, J.S., J.C. Seaman, and P.M. Bertsch. 1999. Immobilization of uranium in contaminated sediments by hydroxyapatite addition. *Environ Sci Technol* 33:337–342.

Barnett, M.O., P. Jardine, and S. Brooks. 2002. U(VI) adsorption to heterogeneous subsurface media: Application of a surface complexation model. *Environ Sci Technol* 36:937–942.

Barnett, M.O., P.M. Jardine, S.C. Brooks, and H.M. Selim. 2000. Adsorption and transport of uranium(VI) in subsurface media. *Soil Sci Soc Am J* 64:908–917.

Basnakova, G., and L.E. Macaskie. 2001. Microbially-enhanced chemisorption of Ni^{2+} ions into biologically-synthesised hydrogen uranyl phosphate (HUP) and selective recovery of concentrated Ni^{2+} using citrate or chloride ion (in English). *Biotechnology Letters* 23:67–70.

Beazley, M.J., R.J. Martinez, P.A. Sobecky, S.M. Webb, and M. Taillefert. 2007. Uranium biomineralization as a result of bacterial phosphatase activity: Insights from bacterial isolates from a contaminated subsurface. *Environ Sci Technol* 41: 5701–5707.

Beazley, M.J., R.J. Martinez, P.A. Sobecky, S.M. Webb, and M. Taillefert. 2009. Nonreductive biomineralization of uranium (VI) phosphate via microbial phosphatase activity in anaerobic conditions. *Geomicrobiology J* 26:431–441.

Beazley, M.J., R.J. Martinez, S.M. Webb, P.A. Sobecky, and M. Taillefert. 2011. The effect of pH and natural microbial phosphatase activity on the speciation of uranium in subsurface soils. *Geochim Cosmochim Acta* 75:5648–5663.

Bernhard, G., G. Geipel, T. Reich, V. Brendler, S. Amayri, and H. Nitsche. 2001. Uranyl(VI) cabonate complex formation: Validation of the $Ca_2UO_2(CO_3)_{3(aq)}$ species. *Radiochem Acta* 89:511–518.

Bernier-Latmani, R., H. Veeramani, E.D. Vecchia, P. Junier, J.S. Lezama-Pacheco, E.I. Suvorova, J.O. Sharp, N.S. Wigginton, and J.R. Bargar. 2010. Non-uraninite products of microbial U(VI) reduction. *Environ Sci Technol* 44:9456–9462.

Bleise, A., P.R. Danesi, and W. Burkart. 2003. Properties, use and health effects of depleted uranium (DU): A general overview. *J Environ Rad* 64:93–112.

Bostick, B.C., S. Fendorf, M.O. Barnett, P.M. Jardine, and S.C. Brooks. 2002. Uranyl surface complexes formed on subsurface media from DOE facilities. *Soil Sci Soc Am J* 66:99–108.

Bostick, W.D., R.J. Stevenson, R.J. Jarabek, and J.L. Conca. 2000. Use of apatite and bone char for the removal of soluble radionuclides in authentic and simulated DOE groundwater. *Adv Environ Res* 3:488–498.

Boult, K.A. 1998. Towards an understanding of the sorption of U(VI) and Se(IV) on sodium bentonite. *J Contam Hydrol* 35:141–150.

Brooks, S.C. 2001. Waste characteristics of the former S-3 ponds and outline of uranium chemistry relevant to NABIR Field Research Center Studies ORNL/TM-2001/27. Oak Ridge National Laboratory, Oak Ridge, TN.

Brooks, S.C., J.K. Fredrickson, S.L. Carroll, D.W. Kennedy, J.M. Zachara, A.E. Plymale, S.D. Kelly, K.M. Kemner, and S. Fendorf. 2003. Inhibition of bacterial U(VI) reduction by calcium. *Environ Sci Technol* 37:1850–1858.

Bruno, J., J.d. Pablo, L. Duro, and E. Figuerola. 1995. Experimental study and modeling of the U(VI)-$Fe(OH)_3$ surface precipitation/coprecipitation equilibria. *Geochim Cosmochim Acta* 59:4113–4123.

Buck, E.C., N.R. Brown, and N.L. Dietz. 1996. Contaminant uranium phases and leaching at the Fernald Site in Ohio. *Environ Sci Technol* 30:81–88.

Burns, P.C., and R. Finch (eds.) 1999. *Uranium: Mineralogy, Geochemistry, and the Environment*. The Mineralogical Society of America, Washington, DC.

Campbell, K.M., J.A. Davis, J. Bargar, D. Giammar, R. Bernier-Latmani, R. Kukkadapu, K.H. Williams, H. Veramani, K.U. Ulrich, J. Stubbs, S. Yabusaki, L. Figueroa, E. Lesher, M.J. Wilkins, A. Peacock, and P.E. Long. 2011. Composition, stability, and measurement of reduced uranium phases for groundwater bioremediation at Old Rifle, CO. *Applied Geochem* 26:S167–S169.

Cerrato, J.M., M.N. Ashner, D.S. Alessi, J.S. Lezama-Pacheco, R. Bernier-Latmani, J.R. Bargar, and D.E. Giammar. 2013. Relative reactivity of biogenic and chemogenic uraninite and biogenic noncrystalline U(IV). *Environ Sci Technol* 47:9756–9763.

Chang, H.-S., S.W. Buettner, J.C. Seaman, P.R. Jaffe, P.G.K. van Groos, D. Li, A.D. Peacock, K.G. Scheckel, and D.I. Kaplan. 2014. Uranium immobilization in an iron-rich rhizosphere of a native wetland plant from the Savannah River Site under reducing conditions. *Environ Sci Technol* 48(16):9270–9278.

Chattanathan, S.A., T.P. Clement, S.R. Kanel, M.O. Barnett, and N. Chatakondi. 2013. Remediation of uranium-contaminated groundwater by sorption onto hydroxyapatite derived from catfish bones. *Water Air & Soil Pol* 224:1429.

Cheng, T., M.O. Barnett, E.E. Roden, and J. Zhuang. 2004. Effects of phosphate on uranium(VI) adsorption to goethite-coated sand. *Environ Sci Technol* 38:6059–6065.

Cheng, T., M.O. Barnett, E. Roden, and J. Zhuang. 2006. Effects of solid-to-solution ratio on uranium(VI) adsorption and its implications. *Environ Sci Technol* 40:3243–3247.

Cheng, T., M.O. Barnett, E.E. Roden, and J. Zhuang. 2007. Reactive transport of uranium(VI) and phosphate in a goethite-coated sand column: An experimental study. *Chemosphere* 68:1218–1223.

Crowley, K., and J.F. Ahearne. 2002. Managing the environmental legacy of U.S. nuclear-weapons production. *American Scientist* 90:514–523.

Davis, J.A., and J.D. Randall. 2003. Application of surface complexation modeling to describe uranium (VI) adsorption and retardation at the Uranium Mill Tailings Site at Naturita, Colorado NUREG/CR-6820. U.S. Nuclear Regulatory Commission Office of Nuclear Regulatory Research, Washington, DC.

Davis, J.A., T.E. Payne, and T.D. Waite. 2002. Simulating the pH and pCO_2 dependence of uranium(VI) adsorption by a weathered schist with surface complexation models, pp. 61–86. *In* P.C. Zhang and P. V. Brady (eds.), *Geochemistry of Soil Radionuclides.* Soil Science Society of America, Madison, WI.

Davis, J.A., J.A. Coston, D.B. Kent, and C.C. Fuller. 1998. Application of the surface complexation concept to complex mineral assemblages. *Environ Sci Technol* 32:2820–2828.

Davis, J.A., D.E. Meece, M. Kohler, and G.P. Curtis. 2004. Approaches to surface complexation modeling of uranium(VI) adsorption on aquifer sediments. *Geochim Cosmochim Acta* 68:3621–3641.

Del Nero, M., C. Galindo, R. Barillon, and B. Madé. 2011. TRLFS evidence for precipitation of uranyl phosphate on the surface of alumina: Environmental implications. *Environ Sci Technol* 45:3982–3988.

Dong, W., and J. Wan. 2014. Additive surface complexation modeling of uranium(vi) adsorption onto quartz-sand dominated sediments. *Environ Sci Technol* 48: 6569–6577.

Dong, W., T.K. Tokumaga, J.A. Davis, and J. Wan. 2012. Uranium(VI) adsorption and surface complexation modeling onto background sediments from the F-Area Savannah River Site. *Environ Sci Technol* 46:1565–1571.
Dong, W., G. Xie, T.R. Miller, M.P. Franklin, T.P. Oxenberg, E.J. Bouwer, W.P. Ball, and R.U. Halden. 2006. Sorption and bioreduction of hexavalent uranium at a military facility by the Chesapeake Bay. *Environ Pollut* 142:132–142.
Duff, M.C., C.F.V. Mason, and D.B. Hunter. 1998. Comparison of acid and base leach for the removal of uranium from contaminated soil and catch-box media. *Can J Soil Sci* 78:675–683.
Elless, M.P., and S.Y. Lee (eds.). 2002. *Radionuclide-Contaminated Soils: A Mineralogical Perspective for their Remediation*. Soil Science Society of America, Madison, WI.
Evans, A.G., L.R. Bauer, J.S. Haselow, D.W. Hayes, H.L. Martin, W.L. McDowell, and J.B. Pickett. 1992. Uranium in the Savannah River Site Environment WSRC-RP-92-315. WSRC.
Ewing, R.C., and L.M. Wang. 2002. Phosphates as nuclear waste forms, pp. 673–699. *In* M.J. Kohn et al. (eds.), *Phosphates—Geochemical, Geobiological, and Materials Importance,* Volume 48. Mineralogical Society of America and Geochemical Society, Washington, DC.
Finn, P.A., J.C. Hoh, S.F. Wolf, S.A. Slater, and J.K. Bates. 1996. The release of uranium, technetium, and iodine from spent fuel under unsaturated conditions. *Radiochim Acta* 75:65–71.
Fox, P.M., J.A. Davis, R. Kukkadapu, D.M. Singer, J. Bargar, and K.H. Williams. 2013. Abiotic U(VI) reduction by sorbed Fe(II) on natural sediments. *Geochim Cosmochim Acta* 117:266–282.
Fuller, C.C., J.R. Bargar, and J.A. Davis. 2003. Molecular-scale characterization of uranium sorption by bone apatite materials for a permeable reactive barrier demonstration. *Environ Sci Technol* 37:4642–4649.
Fuller, C.C., J.R. Bargar, J.A. Davis, and J. Piana. 2002. Mechanisms of uranium interactions with hydroxyapatite: Implications for groundwater remediation. *Environ Sci Technol* 36:158–165.
Galindo, C., M.D. Nero, R. Barillon, E. Halter, and B. Made. 2010. Mechanisms of uranyl and phosphate (co)sorption: Complexation and precipitation at α-Al_2O_3 surfaces. *Journal of Colloid and Interface Science* 347:282–289.
Gavrilescu, M., L.V. Pavel, and I. Cretescu. 2009. Characterization and remediation of soils contaminated with uranium. *J Hazar Mater* 163:475–510.
Goldberg, S. 2005. Equations and models describing adsorption processes in soils, pp. 489–517. *In* M.A. Tabatabai and D.L. Sparks (eds.), *Chemical Processes in Soils.* Soil Science Society of America, Madison, WI.
Goldberg, S., L.J. Criscenti, D.R. Turner, J.A. Davis, and K.J. Cantrell. 2007. Adsorption-desorption processes in subsurface reactive transport modeling. *Vadose Zone J* 6:407–435.
Gorman-Lewis, D., P.C. Burns, and J.B. Fein. 2008. Review of uranyl mineral solubility measurements. *J Chem Thermodyn* 40:335–352.
Gorman-Lewis D., T. Shvarea, K.A. Kubatko, P.C. Burns, D.M. Wellman, B. McNamara, J.E. Szymanowski, A. Navrotsky, and J.B. Fein. 2009. Thermodynamic properties of autunite, uranyl hydrogen phosphate, and uranyl orthophosphate from solubility and calorimetric measurements. *Environ Sci Technol* 43:7416–7422.

Grenthe, I., J. Fuger, R.J.M. Konings, R.J. Lemire, A.B. Muller, C. Nguyen-Trung, and H. Wanner. 1992. *Chemical Thermodynamics of Uranium*. Elsevier, Amsterdam, The Netherlands, 715 pp.
Guillaumont, R., T. Fanghanel, J. Fuger, I. Grenthe, V. Neck, D.A. Palmer, and M.H. Rand. 2003. *Update on the Chemical Thermodynamics of Uranium, Neptunium, Plutonium, Americium and Technetium*. OECD Nuclear Energy Agency. Elsevier, Amsterdam.
Hsi, C., and C. Langmuir. 1985. Adsorption of uranyl onto ferric oxyhydroxides: Application of the surface complexation site-binding model. *Geochim Cosmochim Acta* 49:1931–1941.
Jensen, M.P., K.L. Nash, L.R. Morss, E.H. Appelman, and M.A. Schmidt. 1996. Immobilization of actinides in geomedia by phosphate precipitation. Humic and fulvic acids: Isolation, structure and environmental role. *J Am Chem Soc* 651:272–285.
Jeon, B., B.A. Dempsey, W.D. Burgos, M.O. Barnett, and E.E. Roden. 2005. Chemical reduction of U(VI) by Fe(II) at the solid-water interface using natural and synthetic Fe(III) oxides. *Environ Sci Technol* 39:5642–5649.
Jerden Jr., J.L., and A.K. Sinha. 2006. Geochemical coupling of uranium and phosphorous in soils overlying an unmined uranium deposit: Coles Hill, Virginia. *J Geochemical Explor* 91:56–70.
Jerden Jr., J.L., A.K. Sinha, and L. Zelazny. 2003. Natural immobilization of uranium by phosphate mineralization in an oxidizing saprolite–soil profile: Chemical weathering of the Coles Hill uranium deposit, Virginia. *Chem Geol* 199:129–157.
Jerden Jr., J.L., and A.K. Sinha. 2003. Phosphate based immobilization of uranium in an oxidizing bedrock aquifer. *Appl Geochem* 18:823–843.
Jove-Colon, C.F., C. Sanpawanichakit, H. Xu, R.T. Cygan, J.A. Davis, D.M. Meece, and R.L. Hervig. 2006. A combined analytical study to characterize uranium soil and sediment contamination: The case of the Naturita UMTRA site and the role of grain coatings NUREG/CR-6898. U.S. Nuclear Regulatory Commission, Office of Nuclear Regulatory Research, Washington, DC.
Jroundi, F., M.L. Merroun, J.M. Arias, A. Rossberg, S. Selenska-Pobell, and M.T. González-Muñoz. 2007. Spectroscopic and microscopic characterization of uranium biomineralization in *Myxococcus xanthus*. *Geomicrobiology J* 24:441–449.
Katsenovich, Y.P., D.A. Carvajal, D.M. Wellman, and L.E. Lagos. 2012. Enhanced U(VI) release from autunite mineral by aerobic *Arthrobacter* sp. in the presence of aqueous bicarbonate. *Chem Geol* 308–309:1–9.
Kratz, S., and E. Schnug. 2006. Rock phosphates and P fertilizers as sources of U contamination in agricultural soils, pp. 57–67. Springer-Verlag Berlin, Heidelberg, Federal Republic of Germany.
Landa, E.R. 2004. Uranium mill tailings: Nuclear waste and natural laboratory for geochemical and radioecological investigations. *J Environ Radioact* 77:1–27.
Lenhart, J.J., and B.D. Honeyman. 1999. Uranium (VI) sorption to hematite in the presence of humic acid. *Geochim Cosmochim Acta* 63:2891–2901.
Liger, E., L. Charlet, and P. Van Cappellen. 1999. Surface catalysis of uranium(VI) reduction by iron(II). *Geochim Cosmochim Acta* 63:2939–2955.
Liu, C., Z. Shi, and J.M. Zachara. 2009. Kinetics of uranium(VI) desorption from contaminated sediments: Effect of geochemical conditions and model evaluation. *Environ Sci Technol* 43:6560–6566.

Liu, C., J. Shang, S. Kerisit, J.M. Zachara, and W. Zhu. 2013. Scale-dependent rates of uranyl surface complexation reaction in sediments. *Geochim Cosmochim Acta* 105:326–341.

Lloyd, J.R., and G.M. Gadd. 2011. The geomicrobiology of radionuclides. *Geomicrobiology J* 28:383–386.

Logue, B.A., R.W. Smith, and J.C. Westall. 2004. U(VI) adsorption on natural iron-coated sands: Comparison of approaches for modeling adsorption on heterogeneous environmental materials. *Appl Geochem* 19:1937–1951.

Lovley, D.R., and E.J.P. Phillips. 1992. Bioremediation of uranium contamination with enzymatic uranium reduction. *Environ Sci Technol* 26:2228–2234.

Lovley, D.R., E.J.P. Phillips, Y.A. Gorby, and E.R. Landa. 1991. Microbial reduction of uranium. *Nature* 350:413–416.

Ma, L.Q., and G.N. Rao. 1997. The effect of phosphate rock on Pb distribution in contaminated soils. *J Environ Qual* 26:259–264.

Ma, L.Q., T.J. Logan, and S.J. Traina. 1995. Lead immobilization from aqueous solutions and contaminated soils using phosphate rocks. *Environ Sci Technol* 29:1118–1126.

Ma, Q.Y., T.J. Logan, S.J. Traina, and J.A. Ryan. 1994. Effects of NO_3^-, Cl^-, F^-, SO_4^{2-}, and CO_3^{2-} on Pb^{2+} immobilization by hydroxyapatite. *Environ Sci Technol* 28:408–418.

Macaskie, L.E., K.M. Bonthrone, P. Yong, and D.T. Goddard. 2000. Enzymically mediated bioprecipitation of uranium by a *Citrobacter* sp.: A concerted role for exocellular lipopolysaccharide and associated phosphatase in biomineral formation. *Microbiology* 146:1855–1867.

Macaskie, L.E., R.M. Empson, A.K. Cheetham, C.P. Grey, and A.J. Skarnulis. 1992. Uranium bioaccumulation by a *Citrobacter* sp. as a result of enzymically mediated growth of polycrystalline HUO_2PO_4. *Science* 257:782–784.

Majs, F. 2011. Potential adverse effects of applying phosphate amendments to immobilize soil contaminants. *J Environ Qual* 40:1135–1142.

Marques Fernandes, M., B. Baeyens, R. Dähn, A.C. Scheinost, and M.H. Bradbury. 2012. U(VI) sorption on montmorillonite in the absence and presence of carbonate: A macroscopic and microscopic study. *Geochim Cosmochim Acta* 93:262–277.

Martinez, R.J., M.J. Beazley, M. Taillefert, A.K. Arakaki, J. Skolnick, and P.A. Sobecky. 2007. Aerobic uranium (VI) bioprecipitation by metal-resistant bacteria isolated from radionuclide- and metal-contaminated subsurface soils. *Environ Microbiol* 9:3122–3133.

Mason, C.F.V., W.R.J.R. Turney, B.M. Tomson, N. Lu, P.A. Longmire, and C.J. Chisholm-Brause. 1997. Carbonate leaching of uranium from contaminated soils. *Environ Sci Technol* 31:2707–2711.

Mehta, V.S., F. Maillot, Z. Wang, J.G. Catalano, and D.E. Giammar. 2014. Effect of co-solutes on the products and solubility of uranium(VI) precipitated with phosphate. *Chem Geol* 364:66–75.

Mondani, L., K. Benzerara, M. Carrière, R. Christen, Y. Mamindy-Pajany, L. Février, N. Marmier, W. Achouak, P. Nardoux, C. Berthomieu, and V. Chapon. 2011. Influence of uranium on bacterial communities: A comparison of natural uranium-rich soils with controls. *PLoS ONE* 6:e25771.

Murakami, T., T. Ohnuki, H. Isobe, and T. Sato. 1997. Mobility of uranium during weathering. *Am Mineral* 82:129–157.

N'Guessan, A.L., H.A. Vrionis, C.T. Resch, P.E. Long, and D.R. Lovley. 2008. Sustained removal of uranium from contaminated groundwater following stimulation of dissimilatory metal reduction. *Environ Sci Technol* 42:2999–3004.

Nash, K.L., M.P. Jensen, and M.A. Schmidt. 1998a. Actinide immobilization in the subsurface environment by in-situ treatment with a hydrolytically unstable organophosphorus complexant: Uranyl uptake by calcium phytate. *J Alloys Compd* 271–273:257–261.

Nash, K.L., M.P. Jensen, and M.A. Schmidt. 1998b. In-situ mineralization of actinides for groundwater cleanup: Laboratory demonstration with soil from the Fernald Environmental Management Project, pp. 507–518. *In* W.W. Schulz and N.J. Lombardo (eds.), *Science and Technology for Disposal of Radioactive Tank Waste.* Plenum Press, New York.

Nash, K.L., M.P. Jensen, J.J. Hines, S.A. Friedrich, and M. Redko. 1997. Phosphate mineralization of actinides by measured addition of precipitating anions TTP#: CH2-6-C3-22. Chemistry Division-Argonne National Laboratory, Argonne, IL.

Newsome, L., K. Morris, and J.R. Lloyd. 2014. The biogeochemistry and bioremediation of uranium and other priority radionuclides. *Chem Geol* 363:164–184.

Nico, P.S., B.D. Stewart, and S. Fendorf. 2009. Incorporation of oxididized uranium into Fe (Hydr)oxides during Fe(II) catalyzed remineralization. *Environ Sci Technol* 43:7391–7396.

Nyman, J., T. Marsh, M. Ginder-Vogel, M. Gentile, S. Fendorf, and C. Criddle. 2006. Heterogeneous response to biostimulation for U(VI) reduction in replicated sediment microcosms (in English). *Biodegradation* 17:303–316.

Ohnuki, T., I. Isobe, N. Yanase, T. Nagano et al. 1997. Change in sorption characteristics of uranium during crystallization of amorphous iron minerals. *J Nucl Sci Technol* 34:1153–1158.

Ohnuki, T., N. Kozai, M. Samadfam, R. Yasuda, S. Yamamoto, K. Narumi, H. Naramoto, and T. Murakami. 2004. The formation of autunite ($Ca(UO_2)_2(PO_4)_2 \cdot nH_2O$) within the leached layer of dissolving apatite: Incorporation mechanism of uranium by apatite. *Chem Geol* 211:1–14.

Patterson-Beedle, M., J.E. Readman, J.A. Hriljac, and L.E. Macaskie. 2010. Biorecovery of uranium from aqueous solutions at the expense of phytic acid. *Hydrometallurgy* 104:524–528.

Patterson-Beedle, M., L.E. Macaskie, C.H. Lee, J.A. Hriljac, K.Y. Jee, and W.H. Kim. 2006. Utilisation of a hydrogen uranyl phosphate-based ion exchanger supported on a biofilm for the removal of cobalt, strontium and caesium from aqueous solutions. *Hydrometallurgy* 83:141–145.

Payne, T.E., J.A. Davis, and T.D. Waite. 1996. Uranium adsorption on ferrihydrite—Effects of phosphate and humic acid. *Radiochim Acta* 74:239–234.

Payne, T.E., J.A. Davis, G.R. Lumpkin, R. Chisari, and T.D. Waite. 2004. Surface complexation model of uranyl sorption on Georgia kaolinite. *Appl Clay Science* 26:151–162.

Pickett, J.B., W.P. Colven, and H.W. Bledsoe. 1987. M-Area settling basin and vicinity. Environmental information document. E.I. duPont de Nemours & Co./Savannah River Laboratory, Aiken SC.

Renock, D., M. Mueller, K. Yuan, R.C. Ewing, and U. Becker. 2013. The energetics and kinetics of uranyl reduction on pyrite, hematite, and magnetite surfaces: A powder microelectrode study. *Geochim Cosmochim Acta* 118:56–71.

Riley, R.G., J.M. Zachara, and F.J. Wobber. 1992. Chemical contaminants on DOE lands and selection of contaminant mixtures for subsurface science research. DOE/ER-0547T, Department of Energy.

Rui, X., M.J. Kwon, E.J. O'Loughlin, S. Dunham-Cheatham, J.B. Fein, B. Bunker, K.M. Kemner, and M.I. Boyanov. 2013. Bioreduction of hydrogen uranyl phosphate: Mechanisms and U(IV) products. *Environ Sci Technol* 47:5668–5678.

Salome, K.R., S.J. Green, M.J. Beazley, S.M. Webb, J.E. Kostka, and M. Taillefert. 2013. The role of anaerobic respiration in the immobilization of uranium through biomineralization of phosphate minerals. *Geochim Cosmochim Acta* 106: 344–363.

Sandino, A., and J. Bruno. 1992. The solubility of $(UO_2)_3(PO_4)_2$-$4H_2O(s)$ and the formation of U(VI) phosphate complexes: Their influence in uranium speciation in natural waters. *Geochim Cosmochim Acta* 56:4135–4145.

Schipper, L.A., G.P. Sparling, L.M. Fisk, M.B. Dodd, I.L. Power, and R.A. Littler. 2011. Rates of accumulation of cadmium and uranium in a New Zealand hill farm soil as a result of long-term use of phosphate fertilizer. *Agric Ecosyst Environ* 144:95–101.

Schnug, E., and B.G. Lottermoser. 2013. Fertilizer-derived uranium and its threat to human health. *Environ Sci Technol* 47:2433–2434.

Seaman, J.C. 2000. Thin-foil SEM analysis of soil and groundwater colloids: Reducing instrument and operator bias. *Environ Sci Technol* 34:187–191.

Seaman, J.C., P.M. Bertsch, and R.N. Strom. 1997. Characterization of colloids mobilized from Southeastern Coastal Plain Sediments. *Environ Sci Technol* 31:2782–2790.

Seaman, J.C., T. Meehan, and P.M. Bertsch. 2001a. Immobilization of ^{137}Cs and U in contaminated sediments using soil amendments. *J Environ Qual* 30:1206–1213.

Seaman, J.C., J.S. Arey, and P.M. Bertsch. 2001b. Immobilization of Ni and other metals in contaminated sediments using soil amendments. *J Environ Qual* 30:460–469.

Seaman, J.C., J. Hutchison, B.P. Jackson, and V.M. Vulava. 2003. In situ treatment of metals in contaminated soils using phytate. *J Environ Qual* 32:153–161.

Seaman, J.C., V.J. Vulava, A.G. Sowder, B.P. Jackson, S.A. Aburime, and P.M. Bertsch. 2005. Metal extractability from contaminated SRS sediments: Comparison of column and batch results. *Environ Geosciences* 12:235–242.

Shelobolina, E.S., H. Konishi, H. Xu, and E.E. Roden. 2009. U(VI) Sequestration in hydroxyapatite produced by microbial glycerol 3-phosphate metabolism. *Appl Environ Microbiol* 75:5773–5778.

Sheppard, S.C. 2001. Toxicants in the environment: Bringing radioecology and ecotoxicology together, pp. 63–74. *In* F. Brechignac and B.J. Howard (eds.), *Radioactive Pollutants: Impact on the Environment*. EDP Sciences, Les Ulis Cedex A, France.

Sherman, D.M., C.L. Peacock, and C.G. Hubbard. 2008. Surface complexation of U(VI) on goethite (α-FeOOH). *Geochim Cosmochim Acta* 72:298–310.

Singh, A., K.-U. Ulrich, and D.E. Giammar. 2010. Impact of phosphate on U(VI) immobilization in the presence of goethite. *Geochim Cosmochim Acta* 74:6324–6343.

Singh, A., J.G. Catalano, K.U. Ulrich, and D.E. Giammar. 2012. Molecular-scale structure of uranium(VI) immobilized with goethite and phosphate. *Environ Sci Technol* 46:6594–6603.

Smeaton, C.M., C.G. Weisener, P.C. Burns, B.J. Fryer, and D.A. Fowle. 2008. Bacterially enhanced dissolution of meta-autunite. *Am Mineral* 93:1858–1864.

Sowder, A.G., S.B. Clark, and R.A. Fjeld. 1996. The effect of silica and phosphate on the transformation of schoepite to becquerelite and other uranyl phases. *Radiochimica Acta* 74:45–49.

Sowder, A.G., S.B. Clark, and R.A. Fjeld. 1999. The transformation of uranyl oxide hydrates: The effect of dehydration on synthetic Metaschoepite and its alteration to Becquerelite. *Environ Sci Technol* 33:3552–3557.

Sowder, A.G., P.M. Bertsch, and P.J. Morris. 2003. Partitioning and availability of uranium and nickel in contaminated riparian sediments. *J Environ Qual* 35:885–898.

Stoliker, D.L., D.B. Kent, and J.M. Zachara. 2011. Quantifying differences in the impact of variable chemistry on equilibrium uranium(VI) adsorption properties of aquifer sediments. *Environ Sci Technol* 45:8733–8740.

Stylo, M., D.S. Alessi, P.P. Shao, J.S. Lezama-Pacheco, J.R. Bargar, and R. Bernier-Latmani. 2013. Biogeochemical controls on the product of microbial U(VI) reduction. *Environ Sci Technol* 47:12351–12358.

Tang, Y., and R.J. Reeder. 2009. Enhanced uranium sorption on aluminum oxide pretreated with arsenate. Part I: Batch uptake behavior. *Environ Sci Technol* 43:4446–4451.

Tang, Y., J. McDonald, and R.J. Reeder. 2009. Enhanced uranium sorption on aluminum oxide pretreated with arsenate. Part II: Spectroscopic studies. *Environ Sci Technol* 43:4452–4458.

Thomas, R.A.P., and L.E. Macaskie. 1996. Biodegradation of tributyl phosphate by naturally occurring microbial isolates and coupling to the removal of uranium from aqueous solution. *Environ Sci Technol* 30:2371–2375.

Turner, G.D., J.M. Zachara, J.P. McKinley, and S.C. Smith. 1996. Surface-charge properties and UO_2^{2+} adsorption of a subsurface smectite. *Geochim Cosmochim Acta* 60:3399–3414.

Vandenhove, H., M.V. Hees, K. Wouters, and J. Wannijn. 2007. Can we predict uranium bioavailability based on soil parameters? Part 1: Effect of soil parameters on soil solution uranium concentration. *Environ Pollut* 145:587–595.

Waite, T.D., J.A. Davis, B.R. Fenton, and T.E. Payne. 2000. Approaches to modelling uranium (VI) adsorption on natural mineral assemblages *Radiochim Acta* 88:687–693.

Waite, T.D., J.A. Davis, T.E. Payne, G.A. Waychunas, and N. Xu. 1994. Uranium (VI) adsorption to ferrihydrite: Application of a surface complexation model. *Geochim Cosmochim Acta* 58:5465–5478.

Wang, Z., J.M. Zachara, J.-F. Boily, Y. Xia, T.C. Resch, D.A. Moore, and C. Liu. 2011. Determining individual mineral contributions to U(VI) adsorption in a contaminated aquifer sediment: A fluorescence spectroscopy study. *Geochim Cosmochim Acta* 75:2965–2979.

Wazne, M., G.P. Korfiatis, and X. Meng. 2003. Carbonate effects on hexavalent uranium adsorption by iron oxyhydroxide. *Environ Sci Technol* 37:3619–3624.

Wellman, D.M., J.P. Icenhower, and A.T. Owen. 2006. Comparative analysis of soluble phosphate amendments for the remediation of heavy metal contaminants: Effect on sediment hydraulic conductivity. *Environ Chem* 3:219–224.

Wellman, D.M., E.M. Pierce, and M.M. Valenta. 2007. Efficacy of soluble sodium tripolyphosphate amendments for the in-situ immobilisation of uranium. *Environ Chem* 4:293–300.

Wellman, D.M., Julia N. Glovack, Kent Parker, E.L. Richards, and E.M. Pierce. 2008. Sequestration and retention of uranium(VI) in the presence of hydroxylapatite under dynamic geochemical conditions. *Environ Chem* 5:40–50.

Wu, W., J. Carley, T. Gentry, M. Ginder-Vogel, M. Fienen, T. Mehlhorn, H. Yan, S. Caroll, M. Pace, J. Nyman, J. Luo, M. Gentile, M. Fields, R. Hickey, B. Gu, D. Watson, O. Cirpka, J. Zhou, S. Fendorf, P. Kitanidis, P. Jardine, and C. Criddle. 2006. Pilot-scale in situ bioremediation of uranium in a highly contaminated aquifer. 2. Reduction of U(VI) and geochemical control of U(VI) bioavailability. *Environ Sci Technol* 40:3986–3995.

Zachara, J.M., and J.C. Westall. 1998. Chemical modeling of ion adsorption in soils, pp. 47–95. *In* D.L. Sparks (ed.), *Soil Physical Chemistry*, 2nd Edition. CRC Press, Boca Raton, FL.

Zachara, J.M., S.C. Smith, C.T. Resch, and C.E. Cowan. 1992. Cadmium sorption to soil separates containing layer silicates and iron and aluminum oxides. *Soil Sci Soc Am J* 56:1074–1084.

Zeng, H., A. Singh, S. Basak, K.U. Ulrich, M. Sahu, P. Biswas, and J.G. Catalona. 2009. Nanoscale size effects on uranium(VI) adsorption to hematite. *Environ Sci Technol* 43:1373–1378.

Zhang, P.C., J.L. Krumhansl, and P.V. Brady. 2002. Introduction to properties, sources and characteristics of soil radionuclides, pp. 1–20. *In* P.C. Zhang and P.V. Brady (eds.), *Geochemistry of Soil Radionuclides*. Soil Science Society of America, Madison, WI.

Zhang, S., D.B. Kent, D.L. Elbert, Z. Shi, J.A. Davis, and D.R. Veblen. 2011. Mineralogy, morphology, and textural relationships in coatings on quartz grains in sediments in a quartz-sand aquifer. *J Contam Hydrol* 124:57–67.

4

Influence of Phosphates on Retention and Mobility of Zinc, Cadmium, and Vanadium in Soils

H. Magdi Selim

CONTENTS

Understanding trace elements interactions in the soil–water environment is essential in assessing their bioavailability and potential toxicity. Trace elements include several heavy metals such as zinc (Zn), copper (Cu), arsenic (As), and cadmium (Cd), among others. Several heavy metals such as Zn and Cu are essential micronutrients required in the growth of both plants and animals. Micronutrients are often applied in the form of fertilizers or as supplement in animal feed. Heavy metals are extensively used as fungicides and as bactericides in numerous pharmaceuticals.

The bioavailability of trace elements in the soil–water environment is dependent on an array of soil properties including soil pH, organic matter content, amount and type of dominant clay, and carbonates, among others. In addition, the counterions present in the soil system greatly influence the fate of trace metals in soils. In fact, several studies suggested varied interactions of heavy metals with phosphates in soils. Since phosphorus (P) in the form of phosphate is a commonly used fertilizer, the influence of phosphate on the bioavailability and mobility of heavy metals in soils is of great significance. In this chapter, we focus on three heavy metals: Zn, Cd, and vanadium (V). The findings reported here are based on laboratory experiments where the retention of a heavy metal in the presence of phosphate was carried out. Specifically, two types of experiments were carried out. One type of experiment deals with the transport of a heavy metal in soil in the presence and absence of phosphate. The second type is that where the retention or sorption of a heavy metal in the presence of various P concentrations was investigated over time.

Zinc–Phosphate

Zn availability and mobility in soils may be controlled by several interactions with the soil–water environment. Primary sources of Zn contamination include mining, smelting, and other industrial as well as anthropogenic factors (Adriano 2001). A secondary source of Zn is through phosphate fertilizers, which often contain traces of heavy metals such as Cd, Cu, manganese (Mn), nickel (Ni), lead (Pb), and Zn. Zn is also an essential micronutrient for plants and animals (Adriano 2001; Kabata-Pendias 2000). The understanding of the complex interactions of Zn in the environment is a prerequisite in the effort to predict its behavior in the vadose zone. It is well accepted that several factors influence Zn adsorption, desorption, and equilibrium between the solid and solution phases. These factors include soil pH, clay content, organic matter (OM), cation exchange capacity (CEC), and iron (Fe)/aluminum (Al) oxides (Gaudalix and Pardo 1995), among which soil pH is one of the most important factors (Barrow 1987). Zn sorption increases and Zn desorption reduces with increased pH (Rupa and Tomar 1999; Tagwira et al. 1993). This may occur because increasing pH increases the negative charge of variable-charge soil for Zn adsorption (Saeed and Fox 1979). On the other hand, pH affects the Zn hydrolysis, which would be preferentially sorbed on soil surface (Bolland et al. 1977).

Over the past two decades, phosphate has been observed to increase Zn adsorption and decrease Zn desorption in soils (Agbenin 1998; Rupa and Tomar 1999). Xie and Mackenzie (1989) reported that P sorption increased soil CEC, resulting in increased Zn adsorption on three different studied soils. Saeed and Fox (1979) reported that increased negative charge due to P sorption was responsible for the observed increase in Zn sorption. Xie and Mackenzie (1989) found phosphate sorption enhanced the correlation between Zn sorption and soil OM and Fe content and postulated that enhanced Zn sorption may be a consequence of either an increase in negative surface charge of soil particles, creation of specific sites, or precipitation of hoepite. Ahumada et al. (1997), Pardo (1999), and Sarret (2002) observed significant increases in Zn fractions associated with organic matter and Fe/Al oxides in the presence of P.

Rupa and Tomar (1999) investigated Zn desorption kinetics as influenced by phosphate by alfisol, oxisol, and vertisol. They tested several kinetic models and found that Zn release was best described by the Elovich equation. A primary limitation of simple models such as Elovich is that different rates of reaction are needed to describe Zn adsorption from Zn desorption or release. Multireaction kinetic models are capable of describing adsorption and desorption or release in soils. These models have been successfully used to describe kinetic behavior for several heavy metals and contaminants (Amacher et al. 1988; Barnett et al. 2000; Darland and Inskeep 1997). None of the multireaction-type models have been tested for their capability to describe Zn reactivity in the presence of P in the soil–water environment.

Case Study

Here we present results of a study for Zn–P retention for three surface soils with contrasting properties: a Webster loam, Windsor sand, and Olivier loam. Adsorption of Zn was studied using the batch (mixing) method as described by Selim and Amacher (1997). The duration of adsorption was 1 day. Release or desorption commenced following the 1-day adsorption step using sequential or successive dilutions. To study the influence of P on Zn adsorption as well as release, the Zn batch experiments were also carried out where different levels of P concentrations were added in the solutions.

A family of Zn isotherms is represented in Figure 4.1 for each soil, where different initial P concentrations were added (from 0 to 100 mg L^{-1}). Distinct differences of the extent of the amount of Zn sorbed among the different soils were observed. Highest sorption was observed for the neutral Webster soil. Strong retention of Zn is observed as the soil pH increased (Barrow 1987; Harter and Naidu 2001). Moreover, Webster soil is a fine loamy Haplaquoll with 3.7% calcium carbonates, and the presence of carbonates enhances Zn sorption in soils (see Mesquita and Vieira e Silva 1996). The Olivier and Windsor soils exhibited lower sorption capacities where Windsor soil showed the least sorption. Hinz and Selim (1994) reported stronger retention for Zn by Olivier compared to Windsor, which may be due to higher CEC caused predominantly by smectitic clays. In contrast, Windsor is an entisol and contains parent material that has not been completely weathered to secondary minerals and hence lower sorption capacity for Zn. Increased CEC results in greater negatively charged sites to adsorb Zn (Cavallaro and McBride 1984; Kurdi and Doner 1983).

The influence of P on increased Zn sorption was clearly manifested in the isotherms shown in Figure 4.1 where similar trends were observed for all soils. Saeed and Fox (1979) showed that P fertilization increased Zn adsorption by soils from Hawaii that contained colloids predominantly of the variable charge type. Their results support the hypothesis that phosphate additions to soils increased Zn adsorption by increasing the negative charges on iron and aluminum oxides. Wang and Harrell (2005) reported that Zn sorption was enhanced by $H_2PO_4^-$ as opposed to chloride (Cl^-) or nitrate (NO_3^-) in acid soils.

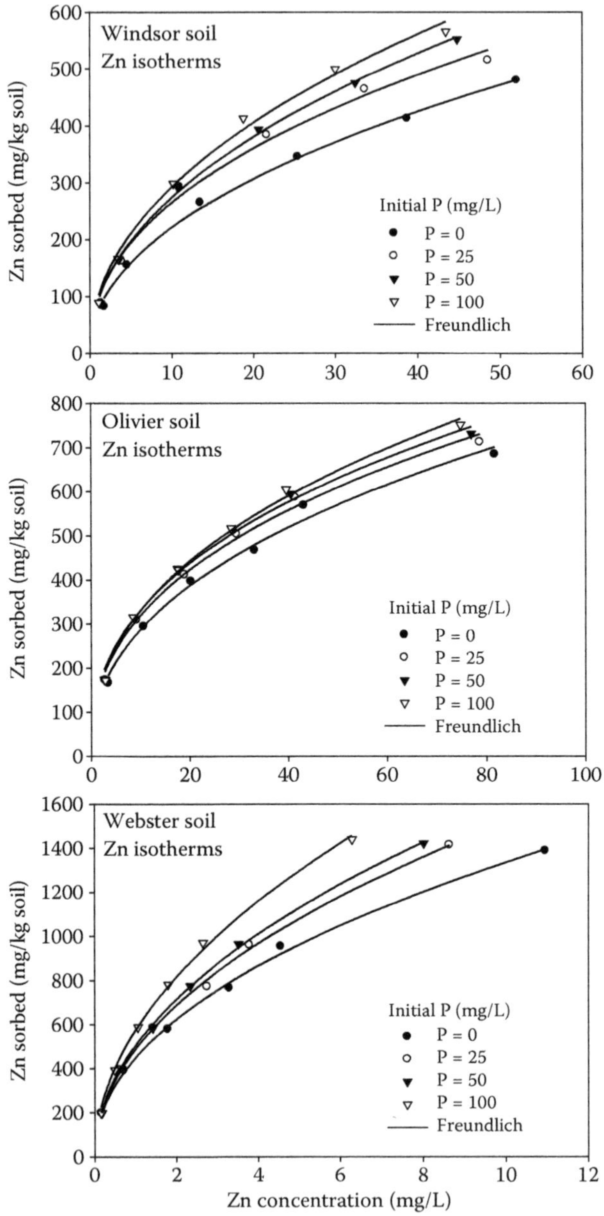

FIGURE 4.1
Zinc adsorption isotherms on different soils after 24-h sorption in the presence of different P concentrations. Solid curves are simulations using the Freundlich equilibrium.

The overwhelming increase in total Zn sorption with P addition was consistent with other studies (Agbenin 1998; Pardo 1999; Rupa and Tomar 1999; Xie and Mackenzie 1989). Moreover, the influence of P increased Zn sorption for Windsor and Webster soils much greater than for Olivier soil.

In Figures 4.2 and 4.3, we present the amount of Zn sorbed versus time to illustrate the kinetics of Zn desorption for the various initial concentrations (C_o) used. As illustrated in the figures, Zn desorption exhibited strong time-dependent behavior as depicted by the continued decrease in the amount sorbed with time. For Windsor and Olivier soils, the rate of Zn desorption was initially rapid and was followed by gradual or slow reactions. In contrast, slow release reactions for Zn were dominant for Webster. The results indicate that a fraction of Zn was weakly sorbed by Windsor and Olivier

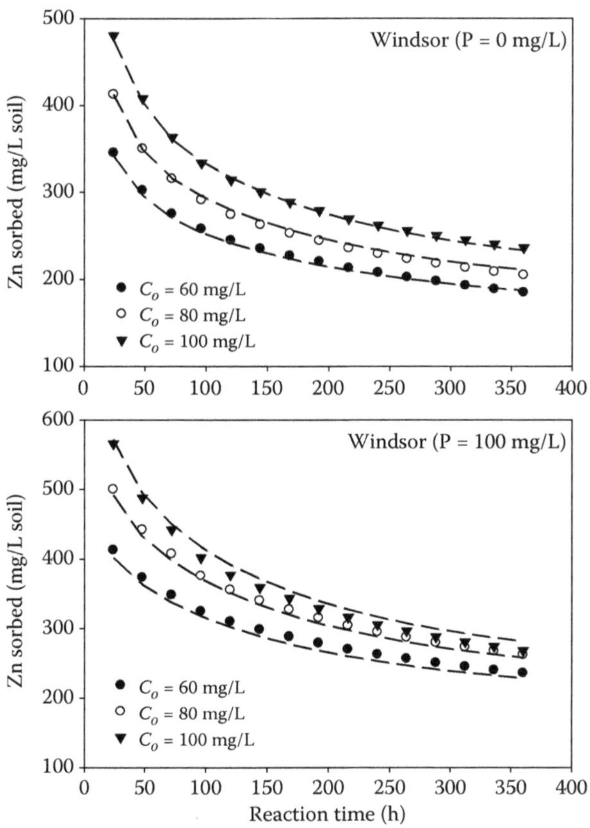

FIGURE 4.2
Zn concentration in Windsor soil versus time during adsorption–desorption for various initial Zn concentrations. Dashed curves are MRM simulations.

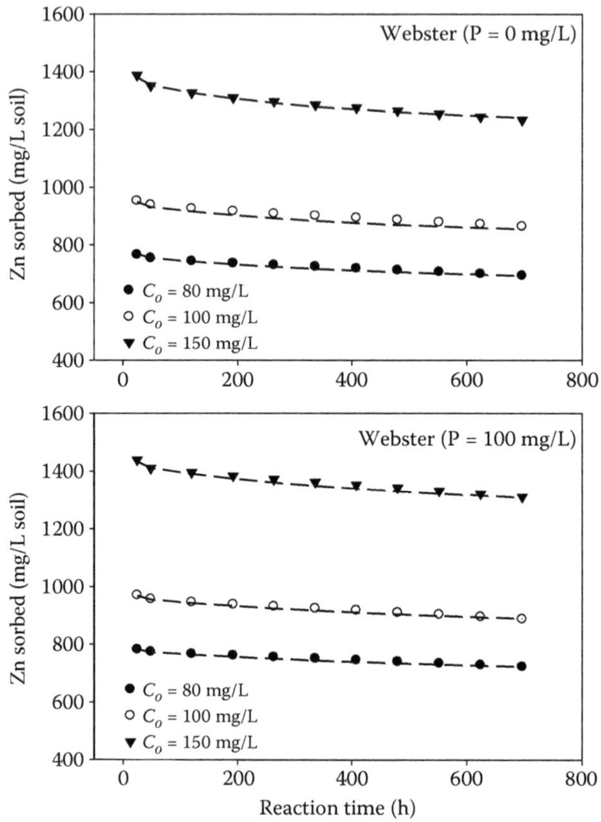

FIGURE 4.3
Zn concentration in Webster soil versus time during adsorption–desorption for various initial Zn concentrations. Dashed curves are MRM simulations.

soils via ion exchange or outer-sphere surface complexation. In contrast, Zn was strongly sorbed in Webster soil and may be bound via inner-sphere surface complexation. Consistent with our results, Zn desorption kinetic results reported by Rupa and Tomar (1999) showed that by increasing the level of applied P to 20 and 40 mg/kg, a subsequent decrease in Zn desorption by 20%–31% and 39%–53% was observed.

Several studies suggested that P-induced Zn sorption may be related to increased surface negative charges and/or creation of specific sorption sites as a result of P sorption on OM, Fe/Al oxides, and clays (Barrow 1987; Bolland et al. 1977; Xie and Mackenzie 1989). Xie and Mackenzie (1989) postulated that sorption of P on OM and oxides can form inner-sphere surface complexes that increase negative charges, resulting in increased surface Zn retention. Kuo and McNeal (1984) proposed that the added P acts as a bridge between soil surfaces and sorbed Zn, or sorbed P alters the soil properties sufficiently to affect Zn sorption. The latter was also proposed by Tagwira et al. (1993). Thus, P application can affect Zn sorption, resulting in increased specific sorption sites for Zn and subsequently reducing Zn release or desorption. Furthermore, P-induced Zn retention may be due in part to precipitation or coprecipitation involving the formation of solid solution of $ZnHPO_4$ as an intermediate product (Agbenin 1998). Al–P precipitation products from reactions of P with Al–OM complexes may form new sites to retain Zn on soil organic matter. Thus, P sorption on the surface increases Zn sorption and restricts Zn desorption, depending on soil pH, surface complexation, and soil precipitation.

Multireaction Model

The multireaction model (MRM) presented here considers several interactions of heavy metals with soil matrix surfaces (Amacher et al. 1988; Hinz and Selim 1994; Selim et al. 1992). Based on soil heterogeneity and observed kinetics of sorption–desorption, the MRM has been proposed to describe the kinetics of reactivities of heavy metals in the soil environment. Basic to the multisite approach is that the soil solid phase is made up of different constituents (clay, silica, organic matter, iron and aluminum oxides, etc.), and that a solute species is likely to react with various constituents (sites) by different mechanisms. The model is multipurpose in nature, which accounts for linear as well as nonlinear reaction processes of the equilibrium and/or kinetic (reversible and irreversible) type. The capability of the model is not only limited to describing commonly measured batch-type sorption data (following a specific reaction time, e.g., 1 day) but also to describing changes in concentration with time of reaction during sorption as well as desorption. The uniqueness of this model is that its aim is to describe the reactivity of solutes with natural systems versus time during sorption or desorption. In contrast, for most models (e.g., simple linear, Freundlich, Langmuir, Elovich, and dual and treble domain reactivity models), two distinct sets of parameters are obtained: one for sorption and one for desorption. Moreover, the use of such models yields a set of parameters that is only applicable for a specific reaction time. On the other hand, the MRM provides a comprehensive accounting of the sorption–desorption processes, where a single set of parameters is sought that is applicable for an entire data set and for a wide range of initial (or input) concentrations. MRM accounts for linear as well as nonlinear reaction processes of equilibrium and/or kinetic (reversible and irreversible) type. The model with reversible as well as irreversible sorption of the concurrent and consecutive type is shown in Figure 4.4. Here S_e represents the amount retained on equilibrium sites (mg/kg), S_1 and S_2 represent the amount retained on reversible kinetic sites (mg/kg), S_{irr} represents the amount retained on consecutive irreversible sites (mg/kg), and S_3 represents the amount retained on concurrent irreversible sites (mg/kg).

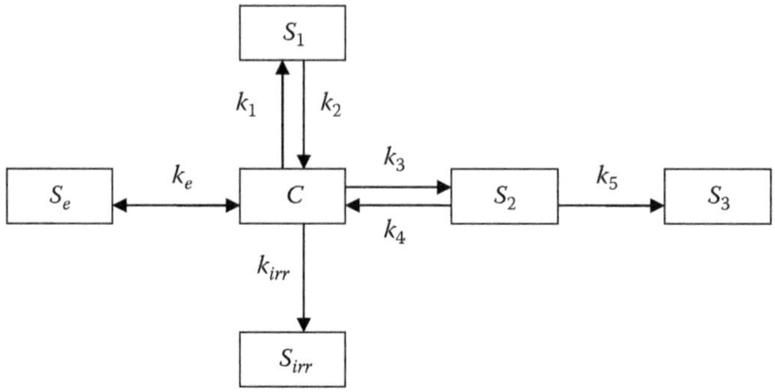

FIGURE 4.4
Schematic of the MRM for solute retention in soils.

The retention reactions associated with the MRM are expressed as

$$S_e = K_e \left(\frac{\theta}{\rho} \right) C^b \tag{4.1}$$

$$\frac{\partial S_1}{\partial t} = k_1 \left(\frac{\theta}{\rho} \right) C^n - k_2 S_1 \tag{4.2}$$

$$\frac{\partial S_2}{\partial t} = \left[k_3 \left(\frac{\theta}{\rho} \right) C^m - k_4 S_2 \right] - k_5 S_2 \tag{4.3}$$

$$\frac{\partial S_3}{\partial t} = k_5 S_2 \tag{4.4}$$

$$S_{irr} = k_{irr} \left(\frac{\theta}{\rho} \right) C \tag{4.5}$$

where t is the reaction time (h), ρ is the soil bulk density (g/cm^3), θ is the water content (cm^3/cm^3), b is the reaction order, and C is the concentration in solution (mg/L). K_e is a dimensionless equilibrium constant (cm^3/g), k_1 and k_2 (h^{-1}) are the forward and backward rate coefficients associated with S_1, respectively, and n is the reaction order; k_3 and k_4 (h^{-1}) are the forward and backward rate coefficients associated with S_2, respectively, and m is the reaction order; k_5 (h^{-1}) is the irreversible rate coefficient; and k_{irr} (h^{-1}) is the rate coefficient associated with S_{irr}. In the above equations, we assumed that $n = m$ since there is no method for estimating n and/or m independently.

Kinetic batch data were fitted to the MRM described here using the nonlinear least squares optimization method. Statistical criteria used often for estimating the goodness-of-fit of the models to the data are the coefficients of determination r^2 and the root mean square error (RMSE). The solid and dashed curves shown in Figures 4.2 and 4.3 are simulations of the time-dependent retention Zn kinetics using our MRM. These simulations illustrate the capability of the model describing Zn release for various initial concentrations (C_o). Moreover, the presence of increasing levels of P on the kinetic behavior of Zn was well described by the MRM. Based on these calculations, it is clear that the MRM was successful in describing the time-dependent behavior of Zn sorption–desorption for all three soils and for the different P concentration.

Desorption or release results that followed adsorption are presented as isotherms in the traditional manner in Figures 4.5 and 4.6 for all soils. Distinct

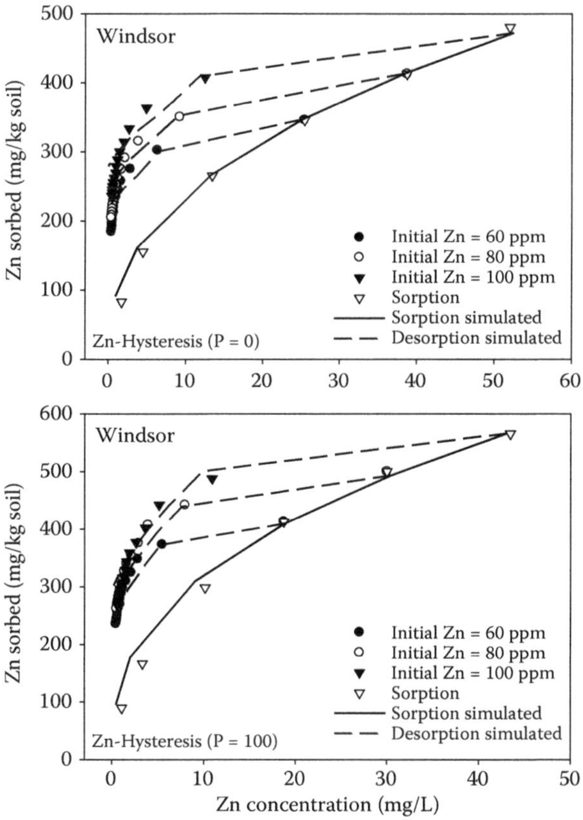

FIGURE 4.5
Adsorption and desorption isotherms for Windsor soil with different P concentration. The solid and dashed curves depict results of the MRM simulation for sorption and desorption.

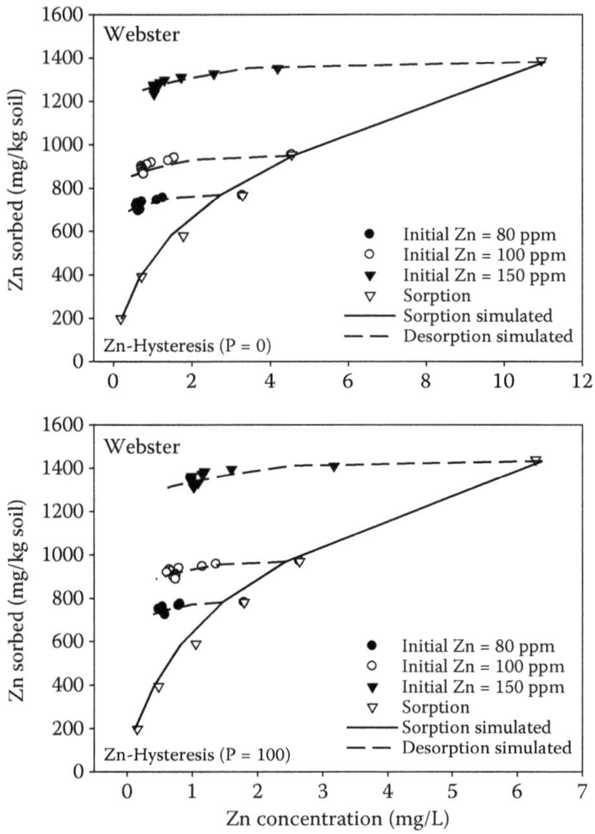

FIGURE 4.6
Adsorption and desorption isotherms for Webster soil with different P concentration. The solid and dashed curves depict results of MRM model simulation for sorption and desorption.

discrepancies between adsorption and successive desorption isotherms are clearly observed in Figures 4.5 and 4.6 and indicate considerable hysteresis for Zn release, the extent of which varied among the three soils. This observed hysteresis was not surprising in view of the kinetic behavior of Zn and is indicative of nonequilibrium behavior of Zn retention mechanisms. Since significant irreversibility of Zn sorbed on mineral surfaces and soils has been extensively reported in the literature, this observed hysteresis may be due in part to slow diffusion, kinetic ion exchange, as well as irreversible mechanisms.

Hysteresis results shown in Figures 4.5 and 4.6 illustrate the extent of kinetics during release of Zn in the three soils. Webster exhibited limited kinetics or very slow release where desorption isotherms exhibited little release over time for all initial Zn concentrations considered. In contrast, Zn release as a percentage of the total sorbed for the two acidic soils was 47%–51% and 42%–49% for Windsor and Olivier soil, respectively. For the neutral Webster, only 9%–11% of the sorbed Zn was released into the soil solution after 28 days of successive desorptions. This is due to its much higher content of clay with smectite as the predominating mineral, organic matter, Fe/Al oxides, and CEC. Soils with high clay, OM, Fe/Al contents, and CEC have more available reactive sites and increased specific sorption and reduce Zn desorption (Xie and MacKenzie 1989). Furthermore, at low initial Zn concentrations, lower proportions of Zn were desorbed, indicating high affinity for Zn in soils, whereas at high concentration, the percentage of Zn desorption increased, indicating lower Zn affinity. Although higher Zn amounts were sorbed as a result of the presence of P, similar release curves were obtained for all P concentrations (Figures 4.5 and 4.6). Thus, the kinetics of Zn retention was not altered by the presence of increased levels of P. However, the application of P resulted in reduced desorption of Zn for all three soils. This is perhaps because the added P increased specific sorption sites on OM and Fe/Al oxide surfaces where Zn was tightly sorbed. Moreover, the effect of P on Zn sorption by Webster soil was more pronounced compared to the acidic soils. The effect of P on Zn sorption is likely due to high clay, OM, Fe/Al contents, and CEC for Webster soil as discussed above.

In order to assess the mobility of Zn in soils, column transport experiments were carried out using the miscible displacement technique. Transport of Zn in soils was quantified using miscible displacement techniques in water-saturated soil columns. A Zn pulse solution with or without P was applied to each column. In Figure 4.7, a Zn breakthrough curve is presented for a Windsor soil column that received a pulse of Zn alone (column 1). In contrast, Figure 4.8 presents results of BTCs where a mixed pulse

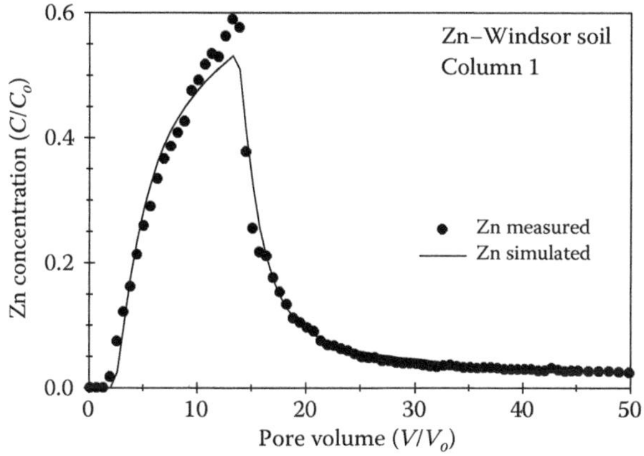

FIGURE 4.7
BTC for Zn form Windsor soil column 1. Solid curve is based on MRM simulation.

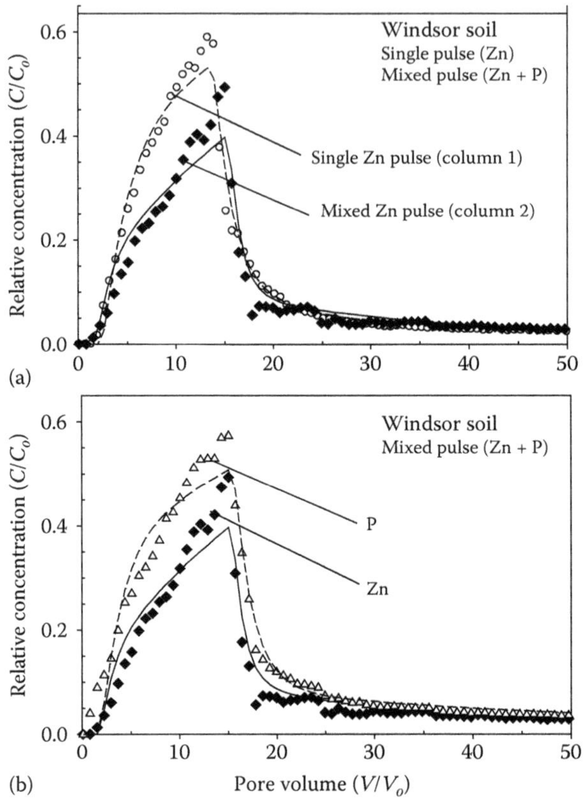

FIGURE 4.8
(a) BTCs for Zn from single (column 1) and mixed pulse (column 2) form Windsor soil. (b) BTCs for Zn and P from mixed pulse (column 2). Solid and dashed curves are based on MRM simulations.

of P and Zn was introduced (column 2). A comparison of the two BTCs clearly shows increased retention of Zn sorption in the presence of P. In fact after some 120 pore volumes, 55% of applied Zn was retained in the presence of P compared to 31% retention without P. Overall Zn concentration in the effluent decreased in the presence of P without a change in the relative position of the BTC (e.g., arrival of a Zn front in the effluent solution). Such behavior is indicative of increased irreversible Zn retention in the presence of P. It is clear that the presence of P resulted in a lower peak concentration and less overall Zn recovery in the effluent when compared to column 1 BTC where P was absent (see Figure 4.9). The transport data are consistent with retention and clearly indicate that the presence of P increases the amount Zn sorbed. Moreover, all BTCs were successfully described by the multireaction model.

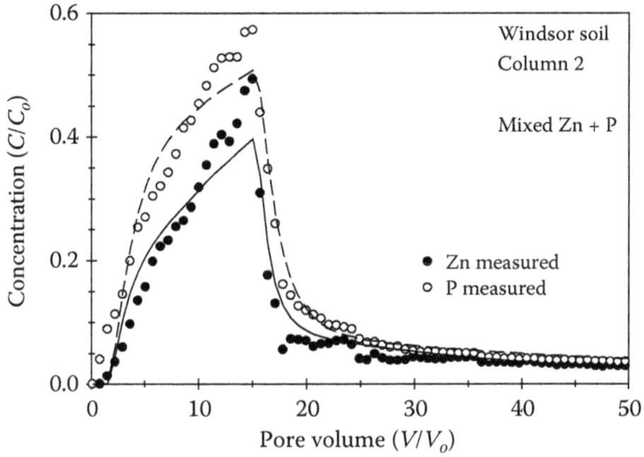

FIGURE 4.9
BTCs for Zn and P form Windsor soil column 2. Solid and dashed curves are based on MRM simulations.

Cadmium–Phosphate

Cd is not known to be an essential nutrient to any plant or animal species, but exposure to it can cause adverse health effects. The availability of Cd to plants depends greatly on the physical and chemical properties of the soil. Symptoms of Cd toxicity in plants include growth retardation, inhibition of photosynthesis, induction and inhibition of enzymes, altered stomata action, water relations, efflux of cations, and generation of free radicals (Prasad 1995). Sources of Cd include combustion of fossil fuels, iron and steel industry, nonferrous metal production, municipal waste combustion, disposal of Cd-containing products, and sewage sludge and phosphate fertilizer application to agricultural fields (Jackson and MacGillivray 1993; Van Assche and Ciarletta 1992). With such high concentrations of Cd accumulating around the world, there is great concern about the ecological impact of its toxicity.

The adsorption of heavy metals such as Cd in soils has been commonly found to be a multistep process involving initial fast adsorption followed by slow adsorption and diffusion into solid particles (O'Reilley et al. 2001; Wang and Xing 2004). Heavy metal interactions in the soil system include metal sorption at surface sites, precipitation, ion exchange with clay minerals, binding by organic colloidal material, or adsorption of metal–ligand complexes (Cantrell et al. 2002; Naidu et al. 1998). In the past several decades, a large amount of research has been conducted to better understand the behavior of Cd in the soil. Cd adsorption–desorption reactions and subsequent transport are greatly affected by soil properties such as CEC, pH, clay content, organic matter content, and the presence of coadsorbing anions (John 1971;

Levi-Minzi et al. 1976). Of these factors, pH is considered the major property controlling metal mobility and availability in the soil (Alloway and Jackson 1991). In a study by Levi-Minzi et al. (1976) in which Cd adsorption by 10 surface soils from Italy was examined, it was found that soils with low CEC, low organic matter content, and high pH had fewer sites for Cd sorption than soils with higher CEC and organic matter, and that were more acidic.

More recent studies have investigated Cd sorption–desorption reactions in the presence of coadsorbing anions such as phosphate. Many of these studies have found that the presence of phosphate increases the sorption and decreases desorption of Cd by soils (Adhikari and Singh 2008; Chen et al. 2006; Pardo 2004; Xiong 1994). A column study by Adhikari and Singh (2008) in which different remediation methods were applied to Cd-contaminated soils found that the presence of phosphate increased the soils' surface net negative charge. When Cd was added along with the phosphate, the increase in negative charge due to phosphate sorption resulted in corresponding increase in Cd sorption. Pardo (2004) also observed that P sorption restricted Cd desorption and attributed this to the need to balance the increase in negative charge induced by the P sorption. However, Pardo (2004) also reported that the increase in negative charge conveyed to the soil surface decreased with increasing P sorption. Pardo (2004) concluded that in acidic soils, most of the effect of P on Cd sorption–desorption reactions was related to variation in the pH of the soil suspensions. Xie and Mackenzie (1989) reported that P sorption increased soil CEC, which would in turn increase Cd adsorption.

Case Study

The influence of the presence of P on Cd sorption in soils is illustrated by the family of Cd isotherms shown in Figure 4.10 where different initial P concentrations were added (0 to 100 mg/L). All Cd isotherms appear nonlinear regardless of the initial P concentration in solution in Webster soil. The influence of P on increased Cd sorption was clearly manifested by the increased Freundlich distribution coefficient K_F from 532 to 601 mL/g for initial P concentrations of 0 and 100 mL/g. In contrast, the nonlinear parameter N did not exhibit much variation (0.50 to 0.54) for all P concentrations.

In Figure 4.11, the BTCs for Cd and P are presented for a Webster soil column that received a Cd pulse followed by a P pulse (i.e., sequential pulses), and in Figure 4.12, BTCs for Cd and P are presented for a Webster soil column that received two consecutive pulses of a mixed P and Cd solution. Cd BTCs indicate strong retention with low Cd concentration in the effluent solution, especially for the mixed column. Peak concentration of the BTC results was 70 mg/L (C/C_o of 0.37), compared to 30 and 60 mg/L associated with the first and second pulses, respectively, of the mixed column. Based on the area under the curve, the amount of Cd recovery was 28% for the sequential column and only 18% for the mixed column, which indicates Cd higher sorption in the presence of P. The observed strong Cd retention exhibited

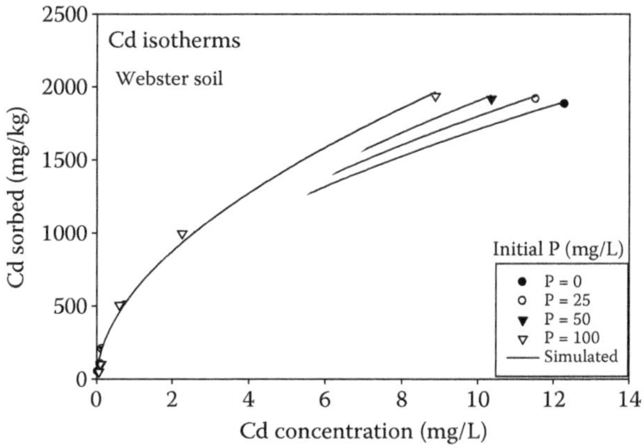

FIGURE 4.10
Cd adsorption isotherms for Webster soil after 24 h of sorption in the presence of various P concentrations. Solid curves are simulations using the Freundlich equation.

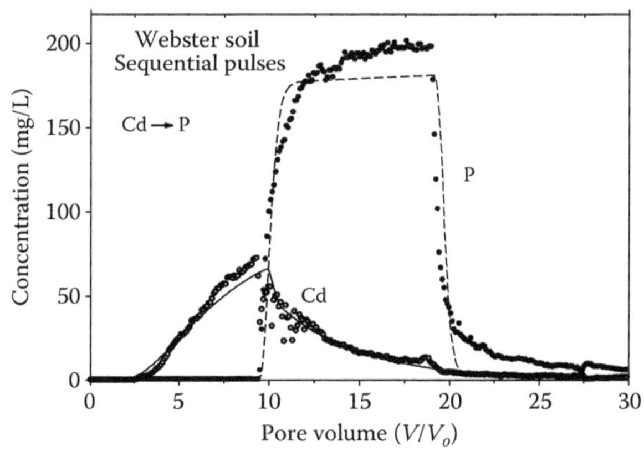

FIGURE 4.11
Cd and P breakthrough results for Webster soil column that received pulses of Cd and P. The solid and dashed curves are MRM predictions.

by the results of Figures 4.11 and 4.12 is supported by the sorption isotherm shown in Figure 4.10, which indicates high Cd sorption that is likely due to high clay content, organic matter, amorphous Fe and Al, and the presence of carbonates. In addition, x-ray diffraction analysis indicates that for Webster soil, smectite is the dominant clay type.

The solid curve shown in Figures 4.12 and 4.13 are MRM simulations for the two Webster columns. For the sequential column, the model provided a good description of Cd as well as P results r^2 of 0.96 for Cd and r^2 of 0.95 for P. For the mixed column, the respective r^2 were 0.85 and 0.86. Overall good model description of the BTCs for Cd and for P was achieved for the sequential

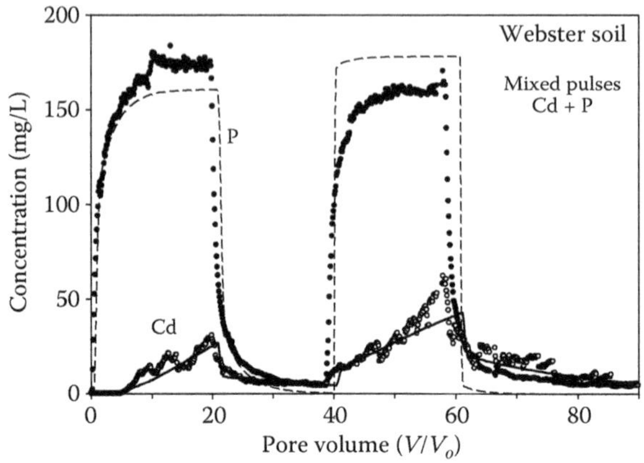

FIGURE 4.12
Cd and P breakthrough results for Webster soil column that received two mixed pulses of Cd and P solution. The solid and dashed curves are MRM predictions.

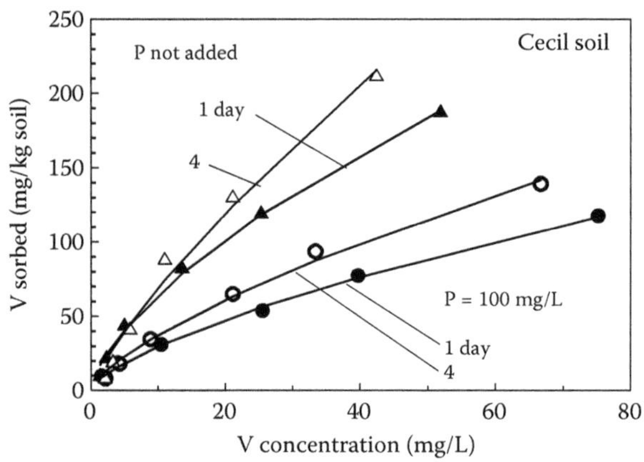

FIGURE 4.13
Vanadium isotherms after 1- and 4-day retention for Cecil soil. Isotherms are also presented in the presence of 100 mg of P in the soil solution.

and mixed columns. Peak P concentration and the tailing of the BTC were not well described by the model, however. The model version that provided the best prediction for Cd was where only kinetic reactions were assumed. Other model versions that account for equilibrium sorption resulted in the finite-difference solution of the convection–dispersion transport equation becoming unstable and convergence was not attained. This is an indication that all Cd retention mechanisms for Webster soil are dominated by kinetic or time-dependent reactions. For P, for the dashed curves shown in Figures 4.11 and 4.12, the multireaction and transport model (MRTM) versions were fully kinetic, which included reversible–irreversible P retention. The breakthrough results for P indicate that for both the sequential and mixed columns, more than 90% P recovery was obtained. These results were not expected, and a significant amount of P sorption was anticipated based on the soil properties of this soil. In fact, the lack of P sorption by the soil columns is in contrast to strong sorption of P for this soil with K_F of 134 and b of 0.682 (figure not shown). These results suggest that the presence of Cd resulted in reduced P retention and enhanced mobility and bioavailability in this soil. Similar findings were obtained for three other soils having a wide range of soil properties.

Vanadate–Phosphate

Vanadium is a ubiquitous trace element in the environment and is an essential trace element for living organisms, but in excessive amounts, it is harmful to humans, animals, and plants (Crans et al. 2004). Vanadium acts as a growth-promoting factor and participates in fixation and accumulation of nitrogen in plants, whereas a high concentration of vanadium may reduce their productivity (Underwood 1977). In soils, vanadium is derived from parental rocks and deposits. The average content of vanadium in the soil in the world has been established at 90 mg kg^{-1} (Reimann and de Caritat 1998); some residual soils derived from rocks enriched in vanadium contain elevated amounts of this metal from about 70 to 100 mg kg^{-1} (Anke 2004).

Environmentally, the occurrence of vanadium in petroleum and coal is of high significance because fuels constitute major sources of vanadium emissions to the atmosphere (Dechaine and Gray 2010). A large fraction of the vanadium-rich atmospheric particles may enter the soil environment as particulate fallout or by dissolving in the rain. Vanadate and vanadyl ions are versatile at forming complexes that inhibit or stimulate activity of many enzymes by specific mechanisms.

Case Study

Here we present vanadium transport and adsorption results for two soils with contrasting properties: a Sharkey clay (very fine, montmorillonitic, nonacid, Vertic Haplaquept) and Cecil soil (clayey, kaolinitic, thermic Typic Kanhapludult). The Sharkey soil has an organic matter content of 1.41‰ and a pH of 5.9, whereas Cecil soil has an organic matter content of 0.74‰ and a pH of 5.6. Miscible displacement experiments were carried out using soil columns under saturated conditions and state flux was slowly saturated upward with 0.01 N NH_4Cl background solution for 4 days prior to V pulse application. A vanadium pulse of 100 mg L^{-1} of NH_4VO_3 in 0.01 N NH_4Cl was applied to each column and was followed by 0.01 N NH_4Cl solution. The effluent solution from each column was collected using a fraction collector. To assess the impact of the presence of P on vanadium mobility in the soil columns, 100 mg L^{-1} of P as meant $NH_4H_2PO_4$ was mixed with the vanadium pulse solution. Moreover, periods of stop flow (or flow interruption) of 2-day duration were implemented to test whether equilibrium or kinetic processes are dominant transport.

Vanadium adsorption by the two soils represents contrasting behavior with regard to intensity as well as kinetics. Sharkey soil exhibited stronger affinity for vanadium than Cecil soil. In addition, Cecil soil, which is predominantly kaolinitic, exhibited an extremely limited kinetic reaction (see Figures 4.13 and 4.14). When compared to 1-day retention, after 4 days of reaction, an increase of only 30% was observed. In contrast, for Sharkey soil, which is predominantly montmorillonitic, a threefold increase in vanadium adsorption was observed when compared to 1-day sorption. The K_d values for Cecil soil were 1.70 and 2.31 L/kg after 1- and 4-day sorption, respectively. The corresponding K_d values for Sharkey soil were 73.03 and 360.91 L/kg after 1- and 21-day sorption, respectively. The presence of P that resulted in decreasing the amount of V sorbed varied among the two soils. In fact, the impact of P on reduced vanadium sorption was greatest for Sharkey where a onefold decrease was observed. In contrast, only a 50% decrease in vanadium sorption was observed for Cecil soil (see Figures 4.13 and 4.14).

The mobility of V as well as P is presented by the BTCs shown in Figures 4.15 through 4.17. These BTCs represent different column scenarios. In column A, a pulse of vanadium was followed by a pulse of phosphate P, which was subsequently followed by a pulse of a mixed solution of P plus V. In column B, a pulse of phosphate P was followed by a pulse of vanadium V, which was subsequently followed by a pulse of a mixed solution of P plus V. For

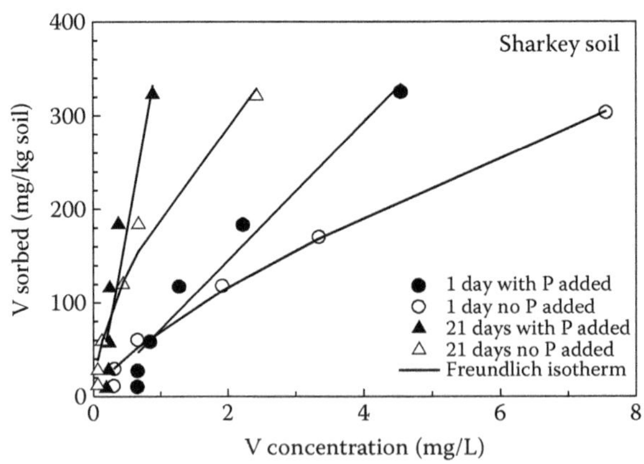

FIGURE 4.14
Vanadium isotherms after 1- and 21-day retention for Sharkey soil. Isotherms are also presented in the presence of 100 mg of P in the soil solution.

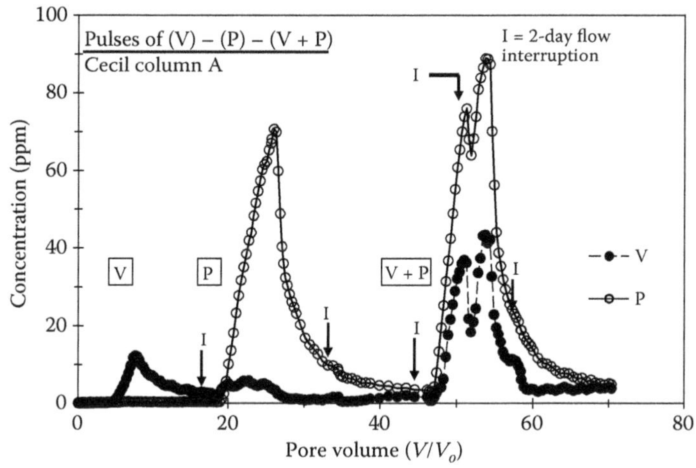

FIGURE 4.15
Breakthrough results (BTCs) of V and P from Cecil soil (column A). The column received three consecutive pulse V solutions P, V, and (P + V), respectively. Arrows indicate where 2-day flow interruptions occurred. Measured V and B results are connected by a straight line.

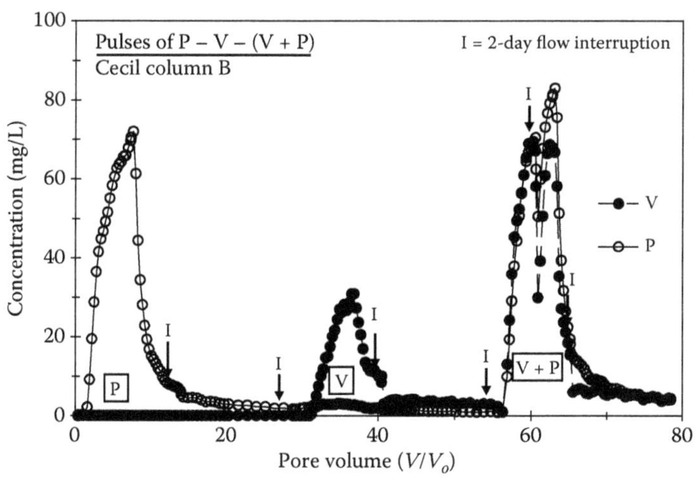

FIGURE 4.16
Breakthrough results (BTCs) of V and P from Cecil soil (column B). The column received three consecutive pulse P solutions V, V, and (P + V), respectively. Arrows indicate where 2-day flow interruptions occurred. Measured V and B results are connected by a straight line.

column C shown in Figure 4.17, the soil column received three consecutive pulses of mixed solution (P pulse V). The BTCs clearly indicate higher affinity for V over P for all cases presented here. This was evident when three consecutive pulses were introduced in the Cecil soil column, as illustrated in Figure 4.16. For the case shown in Figure 4.15, both P and V peaks represent

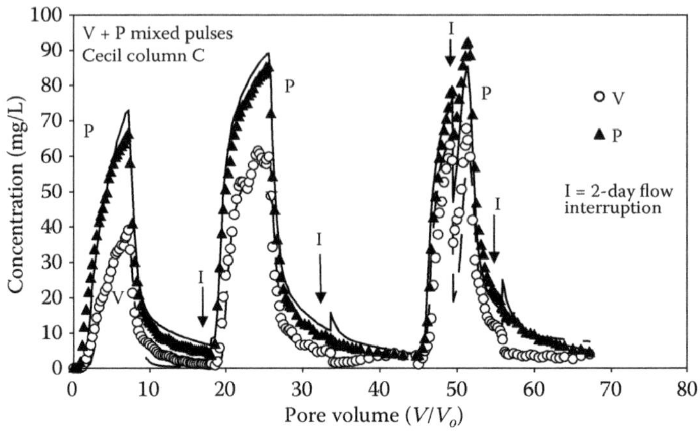

FIGURE 4.17
Breakthrough results (BTCs) of V and P from Cecil soil (column C). The column received three consecutive pulses of mixed solution of P and V. Arrows indicate where 2-day flow interruptions occurred. The solid and dashed curves are simulations for P and V, respectively, based on the competitive MRM.

the input pulse solution during all three pulses. Based on peak concentration and arrival time for each BTC, Cecil soil exhibited higher affinity for V than P. This result was consistent for all three consecutive pulses. The location of BTC maximum was the same for both P and V. This suggests that the retention mechanisms for P as well as V on Cecil soil were similar. However, a major difference is that the amount that appears to be irreversibly sorbed (or slowly released) is larger for V than P. We infer the extent of irreversible sorption (or slow release) based on the area under the BTCs. Moreover, the extensive tailing during leaching or desorption also suggests nonlinear and/or kinetic retention mechanisms for both V and P (Mansell et al. 1992).

Competition between V and P for the same retention sites on soil surfaces is evident when one compares BTCs shown in Figure 4.15 with those of Figure 4.16. The absence of P in the pulse solution resulted in a much higher retention of V with a much lower peak concentration when compared to the case when both P and V were present in the input pulse solution. The extent of retention can be estimated from the area under the BTC for V in the first input pulse where four times the amount of V was found in the effluent solution when P was present. In addition, a subsequent P pulse displaced a significant portion of V already retained by the Cecil soil during the first pulse. However, the opposite was not as apparent when a P pulse followed a V pulse, as shown in Figure 4.15.

Phosphate–soil interactions exhibit various reactions including kinetic, reversible, and nonlinear sorption, often with some degree of irreversibility (Chen et al. 1996; Mansell et al. 1992). The latter primarily occurs due to reactions with Fe and Al oxides in the soil. The term sorption is used in a general sense here to include P adsorption by mineral surfaces and complexation with humic acids (Al and Fe may form bridges between organic ligands and P ions). Irreversible sorption is considered to involve chemisorption and fixation within mineral structures. Scientists generally agree that many types of P retention processes can take place simultaneously. Less is known as to whether these reactions operate in series or in parallel. Experimental techniques are available in the literature to monitor sorption and fixation reactions. Although a mathematical model to completely describe P reactions in soil does not exist, models that approximate these reactions do. Multireaction sorption models constitute a class of such approximations and offer a practical means of approximating P mobility during water flow in soils. In general, the sorption behavior of vanadate in soils is similar to inorganic phosphate. However, V also reduces acid phosphatase activity in soil, which alters the rate of mineralization of organic matter and may reduce phosphate bioavailability (Tyler 1976). The interactions of V with other ions present in the soil solution (e.g., phosphate) and its potential mobility in the soil profile have not been fully investigated (Mikkonen and Tummavuori 1994).

For all three columns with an input pulse containing both P and V, BTCs for P arrived earlier and with higher peak concentrations than those for V. In addition, a decrease in effluent concentration due to flow interruption was observed in all BTCs associated with the third pulse in all three columns, which indicates increases in the amounts of P or V sorbed (Figures 4.15 through 4.17). These observations are consistent for both P and V and show the kinetic nature of the retention mechanisms. For Sharkey clay soil, miscible displacement results illustrated the extent of P as well as V retention in this montmorillonitic soil (results are not shown). For all input V concentrations, the presence resulted in a decrease in the amount of V sorbed at all reaction times. This finding was consistent with that obtained for the kaolinitic Cecil soil as well as by a montmorillonitic Sharkey soil.

Recognizing the competitive nature of V on P, attempts were made to utilize a competitive transport model to describe the data. The results are shown by the solid and dashed curves in Figure 4.17, Cecil column C, which received three consecutive pulse of mixed solution of V plus P. Clearly good agreement of the model-calculated data and measured results was obtained. The model used was originally developed by Sheindorf et al. (1981) and is commonly referred to as the Sheindorf–Rebhun–Sheintuch (SRS) model. Here competitive or multicomponent sorption is assumed to follow that for single-component sorption based on the Freundlich equation

$$S = K_F C^b \tag{4.6}$$

The derivation of the SRS equation was based on the assumption of an exponential distribution of adsorption energies for each component. Specifically, the SRS model was developed to describe competitive equilibrium sorption for multicomponent systems where the sorption isotherms of a single component follow the Freundlich equation. A general form of the SRS equation can be written as

$$S_i = K_i C_i \left(\sum_{j=1}^{l} \alpha_{i,j} C_j \right)^{n_i - 1} \tag{4.7}$$

where the subscripts i and j denote metal components i and j, l is the total number of components, and $\alpha_{i,j}$ is a dimensionless competition coefficient for the adsorption of component i in the presence of component j. The parameters K_i and n_i are the Freundlich parameters representing a single-component system i as described in Equation 4.6. By definition, $\alpha_{i,j}$ equals 1 when $i = j$. If there is no competition (i.e., $\alpha_{i,j} = 0$ for all $j \neq i$), Equation 4.7 yields a single-species Freundlich equation for component i identical to Equation 4.6.

The suitability of the multicomponent SRS equation for describing the competitive adsorption isotherms of trace elements on soil and soil minerals has been investigated by several researchers. A general procedure of applying the SRS equation is first to obtain the Freundlich distribution coefficient K_F and reaction exponent b or n by fitting the single-component isotherms to the Freundlich equation, followed by estimating the competition coefficients $\alpha_{i,j}$ through fitting the experimental isotherms of binary and ternary mixtures to the SRS equation (Roy et al. 1986). The competition coefficients $\alpha_{i,j}$ in SRS represent the relative selectivity of the sorbent to the heavy metal species. It is demonstrated that the SRS equation with competition coefficients estimated through nonlinear least squares optimization successfully described the experimental competitive adsorption isotherms of Ni and Cd on three different soils (Liao and Selim 2010). Gutierrez and Fuentes (1993) employed the SRS equation to represent the competitive adsorption of Sr, Cs, and Co in Ca-montmorillonite suspensions. They found that the SRS competition coefficients $\alpha_{i,j}$ obtained from experiment data of binary mixtures successfully predicted the competitive adsorption of ternary mixture Sr–Cs–Co. Similarly, Bibak (1997) found that values of SRS competitive coefficients obtained from binary sorption experiments successfully predicted sorption data of the ternary solute mixture Cu–Ni–Zn. The SRS equation was successfully used to describe the competitive sorption of Cd, Ni, and Zn on a clay soil by Antoniadis and Tsadilas (2007). In addition, the SRS equation was also used by Wu et al. (2002) in representing the competitive adsorption of molybdate, sulfate, selenate, and selenite on a γ-Al_2O_3 surface where relative affinity coefficients were used instead of competitive coefficients. The relative affinity coefficients were calculated as the ratios of the proton coefficients of competing anions. The simulation result showed that the sorption affinity of anions on the γ-Al_2O_3 surface decreased in the order of $MoO_4^{2-} > SeO_3^{2-} > SeO_4^{2-} > SO_4^{2-}$.

Competetive Multireaction Model (CMRM)

The kinetic (time-dependent) sorption models of the Freundlich type given were developed to simulate the sorption of solutes during their transport in soils and aquifers (Selim 1992). Zhang and Selim (2007) extended the equilibrium Freundlich approach to account for kinetic competitive or multiple component systems. Specifically, the model accounts for equilibrium and kinetic adsorption in a way similar to the SRS equation described previously. This model represents a modification of the MRTM described in Equations 4.1 to 4.5. Zhang and Selim (2007) modified the MRM to account for competitive Freundlich-type retention so that

$$(S_e)_i = K_{e,i} C_i \left(\sum_{j=1}^{l} \alpha_{i,j} C_j \right)^{n_i - 1} \tag{4.8}$$

$$\frac{\partial (S_1)_i}{\partial t} = k_{1,i} \frac{\theta}{\rho} C_i \left(\sum_{j=1}^{l} \alpha_{i,j} C_j \right)^{n_i - 1} - k_{2,i} (S_1)_i \tag{4.9}$$

$$\frac{\partial (S_2)_i}{\partial t} = k_{3,i} \frac{\theta}{\rho} C_i \left(\sum_{j=1}^{l} \alpha_{i,j} C_j \right)^{n_i - 1} - (k_{4,i} + k_{5,i})(S_2)_i \tag{4.10}$$

$$\frac{\partial (S_5)_i}{\partial t} = k_{s,i} (S_5)_i \tag{4.11}$$

$$\frac{\partial (S_{irr})_i}{\partial t} = k_{irr,i} \frac{\theta}{\rho} C_i \tag{4.12}$$

When competition is ignored (i.e., $\alpha_{i,j} = 0$ for all $j \neq i$), Equation 4.8 is reduced to a single species *n*th-order kinetic sorption. Examples of the capability of this approach to describe the transport of the competitive arsenate and phosphate behavior in soil columns are presented in Zhang and Selim (2007) and Ni and Cd by Liao and Selim (2010).

For all three Cecil columns shown in Figures 4.15 through 4.17, with an input pulse containing both P and V, BTCs for P arrived earlier and with higher peak concentrations than those for V. In addition, a decrease in effluent concentration due to flow interruption was observed in all BTCs associated with the third pulse in all three columns, which indicates increases in the amounts of P or V sorbed. These observations are consistent for both P and V and show the kinetic nature of the retention mechanisms. For column

C, which received three mixed pulses, the competitive model well described the effect of flow interruption for all three pulses. Such a response to flow interruption during pulse application and during leaching is indicative of strong kinetic retention behavior for P and V (see Murali and Aylmore 1983). For Sharkey clay soil, miscible displacement results illustrated the extent of P as well as V retention in this montmorillonitic soil (results are not shown). For all input V concentrations, the presence of P resulted in a decrease in the amount of V sorbed at all reaction times. This finding was consistent with that obtained for the kaolinitic Cecil soil as well as by a montmorillonitic Sharkey soil.

References

Adhikari, T. and M. V. Singh. 2008. Remediation of cadmium pollution in soils by different amendments: A column study. *Commun. Soil Sci. Plant Anal.* 39:386–396.

Adriano, D. C. 2001. *Trace Elements in Terrestrial Environments: Biogeochemistry, Bioavailability and Risk of Metals*, Second Edition, Springer, NY.

Agbenin, J. O. 1998. Phosphate-induced zinc retention in a tropical semi-acid soil. *Eur. J. Soil Sci.* 49:693–700.

Ahumada, I. T., A. Bustamante and E. B. Schalscha. 1997. Zinc speciation in phosphate-affected soils. *Commun. Soil Sci. Plant Anal.* 28:989–995.

Amacher, M. C., H. M. Selim and I. K. Iskandar. 1988. Kinetics of chromium(VI) and cadmium retention in soils; a nonlinear multireaction model. *Soil Sci. Soc. Am. J.* 52:398–408.

Anke, M. 2004. Vanadium. In: E. Merian, M. Anke, M. Ihnat and M. Stoeppler (eds.), *Elements and Their Compounds in the Environment*, 2nd Edition, pp. 1171–1191. Weinheim, Germany: Wiley.

Antoniadis, V. and C. D. Tsadilas. 2007. Sorption of cadmium, nickel and zinc in mono- and multimetal systems. *Appl. Geochem.* 22:2375–2380.

Barnett, M. O., P. M. Jardine, S. C. Brooks and H. M. Selim. 2000. Adsorption and transport of uranium(VI) in subsurface media. *Soil Sci. Soc. Am. J.* 64:908–917.

Barrow, N. J. 1987. The effect of phosphate on zinc sorption by soils. *J. Soil Sci.* 38:453–459.

Bibak, A. 1997. Competitive sorption of copper, nickel, and zinc by an oxisol. *Commun. Soil Sci. Plant Anal.* 28:927–937.

Bolland, M. D. A., A. M. Posner and J. P. Quirk. 1977. Zinc adsorption by goethite in the absence or presence of phosphate. *Aust. J. Soil Res.* 15:279–286.

Cantrell, K. J., R. J. Serne and G. V. Last. 2002. *Hanford Contaminant Distribution Coefficient Database and Users Guide.* DOE Pacific Northwest National Lab, PNNL-13895.

Cavallaro, N. and M. B. McBride. 1984. Zinc and copper sorption and fixation by an acid soil clay: Effect of selective dissolutions. *Soil Sci. Soc. Am. J.* 48:1050–1054.

Chen, J. S., R. S. Mansell, P. Nkedi-Kizza and B. A. Burgoa. 1996. Phosphorus transport during transient, unsaturated water flow in an acid sandy soil. *Soil Sci. Soc. Am. J.* 60:42–48.
Chen, X., J. V. Wright, J. L. Conca and L. M. Peurrung. 1997. Evaluation of heavy metals remediation using mineral apatite. *Water Air Soil Pollut.* 98:57–78.
Crans, D. C., J. J. Smee, E. Gaidamauskas and L. Yang. 2004. The chemistry and biochemistry of vanadium and the biological activities exerted by vanadium compounds. *Chem. Rev.* 104:849–902.
Darland, J. E. and W. P. Inskeep. 1997. Effects of pH and phosphate competition on the transport of arsenate. *J. Environ. Qual.* 26:1133–1139.
Dechaine, G. P. and M. P. Gray. 2010. Chemistry and association of vanadium compounds in heavy oil and bitumen, and implications for their selective removal. *Energy Fuels* 24:2795–2808.
Gaudalix, M. E. and M. T. Pardo. 1995. Zinc sorption by acid tropical soils as affected by cultivation. *J. Soil Sci.* 46:317–322.
Gutierrez, M. and H. R. Fuentes. 1993. Modeling adsorption in multicomponent systems using a Freundlich-type isotherm. *J. Contam. Hydrol.* 14:247–260.
Harter, R. S. and R. Naidu. 2001. An assessment of environmental and solution parameter impact on trace-metal sorption by soils. *Soil Sci. Soc. Am. J.* 65:597–612.
Hinz, C. and H. M. Selim. 1994. Transport of zinc and cadmium in soils—Experimental-evidence and modeling approaches. *Soil Sci. Soc. Am. J.* 58:1316–1327.
Jackson, T. and A. MacGillivary. 1993. *Accounting for Cadmium.* Stockholm Environment Institute, London.
John, M. K. 1971. Influence of soil characterization on adsorption and desorption of cadmium. *Environ. Lett.* 2:173.
Kabata-Pendias, A. 2000. *Trace Elements in Soil and Plants*, CRC, Boca Raton, FL.
Kuo, S. and B. L. McNeal. 1984. Effect of pH and phosphate on cadmium sorption by hydrous ferric oxide. *Soil Sci. Soc. Am. J.* 48:1040–1044.
Kurdi, E. and H. E. Doner. 1983. Zinc and copper sorption and interaction in soils. *Soil Sci. Soc. Am. J.* 47:873–876.
Levi-Minzi, R., G. F. Soldatinai and R. Riffaldi. 1976. Cadmium adsorption by soils. *Soil Sci.* 27:10–15.
Liao, L. and H. M. Selim. 2010. Reactivity of nickel in soils: Evidence of retention kinetics. *J. Environ. Qual.* 39:1290–1297.
Mansell, R. S., S. A. Bloom, B. A. Burgoa, P. Nkedi-Krzza and J. S. Chen. 1992. Experimental and simulated P transport in soil using a multireaction model. *Soil Sci.* 153:185–194.
Mikkonen, A. and J. Tummavuori. 1994. Retention of vanadium (V) by three Finnish mineral soils. *Eur. J. Soil Sci.* 45:361–368.
Murali, V. and L. A. G. Aylmore. 1983. Competitive adsorption during solute transport in soils. 3. A review of experimental-evidence of competitive adsorption and an evaluation of simple competition models. *Soil Sci.* 136:279–290.
O'Reilley, S. E., D. G. Strawn and D. L. Sparks. 2001. Residence time effects on arsenate adsorption/desorption mechanisms on goethite. *Soil Sci. Soc. Am. J.* 65:67–77.
Pardo, M. T. 1999. Influence of phosphate on zinc reaction in variable charge soils. *Commun. Soil Sci. Plant Anal.* 30:725–737.
Pardo, M. T. 2004. Cadmium sorption-desorption by soils in the absence and presence of phosphate. *Commun. Soil and Sci. Plant Anal.* 35:1553–1568.

Prasad, M. 1995. Cadmium toxicity and tolerance in vascular plants. *Environ. Exp. Bot.* 35:525–545.

Roy, W. R., J. J. Hassett and R. A. Griffin. 1986. Competitive coefficient for the adsorption of arsenate, molybdate, and phosphate mixture by soils. *Soil Sci. Soc. Am. J.* 50:1176–1182.

Rupa, T. R. and K. P. Tomar. 1999. Zinc desorption kinetics as influenced by pH and phosphorus in soils. *Commun. Soil Sci. Plant Anal.* 30(13–14):1951–1962.

Saeed, M. and R. L. Fox. 1979. Influence of phosphate fertilization on zinc adsorption by tropical soils. *Soil Sci. Soc. Am. J.* 43:683–686.

Sarret, G., P. Saumitou-Laprade, V. Bert, O. Proux, J. Hazemann, A. Traverse, M. A. Marcus and A. Manceau. 2002. Forms of zinc accumulated in the hyperaccumulator *Arabidopsis. Plant Physiol.* 130:1815–1826.

Selim, H. M. 1992. Modeling the transport and retention of inorganics in soils. *Adv. Agron.* 47:331–384.

Selim, H. M. and M. C. Amacher. 1997. *Reactivity and Transport of Heavy Metals in Soils.* CRC Press, Boca Raton, FL.

Selim, H. M., B. Buchter, C. Hinz and L. Ma. 1992. Modeling the transport and retention of cadmium in soils: Multireaction and multicomponent approaches. *Soil Sci. Soc. Am. J.* 56:1004–1015.

Sheindorf, C., M. Rebhun and M. Sheintuch. 1981. A Freundlich-type multicomponent isotherm. *J. Colloid Interface Sci.* 79:136–142.

Tagwira, F., M. Piha and L. Mugwira. 1993. Zinc studies in Zimbabwean soils: Effect of pH and phosphorus on zinc adsorption by two Zimbabwean soils. *Commun. Soil Sci. Plant Anal.* 24:701–716.

Tyler, G. 1976. Influence of vanadium on soil phosphatase activity. *J. Environ. Qual.* 5:216–217.

Underwood, E. J. 1977. *Trace Elements in Human and Animal Nutrition.* Academic Press, New York.

Van Assche, F. J. and P. Ciarletta. 1992. Cadmium in the environment: Levels, trends and critical pathways. Seventh International Cadmium Conference, New Orleans, LA, Cadmium Council, Reston, VA, International Lead Zinc Research Organization Research Triangle Park, NC.

Wang, J. J. and D. L. Harrell. 2005. Effect of ammonium, potassium, and sodium cations and phosphate, nitrate, and chloride anions on zinc sorption and lability in selected acid and calcareous soils. *Soil Sci. Soc. Am. J.* 69:1036–1046.

Wang, K. and B. Xing. 2004. Mutual effects of cadmium and phosphate on their adsorption and desorption by goethite. *Environ. Pollut.* 127:13–20.

Xie, R. J. and A. F. Mackenzie. 1989. Effects of sorbed orthophosphate on zinc status in three soils of eastern Canada. *J. Soil Sci.* 40:49–58.

Xiong, L. M. 1994. Influence of phosphate on cadmium adsorption by soils. *Fert. Res.* 40:31–40.

Zhang, H. and H. M. Selim. 2007. Modeling competitive arsenate-phosphate retention and transport in soils: A multi-component multi-reaction approach. *Soil Sci. Soc. Am. J.* 71:1267–1277.

5

Influence of Phosphate Fertilizer on Cadmium in Agricultural Soils and Crops

Cynthia A. Grant

CONTENTS

Phosphorus fertilizers contain cadmium (Cd) as a contaminant at levels varying from trace amounts to as much as 300 mg Cd kg^{-1} of dry product and therefore can be a major source of Cd input to agricultural systems. Cd will accumulate in the soils if the amount added in fertilizer is greater than what is removed by crop harvest or through other loss pathways such as erosion, bioturbation, or leaching. Phosphate fertilizer may also influence Cd phytoavailability through fertilizer-induced changes in soil pH, osmotic strength of the soil solution, zinc (Zn) availability, soil organic matter, rhizosphere chemistry, soil microbial activity, and plant growth. As both total Cd input into the soil and fertilizer-induced changes in soil properties are a function of the rate of fertilizer application, management practices that improve fertilizer use efficiency should be adopted to minimize fertilizer inputs while maintaining crop yield potential.

Cd accumulation in the food chain is a major international concern. Chronic Cd toxicity may occur from long-term consumption of foods containing excess levels of Cd, possibly resulting in adverse health effects including kidney tubule dysfunction and reduced bone density (Godt et al. 2006). Plants will accumulate Cd from the soil with the amount of uptake depending on a wide range of factors including soil Cd concentration, soil chemical, physical, and biological characteristics, weather, agricultural management practices, and crop genetics (Grant et al. 1999, 2008; Sheppard et al.

2007, 2009a). Excess exposure to Cd may also harm soil organisms, affecting soil ecology and health (McGrath 1999). Therefore, it is important to ensure that Cd availability in soils does not exceed levels that can adversely affect soil or crop quality.

Cd is present in soils naturally, but can also be added to the soil from industrial contamination, atmospheric deposition, biosolid application, irrigation water, and agricultural inputs such as fertilizers, manures, and soil amendments (Alloway and Steinnes 1999; De Vries and McLaughlin 2013; Keller and Schulin 2003b; Sheppard et al. 2009b). Cd loss from agricultural soils occurs primarily from crop removal, but erosion, leaching, and bioturbation can also lead to Cd removal from the crop rooting zone (Sheppard et al. 2009b). Cd input and off-take are normally substantially smaller than the pool of total soil Cd (Christensen and Huang 1999). However, Cd concentration in soils can increase over time if input exceeds removal (Sheppard et al. 2009b). A gradual increase in Cd concentration has been reported in long-term studies in many parts of the world, including Sweden (Andersson 1977), Denmark (Dam Kofoed and Sondergard-Klausen 1983), England (Jones et al. 1987), and New Zealand (Schipper et al. 2011), reflecting a positive balance between Cd input and removal.

Phosphate fertilizer is a major source of Cd input in agricultural systems. Phosphorus (P) is an essential nutrient for plant growth, and crop productivity will decrease if an adequate supply of P is not available for plant uptake (Grant et al. 2001). If soils are deficient in P, application of organic or inorganic fertilizer is needed for optimum crop yield. However, inorganic P fertilizers and organic P sources such as manures and biosolids contain varying amounts of Cd (Syers et al. 1986) and can serve as a major anthropogenic source of Cd input into agricultural soils (De Vries and McLaughlin 2013; Jiao et al. 2012; Keller and Schulin 2003a; Sheppard et al. 2009b). The effect of phosphate on the risk of Cd movement into the food chain will depend not only on the total amount of Cd added to the soil but also on the influence of phosphate on the availability of Cd in the soil–plant system through indirect effects on soil chemistry, crop growth, and microbial interactions. Therefore, management of phosphate fertilizer, both in the short and long term, can directly and indirectly influence Cd concentration in crops and its movement into the food chain.

Direct Addition of Cadmium in Phosphate Fertilizer

Phosphate fertilizers contain Cd at concentrations ranging from only trace amounts to greater than 300 mg kg^{-1} on a total dry weight basis (Syers et al. 1986). The variation in Cd concentration of P fertilizers reflects the concentration of Cd in the phosphate rock used for its manufacture (Table 5.1) (Syers

TABLE 5.1
Origin and Concentration of P and Cd in Selected Phosphate Deposit

Phosphate Rock	Origin	P g kg^{-1}	Cd mg kg^{-1}
Gafsa	Tunisia	134	38
North Florida	North Florida, United States	133	3
Jordanian	Jordan	134	4
North Carolina	North Carolina, United States	127	48
Sechura	Sechura Desert, Peru	131	11
Mexican	Mexico	140	8
Nauru Island	Nauru Island, South Pacific	156	100
Arad	Negev Desert, Israel	141	12
Makatea Island	Makatea Island, South Pacific	130	10
Chatham Rise phosphorite	Chatham Rise, New Zealand	89	2

Source: Compiled from data presented in Syers, J. K. et al., *J. Sci. Food Agric.* 37:1057–1064, 1986.

et al. 1986). Cd concentration of sedimentary rocks is normally higher than that of igneous rocks because of differences in the amount of Cd coprecipitated as a substitute for calcium (Ca) in the phosphate compounds during geological formation of the deposits (Traina 1999). Cd present in the rock is retained and concentrated during the fertilizer production process, resulting in widely varying Cd concentrations in fertilizers produced from different sources of phosphate rock (Chien et al. 2003, 2011; Molina et al. 2009; Syers et al. 1986). Cd present in phosphate fertilizer will enter the soil–plant system when the fertilizer is used for crop production.

Cd added to agricultural systems in phosphate fertilizer can accumulate over time if application is in excess of removal (Schipper et al. 2011; Sheppard et al. 2009b). Cadmium added to soils from P fertilizer application is a function of the rate of application, the frequency of application, and the concentration of Cd in the fertilizer material. Over the long term, Cd concentration in soils reflects the balance between Cd input and off-take (Schipper et al. 2011; Sheppard et al. 2009b). Total input and net Cd balance will vary widely depending on the fertilizer source and the crops grown, although both input and off-take of Cd tend to be low relative to the total amount of Cd present in the soil (Christensen and Huang 1999; Christensen and Tjell 1991; Sheppard et al. 2009b). A number of assessments have shown that the Cd balance related to P fertilization calculated as input minus off-take is positive in many agricultural systems, indicating the potential for net accumulation of Cd in the soil (De Vries and McLaughlin 2013; Keller and Schulin 2003a,b; Keller et al. 2001; McDowell et al. 2013; Sheppard et al. 2009b). As a result, gradual increases in soil Cd due to long-term P fertilization have been observed in studies from many regions of the world, including the United States (Mulla et al. 1980), Denmark (Christensen and Tjell 1991; Dam Kofoed and Sondergard-Klausen 1983), Norway (Baerug and Singh 1990), Finland (Mäkelä-Kurtto et al. 1991), Britain (Nicholson and Jones 1994), and New Zealand (Schipper et al. 2011), reflecting a surplus Cd balance. In pasture systems in Australia (Williams and David 1976) and New Zealand (Andrewes et al. 1996; Gray et al. 1999a; Loganathan et al. 1995, 1997, 2003; Roberts et al. 1994), applications of phosphate fertilizers that were estimated to contain in the range of 20 to 50 mg Cd kg^{-1} led to significant increases in the Cd concentration in the surface soils. Similarly, in a pasture system in Ireland, 31 years of application of 30 kg P ha^{-1} as triple superphosphate containing approximately 39 mg Cd kg^{-1} significantly increased Cd concentration in the surface 10 cm of soil, but the increase was <0.1 mg kg^{-1} (McGrath and Tunney 2010). Correlations between phosphate and Cd in agricultural soils indicate that Cd addition in phosphate fertilizers has been a major driver of increased Cd concentration in soils over time (De Vries and McLaughlin 2013; Jones et al. 1987; Pan et al. 2010; Roberts et al. 1994).

Since input of Cd reflects the total amount of Cd applied, accumulation in the soil will be related both to the phosphate application rate and the concentration of Cd present in the fertilizer (Figure 5.1) (Lambert et al. 2007). The risk of excessive Cd accumulation in soil can therefore be reduced by the use of low-Cd phosphate sources (Sheppard et al. 2009b). For example, evaluation of nine long-term soil fertility experiments in the United States showed that more than 50 years of application of recommended rates of P fertilizers containing an estimated 5 mg Cd kg^{-1} increased soil Cd concentration only marginally by between 0 and 0.5% per year (Mortvedt 1987). In Sweden,

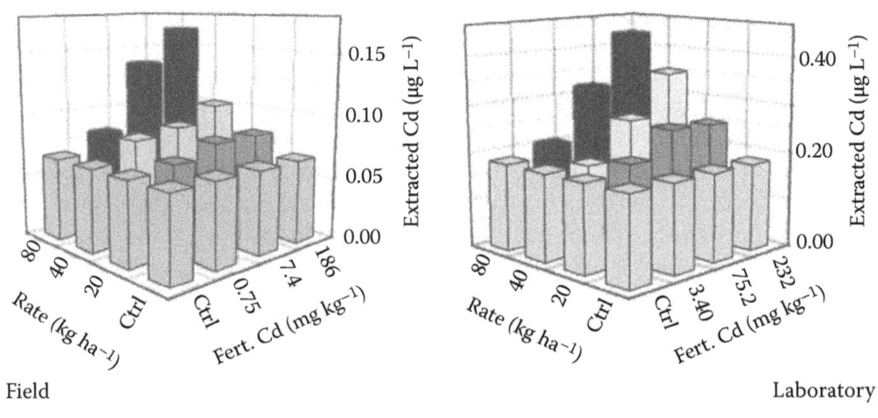

FIGURE 5.1
Mean Cd concentration in soil extracts as affected by the rate of application of monoammonium phosphate (MAP) fertilizer and Cd concentration of the fertilizer averaged over five soils. The field experiment received fertilizer at the specified rate of application for three consecutive years. The laboratory experiment received the equivalent of 15 years of application of the specified rate of application. (From Lambert, R. et al., *Sci. Total Environ.* 378(3):293–305, 2007.)

TABLE 5.2

Estimated Input of Trace Metals in Arable Soils via Phosphate Fertilizers Based on the Average Fertilizer Use in Arable Land and on Utilized Agricultural Areas in 1999/2000 and on Country-Averaged Values of Trace Element Concentration

Country	P_2O_5 Input (kg ha^{-1} yr^{-1})	Cd Input from P Fertilizer (g ha^{-1} yr^{-1})	Cd Input from Atmospheric Deposition[a] (g ha^{-1} yr^{-1})
Austria	37	1.3	2.7
Belgium	31	1.2	<0.1
Denmark	20	0.5	0.3
Finland	29	0.1	0.2
France	49	2.5	–
Germany	33	1.3	2.5
Ireland	78	1.8	0.6
Italy	50	2	3.3
The Netherlands	45	1.9	1.3
Portugal	50	3.1	–
Spain	48	1.9	–
Sweden	20	0.1	0.8
European Average	43	1.6	1.9

Source: Nziguheba, G. and Smolders, E., *Sci. Total Environ.* 390(1):53–57, 2008.

[a] Atmospheric input was based on a compilation made in Nicholson et al. (2003).

application of 25 kg P ha^{-1} over a 15-year period increased soil Cd by 0.33% to 1.1% per year in the 0- to 20-cm depth, a minor effect relative to variation caused by uncontrolled factors (Andersson and Hahlin 1981). Application of agronomic rates of P for 29 years did not increase soil Cd concentration in a barley trial in England (Richards et al. 1998), while neither total nor available Cd concentration in the soils was increased significantly by the Cd added in fertilizer in a 70-year fertility study in Norway (Jeng and Singh 1995). Recent estimates of Cd inputs with fertilizers, manure, and atmospheric deposition and outputs of Cd with removal of crop harvest suggest that the soil Cd in European soils is relatively low (Table 5.2) and mass balance is close to a steady state (Nziguheba and Smolders 2008). Net accumulation can be minimized by reducing the difference between Cd input and removal.

Indirect Effects of Phosphate Fertilizers on Cadmium Availability in Soils

While phosphate fertilizers can contribute to increased Cd concentration in soils if the Cd balance is positive, the risk of Cd toxicity to the soil ecosystem or of increased Cd accumulation in crops will also depend on the influence of the phosphate on the bioavailability of the Cd in the soil (Carrillo-Gonzalez et al. 2006; Hamon et al. 1998; McLaughlin et al. 2000). Cd in the soil undergoes a range of reactions, including direct sorption to soil colloids, reaction with inorganic and organic ligands with subsequent adsorption of the ligand complexes to soil colloids, and precipitation as compounds with varying solubility (Loganathan et al. 2012). Sorption is the dominating reaction with low soil Cd concentrations, while precipitation occurs only at high concentrations, such as in contaminated sites (Carrillo-Gonzalez et al. 2006; Loganathan et al. 2012). Thus, Cd availability will be influenced by factors affecting adsorption or precipitation reactions, including soil pH, ionic strength, soil organic matter content, cation exchange capacity, and the presence of ions in the soil solution that may form Cd complexes (Carrillo-Gonzalez et al. 2006; Naidu et al. 1994a; Smolders et al. 1998; Smolders and McLaughlin 1996). Soil microbiological activity and rhizosphere dynamics can also influence Cd availability (Rivera-Becerril et al. 2002). Thus, phosphate fertilizer can have indirect effects on Cd availability by affecting soil chemical reactions, soil biological process, and plant growth. Effects of phosphate application on Cd can range from immobilization to increased availability depending on the Cd concentration in the soil, the amount and source of phosphate applied, and the soil characteristics.

Where Cd concentration in the soil is very high, such as on contaminated sites, application of high rates of phosphate may be used as a remediation practice to reduce Cd availability, with effects depending greatly on the concentration of Cd in the soil and the level of P applied (Basta and McGowen 2004; Basta et al. 2005; Hong et al. 2010; McGowen et al. 2001). Phosphate-mediated Cd remediation has been discussed in reviews by Basta and McGowen (2004) and Hong et al. (2010). Immobilization may occur through formation of relatively insoluble metal–phosphate minerals or precipitates such as Cd as $Cd_3(PO_4)$ and $Cd(OH)_2$ (Bolan and Duraisamy 2003; Bolan et al. 2003, 2013). However, as the free activity of Cd^{2+} in the soil solutions is generally much below the potential chemical equilibrium for phosphate minerals, the solubility and mobility of Cd in soils are commonly controlled by sorption or coprecipitation mechanisms rather than chemical solubility equilibria with the mineral phase (Carrillo-Gonzalez et al. 2006). Phosphate application may immobilize Cd by enhanced surface complex formation on the P compound, increased surface charge, or coadsorption of P and Cd as an ion pair (Bolan and Duraisamy 2003; Carrillo-Gonzalez et al. 2006; McGowen et al. 2001). Specific sorption of PO_4^{3-} can increase the negative-charged sites available for Cd sorption in both variable charge and permanent charge soils (Hong et al. 2008, 2010; Loganathan et al. 2012; Naidu et al. 1994b).

For phosphate stabilization of metal-contaminated soils to occur, both the soluble, metal-bearing phase and the phosphate source must be in solution (Munksgaard and Lottermoser 2011). Therefore, the effectiveness of phosphate addition in reducing Cd availability tends to be greater with more soluble sources (Basta and McGowen 2004). In leaching studies using various types of phosphate fertilizers, potassium orthophosphate fertilizer was the most effective amendment for Cd stabilization (Munksgaard and Lottermoser 2011). In pot studies, high application rates of P as finely ground rock phosphate or monopotassium phosphate (KH_2PO_4) to an industrially contaminated soil reduced Cd concentrations in sorghum-sudan grass (*Sorghum bicolor* L. Moench) (Zwonitzer et al. 2003). Similarly, immobilization of Cd by application of bone char as a Cd-free phosphate source was related to the dissolution dynamics of the P from the bone char (Siebers and Leinweber 2013). In other studies, applications of up to 400 mg P kg^{-1} as fused superphosphate increased Cd availability to radish (*Raphanus sativus*, L.) while increasing the rate of application to over 800 mg P kg^{-1} decreased availability (Hong et al. 2008). Both the release of Cd in sequential extractions and the rate of Cd release as measured by diethylenetriamine pentaacetic acid (DTPA) extraction after incubation of the fertilizer in the soil were higher from soluble triple superphosphate than from relatively insoluble rock phosphate (Kuo et al. 2007).

Soil pH is a major factor influencing Cd solubility and mobility in soils (Carrillo-Gonzalez et al. 2006; Gray et al. 1999b; Rieuwerts et al. 2006), and phosphate applications may influence Cd availability through effects on soil pH. As soil pH increases, Cd solubility and ion activity decrease while sorption increases (Carrillo-Gonzalez et al. 2006; Naidu et al. 1994a). Phosphate fertilization will affect soil pH, with the effect depending on the type and rate of phosphate compound applied, the buffering capacity of the soil, and the extent of P adsorption. Phosphorus sources that do not contain ammonium may have a liming effect on low pH soils, with the phosphate ions replacing surface hydroxyl ions, thus raising soil pH and increasing the negative charge on the soil colloids, which would increase the adsorption capacity for Cd (Basta et al. 2005). In pot studies conducted with Cd-contaminated soil, application of a high rate of nitrogen-free commercial phosphate fertilizer increased soil pH and reduced extractable soil Cd (Siebers et al. 2013b). Similarly, pot studies with rice grown on contaminated soils showed decreased Cd availability in soils treated with calcium magnesium phosphate and KH_2PO_4 related to an increase in soil pH (Liu et al. 2012). Levi-Minzi and Petruzzelli (1984) reported that the effect of P on pH and Cd solubility was minimal in an organic soil with high pH-buffering capacity, but in a poorly buffered mineral soil, the addition of diammonium phosphate (DAP) increased soil pH, thereby reducing the solubility of Cd. However, other pot studies with high rates of phosphate fertilizers showed that phosphate sources that did not influence soil pH tended to reduce Cd availability on contaminated soils (Chen et al. 2011).

While applications of high rates of phosphate fertilizer that increase pH may reduce the Cd availability on contaminated sites, effects may differ with the use of phosphate sources that lower rather than raise soil pH. Decreasing soil pH increases the solubility of Cd compounds, increasing the soil solution concentration. In addition, decreasing pH decreases the number of hydroxyl species that are preferentially adsorbed. Soil acidification not only reduces Cd sorption but also shifts Cd sorption from specific to nonspecific ion exchange sorption (Loganathan et al. 2012). Finally, in variable-charge soils, decreasing pH decreases the net surface negative charge, reducing Cd adsorption (Naidu et al. 1994a). The soil distribution coefficient (Kd), calculated as the ratio of the concentration of Cd on the soils to the concentration in the liquid phase, decreases and phytoavailability increases as pH declines (Anderson and Christensen 1988; Holm et al. 1995; Rieuwerts et al. 2006; Sauvé et al. 2000a). There was about a 10-fold decline in the Kd of 112 Canadian soil as soil pH decreased from pH 7.8 to 3.5 (Sheppard et al. 2007), while in a study of 97 soil samples from north Wales, calcium chloride ($CaCl_2$) extractability of Cd increased by about 10-fold as pH decreased from 6 to 4, although Kd was not strongly affected (Rieuwerts et al. 2006).

Pot studies with high rates of phosphate fertilizers showed that phosphate sources that decreased the pH tended to increase available Cd on contaminated soils (Chen et al. 2011). Applications of nitrogen (N)- and Ca-containing P fertilizers generally decrease the soil pH due to acidification reactions, potentially increasing the availability of Cd (Lambert et al. 2007; Levi-Minzi and Petruzzelli 1984; Loganathan et al. 2012). In leaching studies, superphosphate, triple superphosphate, and KH_2PO_4 applications decreased the pH and led to a release of Cd in leachate water in contaminated soils treated with high amounts of fertilizer (Munksgaard and Lottermoser 2011). Similarly, application of ammonium phosphate at high rates of 500 kg P kg^{-1} soil increased leaching of Cd in soil columns (Zhang and Zhang 2010). The effect of P applications on soil pH would be enhanced by band application of fertilizers, as the high concentration of fertilizers associated with the microregion of the fertilizer band would accentuate pH effects. In addition, repeated fertilizer applications over time would have a cumulative effect on soil pH and Cd availability, particularly on poorly buffered soils.

The decrease in pH due to phosphate application may not always lead to a major increase in Cd availability. After 40 years of application of MAP fertilizer at rates of 26, 39, and 52 kg P ha^{-1}, soil pH decreased, but acid-extractable Cd increased only slightly with an increasing P rate presumably due to low concentration of Cd in the fertilizer (Cakmak et al. 2010). In incubation studies evaluating the effect of triple superphosphate on Cd speciation, application of 1000 mg P kg^{-1} soil reduced soil pH immediately after application due to the release of phosphoric acid during fertilizer dissolution (Figure 5.2) (Siebers et al. 2013a). In this case, the application of high rates of triple superphosphate to a contaminated soil led to a reduction in easily mobilizable Cd, likely because the decrease in pH led to a dissolution of Cd carbonates (Siebers et al. 2013a).

Applications of soluble phosphate sources can also influence Cd dynamics by increasing the ionic strength of the soil solution. Cadmium sorption decreases with increasing ionic strength of the soil solution, with effects being greatest in soils with a low sorption capacity and being minimal at pH

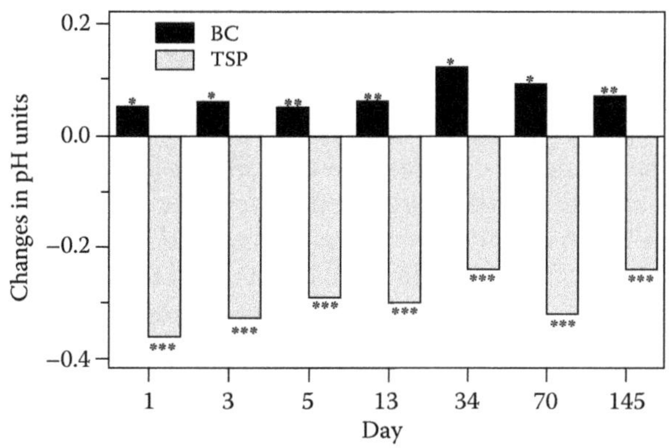

FIGURE 5.2
Change in soil pH units after fertilizer addition in relation to the control. BC = bone char, TSP = triple superphosphate. Significant at $^{*}p < 0.05$; $^{**}p < 0.01$; $^{***}p < 0.001$, $n = 3$.

values exceeding 6, where sorption is high (Naidu et al. 1994a). The decreased sorption with increasing ionic strength may relate to increased competition of cations in solution with Cd for binding sites, decreased activity of Cd in solution, formation of ion pairs such as $CdCl^{+}$ that have a lower sorption affinity, pH decrease due to proton desorption, or changes in the electrostatic potential in the adsorption plane (Loganathan et al. 2012). Application of phosphate fertilizer will increase the ionic strength of the soil solution, with the effect being the greatest with soluble fertilizer sources and in the region of high concentration around the fertilizer granule.

In addition to Cd, phosphate fertilizers also contain Zn as a contaminant at concentrations that can range from less than 50 to in excess of 1500 mg kg^{-1} (Alloway and Steinnes 1999). Cd and Zn compete for the same binding sites in the soil; therefore, Zn can desorb Cd from the cation exchange, increasing Cd concentration in the soil solution (Christensen 1984; François et al. 2009). The ammonium from MAP and diammonium phosphate may also lead to displacement of Cd from adsorption sites, increasing Cd availability (Lambert et al. 2007; Lorenz et al. 1994; Mitchell et al. 2000).

Impacts of Phosphate Fertilizer on Cadmium Accumulation in Crops

Effects of Long-Term Cd Accumulation

Cadmium concentration of crops is strongly affected by soil Cd concentration (Adams et al. 2004); therefore, long-term increases in Cd in the soil from P application will commonly result in higher crop Cd accumulation

(Brennan and Bolland 2004; Hamon et al. 1998; Jones and Johnston 1989; Jones et al. 1987; Kashem and Singh 2002; Mulla et al. 1980; Williams and David 1973, 1976). For example, in studies with wheat and canola on a sandy soil in Australia, both soil and crop Cd increased due to phosphate fertilization with the crop Cd concentration being highly correlated with soil test Cd (Brennan and Bolland 2004). Similarly, in field studies conducted on soils across the Canadian prairies, concentration of Cd in durum wheat grain increased with both P fertilizer rate and Cd concentration, reflecting the total input of Cd to the soil over time (Grant et al. 2013) (Figure 5.3). In contrast, in trials where long-term application of low-Cd fertilizer did not increase soil Cd, the Cd concentration of wheat (*Triticum aestivum*, L.), corn (*Zea mays*, L.), soybean (*Glycine max*, L.), and timothy (*Phleum pratense*, L.) was also unaffected (Mortvedt 1987).

An increase in soil Cd does not always result in a change in Cd proportional to the Cd addition. The availability of Cd added to the soil from fertilizers will vary substantially with soil properties, with availability tending to be higher on light-textured, low pH soils with low cation exchange capacity (Grant et al. 2013; Lambert et al. 2007). As well as adding Cd accumulation to soils, long-term P fertilization can have other effects on the soil–plant system that may influence Cd availability, including impacts on soil pH, presence of other competing ions, type and content of organic matter, cation exchange capacity (CEC), complex formation, root growth, and mycorrhizal activity. Over time, Cd applied to soils may convert to less available forms. Using isotopic exchange techniques, Hamon et al. (1998) reported that in Australian soils that received Cd inputs from superphosphate over a 45-year period, from 1.0% to 1.5% of the total Cd present in the soil was converted to nonavailable forms each year. Therefore, the effect of long-term phosphate

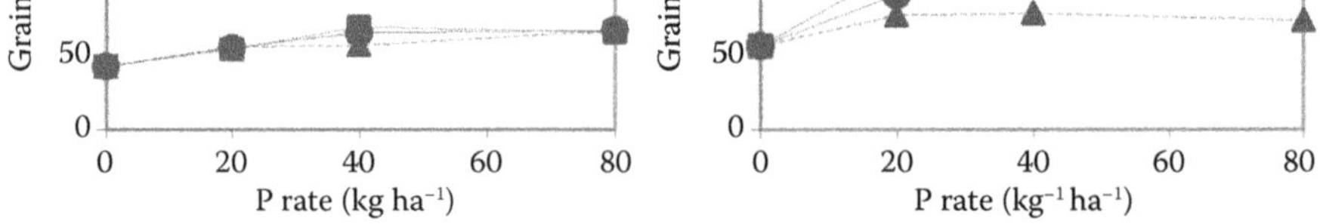

FIGURE 5.3
Effect of 7 years of application of MAP fertilizer with varying Cd concentration on Cd concentration in durum wheat grain, on two contrasting soils on the Canadian prairies. Low, medium, and high Cd fertilizers contained 0.38, 71, and 211 mg Cd kg^{-1}, respectively. The Brandon soil was higher in pH and cation exchange capacity than the Sylvania soil. (From Grant, C. et al., *Plant Soil* 372(1–2):221–233, 2013.)

fertilization will be affected by the interaction of Cd input in the fertilizer with effects on other components of the soil–plant system.

In studies in Norway, soil Cd concentrations were higher in long-term cultivated soils than in newly cultivated soils that had never received Cd inputs through P fertilizers, but the crop Cd accumulation was lower (Andersson and Hahlin 1981; Baerug and Singh 1990; He and Singh 1993b,c). Phytoavailability of Cd may have been reduced by the higher soil pH that was associated with long-term cultivation in this study. A similar effect was noted in trials in England where the Cd concentration of grasses that had received long-term application of phosphate fertilizer was higher than that of unfertilized grasses in the absence of lime application, but did not differ if lime was added to control acidification (Nicholson and Jones 1994).

Phosphate fertilizers may also influence Cd availability to crops by effects on Zn supply. In addition to Cd, phosphate fertilizers also contain Zn as an impurity, at concentrations ranging from less than 50 to more than 1500 kg Zn kg^{-1} (Alloway and Steinnes 1999). Over time, the Zn may accumulate in the soil if input is less than off-take (Lambert et al. 2007). For example, in field studies conducted across western Canada, repeated applications of P fertilizers increased the soil solution concentration of both Zn and Cd (Lambert et al. 2007). While Zn will compete with Cd for sorption sites in the soil, increasing Cd concentration in the soil solution as discussed previously (Christensen 1984), Zn will also compete with Cd for uptake and translocation by the plant. Therefore, the impact of Cd added to soils from long-term phosphate application on Cd accumulation in crops may be attenuated to some extent by the presence of Zn in the fertilizer (Cakmak et al. 2000; Jiao et al. 2004; Oliver et al. 1994; Williams and David 1976).

If phosphate fertilizers are effective at increasing crop yield on P-deficient soils, they will also serve to increase the return of root material, exudates, and crop residue to the soil. Over time, effective fertilization can be associated with higher soil organic matter content relative to unfertilized soil due to the increased amount of residue return. For example, a 31-year pasture study in Ireland showed increased soil organic matter content with phosphate fertilization, presumably because P application increased pasture yield and input of plant residues to the soil (McGrath and Tunney 2010). In Denmark, application of N, P, and K fertilizer in a 90-year field study increased soil organic matter content as compared to an unfertilized control (Schjønning et al. 1994). Limited direct information is available on the influence of changes in organic matter content caused by P fertilization on Cd accumulation in crops. However, increased soil organic matter content and the addition of organic amendments to soil have shown varying effects on Cd availability. Dissolved organic carbon can form soluble Cd complexes, thus increasing Cd availability (Rieuwerts et al. 2006; Sauvé et al. 2000a,b). However, increasing soil organic matter content will also increase cation exchange capacity, increasing Cd adsorption (Gray et al. 1999b; Rieuwerts et al. 2006). Although effects may vary, Cd availability generally appears to decrease or be unaffected by increased organic matter content or by application of organic materials such as biosolids or manures (Bolan and Duraisamy 2003; He and Singh 1993a; Willaert et al. 1990).

Short-Term Effects of Agronomic Phosphate Applications

In addition to long-term effects, agronomic rates of phosphate fertilizer can have immediate effects on crop Cd concentration in the year of application (Adams et al. 2004; François et al. 2009; Gao et al. 2010a,b; Grant 2011; Grant and Bailey 1997, 1998; Grant et al. 1996, 1999, 2002, 2009, 2010; Grant and Sheppard 2008; He and Singh 1993c, 1994; Singh 1990; Singh and Myhr 1998). Short-term availability of the Cd from P fertilizers for crop uptake will depend in part on the Cd concentration and the solubility of the fertilizer being used. Effects of phosphate fertilization have been frequently observed in pot studies under greenhouse and growth chamber conditions. In greenhouse studies, application of nitrogen, phosphorus, and potassium (NPK) fertilizer containing 417 mg Cd kg^{-1} P increased Cd concentration in carrots (*Daucus carota*, L.), lettuce (*Latuca sativa*, L.), oat (*Avena sativa*, L.), and ryegrass (*Lolium multiflorum*, L.), but addition of a low-solubility rock phosphate containing a similar Cd concentration had no significant effect (He and Singh 1995). In pot studies, the uptake of Cd by rape and oats was greater with the use of single superphosphate than with a NPK source having a lower Cd concentration (Singh 1990). Pot studies with flax showed greater accumulation of Cd with the use of commercial triple superphosphate or MAP containing Cd than with reagent-grade sources containing only trace amounts of Cd (Jiao et al. 2004).

It is important to recognize that effects of the Cd added in annual agronomic applications of P may be lower under field conditions than observed in pot studies (Guttormsen et al. 1995). Under field conditions, plant roots can explore a larger soil volume that is less affected by the phosphate addition than in confined pots. Annual additions of phosphate fertilizer increased Cd concentration in a range of crops grown in pot studies (He and Singh 1995), but under field conditions, application of phosphate fertilizers containing a similar concentration of Cd did not influence the Cd concentration of cereal crops (Singh and Myhr 1998) or vegetables (Guttormsen et al. 1995). Under field conditions, Cd concentration of potatoes (*Solanum tuberosum*, L.) in the year of fertilizer application was not affected by the concentration of Cd in superphosphate (McLaughlin et al. 1995; Sparrow et al. 1993a,b), MAP, diammonium phosphate, or rock phosphate (McLaughlin et al. 1995). Similarly, in field studies with durum wheat (*Triticum turgidum*, L.) on the Canadian prairies, although crop Cd concentration increased with application of MAP fertilizer, it was generally unaffected by the concentration of Cd present in fertilizers (Table 5.3) (Grant et al. 2002).

The short-term impact of phosphate fertilizer on Cd concentration in crops is not solely a function of the Cd content of the fertilizer, but may also be

TABLE 5.3

Cadmium Concentration in Durum Wheat as a Function of Addition of MAP Containing Low, Moderate, or High Concentrations of Cadmium at Three Locations over 3 Years

		1994				1995				1996		
	P	Dalemede	Heeny	Irricana	Jury	Heeney	Irr-A	Irr-L	Jury	Heeney	Irricana	Jury
Source	kg ha^{-1}	μg kg^{-1}										
None	0	157	71	117	86	102	91	99	81	87	120	74
Russia[a]	9	168	96	145	97	99	117	116	74	99	122	95
Russia	18	205	117	163	97	121	133	131	81	114	147	150
Florida	9	174	100	144	106	106	110	128	75	98	126	102
Florida	18	199	97	174	95	122	116	120	94	115	143	105
Idaho	9	169	82	142	99	115	100	123	69	96	129	89
Idaho	18	205	89	178	93	120	114	144	91	119	155	106
Contrast												
Control vs P		0.0111	0.0023	0.0001	ns	0.081	0.0065	0.0001	ns	0.0398	0.0107	0.016
Low vs high P		0.0004	ns	0.0001	ns	0.0039	0.0696	0.0599	0.0333	0.0138	0.0001	0.0187
Russia vs Florida		ns	ns	ns	ns	ns	ns	ns	ns	ns	ns	ns
Russia vs Idaho		ns	ns	ns	ns	ns	ns	ns	ns	ns	ns	ns
Florida vs Idaho		ns	ns	ns	ns	ns	ns	ns	ns	ns	0.0419	ns
Russian vs Idaho (high P)		ns	0.0118	ns	ns	ns	ns	ns	ns	ns	ns	0.0154
SE		10.52	7.74	5.59	2.86	5.94	7.8	6.55	8.95	8.79	5.86	12.52

Source: Adapted from Grant, C. A. et al., *J. Sci. Food. Agric.* 82(3):301–308, 2002.

[a] Concentration of Cd in fertilizers was 0.2 μg g^{-1} in the Russian source, 7.8 μg g^{-1} in the Florida source, and 186.0 μg g^{-1} in the Idaho source.

affected by fertilizer effects on soil and plant dynamics. Under growth chamber conditions, application of reagent-grade MAP that contained only trace amounts of Cd increased Cd concentration of durum wheat (Choudhary et al. 1994). Similarly, in pot studies with flax and durum wheat, reagent-grade triple superphosphate and MAP both increased Cd with no significant difference occurring in durum wheat between reagent- and commercial-grade phosphate sources (Jiao et al. 2004). Effects of phosphate fertilizer in the absence of significant Cd input may be related to short-term effects of the fertilizer on soil pH, ionic strength, Zn concentration, and plant growth that can affect the Cd accumulation by the plant.

As discussed the section "Indirect Effects of Phosphate Fertilizers on Cadmium Availability in Soils," soluble phosphate fertilizer may influence pH with the addition of nitrogen and calcium-containing P fertilizers generally decreasing soil pH (Lambert et al. 2007; Levi-Minzi and Petruzzelli 1984; Loganathan et al. 2012). While a single agronomic rate of application of phosphate fertilizer is unlikely to affect the pH of the bulk soils, it can have a significant effect in the fertilizer reaction zone. Triple superphosphate and MAP have been shown to significantly decrease pH in reaction zones around the fertilizer granule (Du et al. 2010; Fan and MacKenzie 1993). In addition, many crops can increase rooting when they contact a high concentration of phosphate as occurs in the reaction zone of the fertilizer granule (Strong and Soper 1974a,b). The combination of lower pH and high root proliferation may increase the Cd uptake by the plant from the fertilizer reaction zone. The increased Cd availability may be accentuated by increased ionic strength of the soil solution caused by the high salt concentration around the granule of a highly soluble phosphate source (Loganathan et al. 2012; Naidu et al. 1994b).

Phosphate fertilizer may influence Cd accumulation in plants by affecting the Zn–Cd interaction. (Moraghan 1984). Studies support the concept that Zn and Cd compete for transport sites on the root cell plasma membrane (Hart et al. 2002) or for transmembrane movement, thus affecting Cd translocation processes into and within the plant, including Cd transport across the cell membrane from xylem parenchyma cells into the xylem transport system or membrane transport in and out of the phloem cells (Hart et al. 1998, 2002, 2005, 2006; Mahdy 2012; Mills et al. 2012; Zhao et al. 2005). Agronomic rates of Zn fertilization reduced Cd concentration in a range of crops in both field and pot studies (Choudhary et al. 1994; Grant and Bailey 1997; Hart et al. 2005; Jiao et al. 2004; McKenna et al. 1993; Oliver et al. 1994). Therefore, Cd accumulation in crops may be reduced by the presence of Zn contamination in phosphate fertilizer (Cakmak et al. 2000; Jiao et al. 2004; Oliver et al. 1994; Williams and David 1976). However, there are also numerous studies demonstrating that phosphate applications can reduce Zn concentrations in plants, with the effect possibly being due to P–Zn interaction in soil (Lambert et al. 2007), dilution effects from a large P yield response, P toxicity from restricted Zn supply, or a metabolic disorder (Cakmak and Marschner 1986, 1987). For example, application of phosphate fertilizer in growth chamber studies decreased Zn concentration and increased Cd and P concentrations in both the root and shoot of durum wheat (Gao et al. 2011). Therefore, if P fertilizer reduces Zn availability, it may indirectly increase Cd accumulation by allowing greater Cd uptake and translocation in the plant (Grant and Bailey 1997; Grant et al. 2010).

In P-deficient soils, application of phosphate fertilizers may increase crop biomass production, potentially affecting plant Cd concentration and total Cd accumulation. If the soil supply of Cd is limited, higher biomass production can decrease Cd concentration in the plant through biological dilution. Dilution occurs when the uptake of Cd from the soil remains relatively constant but the amount of plant material through which the Cd is distributed increases. In contrast, where Cd availability in the soil is relatively high, the plant sink for Cd serves as a greater limitation to uptake than the soil supply. Consequently, increasing the biomass production increases the sink available for Cd storage, and the high availability of Cd in the soil allows increased crop accumulation; total uptake of Cd by the plant would therefore increase while Cd concentration could remain constant or decline only slightly. Over time, if the Cd concentration of the fertilizer was low and the biomass production and Cd concentration in the tissue were high, the greater off-take in the plant material encouraged by the P uptake could increase Cd removal from the soil, which could be of significance where phytoremediation was being practiced (Li et al. 2013).

Phosphorus fertilization can affect biological processes that influence the phytoavailability of Cd. Many plants will increase root formation and exudation of carboxylate protons such as citrate and malate under P-deficient conditions as a means of increasing P availability (Pearse et al. 2006; Raynaud et al. 2008; Richardson et al. 2009a,b). In studies using a normal and a mutant form of barley where the mutant did not produce root hairs, presence of root hairs increased the shoot Cd concentration and uptake efficiency of Cd by the plant (Figure 5.4) (Zheng et al. 2011). Addition of P reduced root length and the length and density of root hair production in the nonmutant barley

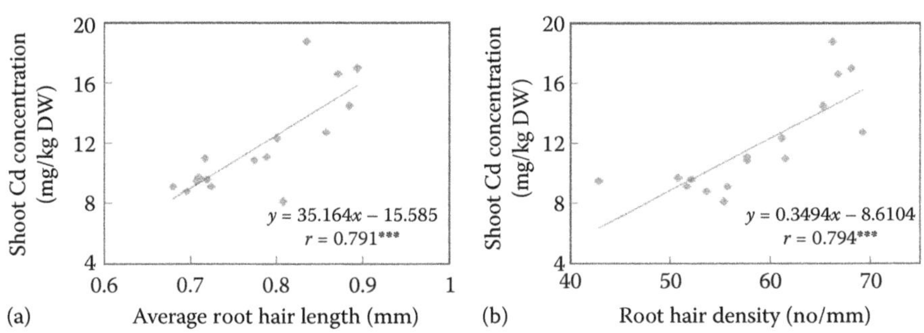

FIGURE 5.4
Relationships between average root hair length and shoot Cd concentration (a) and average root hair density and shoot Cd concentration (b) in a wild-type barley. Note: *** = significance at the probability level of $p < 0.001$. (From Zheng, R. et al., *Environ. Pollut.* 159(2):408–415, 2011.)

TABLE 5.4

Effect of Preceding Crop, Tillage System, P Application to Preceding Crop, and P Application to Flax on Concentration of Cd and Zn in Flaxseed at Two Sites over 3 Years

	Cd Concentration in Flaxseed					
	BRC-North			MZTRF		
	2000	2001	2002	2000	2001	2002
	μg kg^{-1}					
Preceding Crop						
Canola	321	337	348	254	246	226
Wheat	268	273	345	209	206	199
Standard error (error c df = 36)	8.84	63.3	15.4	9.97	10.2	10.9
P Application to Flax						
0	294	286	325	224	218	208
11	295	323	368	238	235	217
Standard error (error b df = 30)	8.07	63.3	14.5	9.63	10.4	10.9
P Application to Preceding Crop						
0	282	300	341	218	219	210
11	296	303	336	235	230	208
22	305	312	363	241	229	219
Standard error (error b df = 30)	9.77	64.6	18.3	10.3	10.4	11.8
Tillage						
CT	285	370	321	233	240	200
RT	304	239	372	230	212	225
Standard error (error a df = 3)	13.22	70.5	52.8	12.1	16.3	16.7

		Contrast Analysis					
		BRC-North			MZTRF		
Treatment	**DF**	**2000**	**2001**	**2002**	**2000**	**2001**	**2002**
Crop	1	0.0001	0.0179	ns	0.0001	0.0001	0.0002
Pflax	1	ns	ns	0.0157	0.0127	0.0105	ns
Pprev	2	ns	ns	ns	0.0056	ns	ns
Tillage	1	ns	ns	ns	ns	0.0297	ns
Pprev × Pflax	2	0.0428	ns	ns	ns	ns	ns
Pflax × tillage	1	ns	ns	0.0367	ns	ns	ns
Crop × tillage	1	ns	ns	ns	ns	ns	ns
Crop × Pprev × Pflax × tillage	2	0.0031	ns	ns	ns	ns	ns

Source: Adapted from Grant, C. A. et al., *Plant Soil* 333(1):337–350, 2010.
Note: Pflax = P application to flax, Pprev = P application to preceding crop.

and decreased Cd concentration in the plant. Foliar application of P also increased plant P status, reduced root hair production, and decreased Cd uptake (Zheng et al. 2011). The reduction in rooting that occurs in response to improved plant P sufficiency may contrast with the more localized increase in rooting that can occur when a plant root contacts a region of high P concentration (Strong and Soper 1974a,b). Modification of the chemistry and biology of the rhizosphere by release of root exudates and acidification through proton extrusion can also increase the availability of trace elements such as Zn and Cd (Raynaud et al. 2008).

Phosphorus status of the plant is also a controlling factor in the development of mycorrhizal associations that may influence Zn and Cd accumulation by crops (Grant et al. 2005). Arbuscular mycorrhizal inoculation can either increase or decrease Cd phytoavailability and accumulation in tobacco (*Nicotiana tabacum,* L.) plants, possibly due to enhanced root growth or changes in root exudation (Janoušková et al. 2007). Mycorrhizal colonization in pea (*Pisum sativum,* L.) increased Cd accumulation in shoots and decreased Cd concentration in pods under high Cd conditions where 100 mg Cd kg^{-1} was added to the soil as $CaCl_2$ but decreased Cd accumulation in pea shoots when no Cd was added to the soil (Rivera-Becerril et al. 2002). In flax (*Linum usitatissimum,* L.) application of P fertilizer or sequencing of the crop after a nonmycorrhizal canola (*Brassica napus,* L.) crop rather than after a mycorrhizal spring wheat (*Triticum aestivum,* L.) crop led to higher Cd concentration, possibly as a result of reduced mycorrhizal colonization (Table 5.4) (Grant et al. 2010). However, in other studies, no relationship was found between mycorrhizal colonization and Cd accumulation in durum wheat (Gao et al. 2010a). Therefore, while application of P fertilizers may possibly affect Cd accumulation through suppression of mycorrhizal colonization, further research is required to more clearly understand these interactions.

Conclusions

Phosphorus is an essential nutrient for crop production and an adequate supply is required for optimum crop yield. Both inorganic P fertilizers and organic P sources such as biosolids and manure contain Cd, and P fertilizers are a major anthropogenic input of P to agricultural soils. Cadmium addition to the soil increases with the rate of applied P, Cd concentration in the fertilizer, and frequency of application. Phosphate applications can influence Cd availability in the soil directly by the addition of Cd to the soil and indirectly by fertilizer-induced effects on soil chemical and biological properties such as soil pH, organic matter content, ionic strength of the soil solution, Zn availability, rhizosphere chemistry, and mycorrhizal associations. Long-term accumulation of Cd in the soil can be reduced by using fertilizer with low Cd concentration, while both Cd input and effects on availability may be restricted by limiting fertilizer inputs through adoption of management practices that optimize fertilizer use efficiency.

References

Adams, M. L., Zhao, F. J., McGrath, S. P., Nicholson, F. A. and Chambers, B. J. 2004. Predicting cadmium concentrations in wheat and barley grain using soil properties. *J. Environ. Qual.* 33:532–541.

Alloway, B. J. and Steinnes, E. 1999. Anthropogenic additions of cadmium to soils. *In* M. J. McLaughlin, B. R. Singh, eds. *Cadmium in Soils and Plants*, 97–123. Kluwer Academic Publishers, Dordrecht, Netherlands.

Anderson, P. R. and Christensen, T. H. 1988. Distribution coefficients of Cd, Co, Ni and Zn in soils. *J. Soil Sci.* 39:15–22.

Andersson, A. 1977. Heavy metals in Swedish soils: On their retention, distribution and amounts. *Swed. J. Agric. Res.* 7:7–20.

Andersson, A. and Hahlin, M. 1981. Cadmium effects from phosphorus fertilization in field experiments. *Swed. J. Agric. Res.* 11(1):3–10.

Andrewes, P., Town, R. M., Hedley, M. J. and Loganathan, P. 1996. Measurement of plant-available cadmium in New Zealand soils. *Aust. J. Soil Res.* 34(3):441–452.

Baerug, R. and Singh, B. R. 1990. Cadmium levels in soils and crops after long-term use of commercial fertilizers. *Nor. J. Agric. Sci.* 4:251–260.

Basta, N. T. and McGowen, S. L. 2004. Evaluation of chemical immobilization treatments for reducing heavy metal transport in a smelter-contaminated soil. *Environ. Pollut.* 127(1):73–82.

Basta, N. T., Ryan, J. A. and Chaney, R. L. 2005. Trace element chemistry in residual-treated soil: Key concepts and metal bioavailability. *J. Environ. Qual.* 34(1):49–63.

Bolan, N. S. and Duraisamy, V. P. 2003. Role of inorganic and organic soil amendments on immobilisation and phytoavailability of heavy metals: A review involving specific case studies. *Aust. J. Soil Res.* 41:533–555.

Bolan, N. S., Adriano, D. C. and Naidu, R. 2003. Role of phosphorus in (im)mobilization and bioavailability of heavy metals in the soil-plant system. *Reviews of Environmental Contamination and Toxicology*, 1–44. Springer, New York.

Bolan, N. S., Makino, T., Kunhikrishnan, A., Kim, P. J., Ishikawa, S., Murakami, M., Naidu, R. and Kirkham, M. B. 2013. Cadmium contamination and its risk management in rice ecosystems. *Adv. Agron.* 119:183–273.

Brennan, R. F. and Bolland, M. D. A. 2004. Wheat and canola response to concentrations of phosphorus and cadmium in a sandy soil. *Aust. J. Exp. Agr.* 44:1025–1029.

Cakmak, D., Saljnikov, E., Mrvic, V., Jakovljevic, M., Marjanovic, Z., Sikiric, B. and Maksimovic, S. 2010. Soil properties and trace elements contents following 40 years of phosphate fertilization. *J. Environ. Qual.* 39(2):541–547.

Cakmak, I. and Marschner, H. 1986. Mechanism of phosphorus-induced zinc deficiency in cotton. I. Zinc deficiency-enhanced uptake rate pf phosphorus. *Physiol. Plant.* 68:483–490.

Cakmak, I. and Marschner, H. 1987. Mechanism of phosphorus-induced zinc deficiency in cotton. III. Changes in physiological availabiity of zinc in plants. *Physiol. Plant.* 70:13–20.

Cakmak, I., Welch, R. M., Erenoglu, B., Römheld, V., Norvell, W. A. and Kochian, L. V. 2000. Influence of varied zinc supply on re-translocation of cadmium (^{109}Cd) and rubidium (^{86}Rb) applied on mature leaf of durum wheat seedlings. *Plant Soil* 219(1–2):279–284.

Carrillo-Gonzalez, R., Simunek, J., Sauve, S. and Adriano, D. 2006. Mechanisms and pathways of trace element mobility in soils. *Adv. Agron.* 91:111–178.

Chen, Q., Hu, C., Tan, Q. and Sun, X. 2011. Effect of different phosphate sources on availability of cadmium in soil. *Huanjing Kexue Xuebao/Acta Scientiae Circumstantiae* 31(10):2254–2259.

Chien, S. H., Carmona, G., Prochnow, L. I. and Austin, E. R. 2003. Cadmium availability from granulated and bulk-blended phosphate-potassium fertilizers. *J. Environ. Qual.* 32(5):1911–1914.

Chien, S. H., Prochnow, L. I., Tu, S. and Snyder, C. S. 2011. Agronomic and environmental aspects of phosphate fertilizers varying in source and solubility: An update review. *Nutr. Cycling Agroecosyst.* 89(2):229–255.

Choudhary, M., Bailey, L. D. and Grant, C. A. 1994. Effect of zinc on cadmium concentration in the tissue of durum wheat. *Can. J. Plant Sci.* 74(3):549–552.

Christensen, T. 1984. Cadmium soil sorption at two conentrations: II. Reversibility, effect of changes in solute composition, and effects of soil aging. *Water Air Soil Pollut.* 21:115–125.

Christensen, T. H. and Huang, P. M. 1999. Solid phase cadmium and the reactions of aqueous cadmium with soil surfaces. *In* M. J. McLaughlin, B. R. Singh, eds. *Cadmium in Soils and Plants,* 65–96. Kluwer Academic Publishers, Dordrecht, Netherlands.

Christensen, T. H. and Tjell, J. C. 1991. Sustainable management of heavy metals in agriculture. Example: Cadmium. *In* J. G. Farmer, ed. *Heavy Metals in the Environment,* Volume 1, 40–49. CEP Consultants, Edinburgh, Scotland.

Dam Kofoed, A. and Sondergard-Klausen, P. 1983. Effect of fertilization on Cd content of soil and plants (in Danish). *Tidsskr. Planteavl* 87:23–32.

De Vries, W. and McLaughlin, M. J. 2013. Modeling the cadmium balance in Australian agricultural systems in view of potential impacts on food and water quality. *Sci. Total Environ.* 461–462:240–257.

Du, Z., Zhou, J., Wang, H., Chen, X. and Wang, Q. 2010. Soil pH changes from fertilizer site as affected by application of monocalcium phosphate and potassium chloride. *Commun. Soil Sci. Plant Anal.* 41(15):1779–1788.

Fan, M. X. and MacKenzie, A. F. 1993. Interaction of urea with triple superphosphate in a simulated fertilizer band. *Fert. Res.* 36:35–44.

François, M., Grant, C., Lambert, R. and Sauvé, S. 2009. Prediction of cadmium and zinc concentration in wheat grain from soils affected by the application of phosphate fertilizers varying in Cd concentration. *Nutr. Cycling Agroecosyst.* 83(2):125–133.

Gao, X., Akhter, F., Tenuta, M., Flaten, D. N., Gawalko, E. J. and Grant, C. A. 2010a. Mycorrhizal colonization and grain Cd concentration of field-grown durum wheat in response to tillage, preceding crop and phosphorus fertilization. *J. Sci. Food Agric.* 90(5):750–758.

Gao, X., Brown, K. R., Racz, G. J. and Grant, C. A. 2010b. Concentration of cadmium in durum wheat as affected by time, source and placement of nitrogen fertilization under reduced and conventional-tillage management. *Plant Soil* 337(1):341–354.
Gao, X., Flaten, D. N., Tenuta, M., Grimmett, M. G., Gawalko, E. J. and Grant, C. A. 2011. Soil solution dynamics and plant uptake of cadmium and zinc by durum wheat following phosphate fertilization. *Plant Soil* 338(1):423–434.
Godt, J., Scheidig, F., Grosse-Siestrup, C., Esche, V., Brandenburg, P., Reich, A. and Groneberg, D. A. 2006. The toxicity of cadmium and resulting hazards for human health. *J. Occup. Med. Toxicol.* 1:22.
Grant, C., Bittman, S., Montreal, M., Plenchette, C. and Morel, C. 2005. Soil and fertilizer phosphorus: Effects on plant P supply and mycorrhizal development. *Can. J. Plant Sci.* 85(1):3–14.
Grant, C., Flaten, D., Tenuta, M., Malhi, S. and Akinremi, W. 2013. The effect of rate and Cd concentration of repeated phosphate fertilizer applications on seed Cd concentration varies with crop type and environment. *Plant Soil* 372(1–2):221–233.
Grant, C. A. 2011. Influence of phosphate fertilizer on cadmium in agricultural soils and crops. *Pedologist* 54(3):143–155.
Grant, C. A. and Bailey, L. D. 1997. Effects of phosphorus and zinc fertiliser management on cadmium accumulation in flaxseed. *J. Sci. Food. Agric.* 73(3):307–314.
Grant, C. A. and Bailey, L. D. 1998. Nitrogen, phosphorus and zinc management effects on grain yield and cadmium concentration in two cultivars of durum wheat. *Can. J. Plant Sci.* 78(1):63–70.
Grant, C. A., Bailey, L. D., Harapiak, J. T. and Flore, N. A. 2002. Effect of phosphate source, rate and cadmium content and use of *Penicillium bilaii* on phosphorus, zinc and cadmium concentration in durum wheat grain. *J. Sci. Food. Agric.* 82(3):301–308.
Grant, C. A., Bailey, L. D., McLaughlin, M. J. and Singh, B. R. 1999. Management factors which influence cadmium concentration in crops. *In* M. J. McLaughlin, B. R. Singh, eds. *Cadmium in Soils and Plants,* 151–198. Kluwer Academic Publishers, Dordrecht, Netherlands.
Grant, C. A., Bailey, L. D. and Themen, M. C. 1996. The effect of N, P and KCl fertilizers on grain yield and Cd concentration of malting barley. *Fert. Res.* 45(2):153–161.
Grant, C. A., Clarke, J. M., Duguid, S. and Chaney, R. L. 2008. Selection and breeding of plant cultivars to minimize cadmium accumulation. *Sci. Total Environ.* 390(2–3):301–310.
Grant, C. A., Flaten, D. N., Tomasiewicz, D. J. and Sheppard, S. C. 2001. The importance of early season phosphorus nutrition. *Can. J. Plant Sci.* 81(2):211–224.
Grant, C. A., Monreal, M. A., Irvine, R. B., Mohr, R. M., McLaren, D. L. and Khakbazan, M. 2009. Crop response to current and previous season applications of phosphorus as affected by crop sequence and tillage. *Can. J. Plant Sci.* 89(1):49–66.
Grant, C. A., Monreal, M. A., Irvine, R. B., Mohr, R. M., McLaren, D. L. and Khakbazan, M. 2010. Preceding crop and phosphorus fertilization affect cadmium and zinc concentration of flaxseed under conventional and reduced tillage. *Plant Soil* 333(1):337–350.
Grant, C. A. and Sheppard, S. C. 2008. Fertilizer impacts on cadmium availability in agricultural soils and crops. *Hum. Ecol. Risk Assess.* 14(2):210–228.
Gray, C. W., McLaren, R. G., Roberts, A. H. C. and Condron, L. M. 1999a. The effect of long-term phosphatic fertiliser applications on the amounts and forms of cadmium in soils under pasture in New Zealand. *Nutr. Cycling Agroecosyst.* 54(3):267–277.

Gray, C. W., McLaren, R. G., Roberts, A. H. C. and Condron, L. M. 1999b. Solubility, sorption and desorption of native and added cadmium in relation to properties of soils in New Zealand. *Eur. J. Soil Sci.* 50(1):127–137.
Guttormsen, G., Singh, B. R. and Jeng, A. S. 1995. Cadmium concentration in vegetable crops grown in a sandy soil as affected by Cd levels in fertilizer and soil pH. *Fert. Res.* 41(1):27–32.
Hamon, R. E., McLaughlin, M. J., Naidu, R. and Correll, R. 1998. Long-term changes in cadmium bioavailability in soil. *Environ. Sci. Technol.* 32(23):3699–3703.
Hart, J. J., Norvell, W. A., Welch, R. M., Sullivan, L. A. and Kochian, L. V. 1998. Characterization of zinc uptake, binding, and translocation in intact seedlings of bread and durum wheat cultivars. *Plant Physiol.* 118:219–226.
Hart, J. J., Welch, R. M., Norvell, W. A., Clarke, J. M. and Kochian, L. V. 2005. Zinc effects on cadmium accumulation and partitioning in near-isogenic lines of durum wheat that differ in grain cadmium concentration. *New Phytol.* 167(2):391–401.
Hart, J. J., Welch, R. M., Norvell, W. A. and Kochian, L. V. 2002. Transport interactions between cadmium and zinc in roots of bread and durum wheat seedlings. *Physiol. Plant.* 116:73–78.
Hart, J. J., Welch, R. M., Norvell, W. A. and Kochian, L. V. 2006. Characterization of cadmium uptake, translocation and storage in near-isogenic lines of durum wheat that differ in grain cadmium concentration. *New Phytol.* 172(2):261–271.
He, Q. B. and Singh, B. R. 1993a. Cadmium distribution and extractability in soils and its uptake by plants as affected by organic matter and soil type. *J. Soil Sci.* 44:641–650.
He, Q. B. and Singh, B. R. 1993b. Plant availability of Cd in soils. I. Extractability of Cd in newly- and long-term cultivated soils. *Acta Agric. Scand. B. Soil Plant Sci.* 43:134–141.
He, Q. B. and Singh, B. R. 1993c. Plant availability of Cd in soils. II. Factors affecting the extractability and plant uptake of Cd in cultivated soils. *Acta Agric. Scand. B. Soil Plant Sci.* 43:142–150.
He, Q. B. and Singh, B. R. 1994. Crop uptake of cadmium from phosphorus fertilizers. I. Yield and cadmium content. *Water Air Soil Pollut.* 74(3/4):251–265.
He, Q. B. and Singh, B. R. 1995. Cadmium availability to plants as affected by repeated applications of phosphorus fertilizers. *Acta Agric. Scand. B. Soil Plant Sci.* 45(1):22–31.
Holm, P. E., Christensen, T. H., Tjell, J. C. and McGrath, S. P. 1995. Speciation of cadmium and zinc with application to soil solutions. *J. Environ. Qual.* 24(1):183–190.
Hong, C. O., Chung, D. Y., Lee, D. K. and Kim, P. J. 2010. Comparison of phosphate materials for immobilizing cadmium in soil. *Arch. Environ. Con. Tox.* 58(2):268–274.
Hong, C. O., Lee, D. K. and Kim, P. J. 2008. Feasibility of phosphate fertilizer to immobilize cadmium in a field. *Chemosphere* 70(11):2009–2015.
Jeng, A. S. and Singh, B. R. 1995. Cadmium status of soils and plants from a long-term fertility experiment in southeast Norway. *Plant Soil* 175(1):67–74.
Jiao, W., Chen, W., Chang, A. C. and Page, A. L. 2012. Environmental risks of trace elements associated with long-term phosphate fertilizers applications: A review. *Environ. Pollut.* 168:44–53.
Jiao, Y., Grant, C. A. and Bailey, L. D. 2004. Effects of phosphorus and zinc fertilizer on cadmium uptake and distribution in flax and durum wheat. *J. Sci. Food. Agric.* 84(8):777–785.

Jones, K. C. and Johnston, A. E. 1989. Cadmium in cereal grain and herbage from long-term experimental plots at Rothamsted, UK. *Environ. Pollut.* 57(3):199–216.
Jones, K. C., Symon, C. J. and Johnston, A. E. 1987. Long-term changes in soil and cereal grain cadmium: Studies at Rothamsted Experimental Station. *In* D. D. Hemphill, ed., 450–460. *Trace Substances and Environmental Health-XXI: Proceedings of the University of Missouri's 21st Annual Conference on Trace Substances Environmental Health.* Curators of the University of Missouri, Columbia, MO.
Kashem, M. A. and Singh, B. R. 2002. The effect of fertilizer additions on the solubility and plant-availability of Cd, Ni and Zn in soil. *Nutr. Cycling Agroecosyst.* 62:287–296.
Keller, A. and Schulin, R. 2003a. Modelling heavy metal and phosphorus balances for farming systems. *Nutr. Cycling Agroecosyst.* 66(3):271–284.
Keller, A. and Schulin, R. 2003b. Modelling regional-scale mass balances of phosphorus, cadmium and zinc fluxes on arable and dairy farms. *Eur. J. Agron.* 20(1–2):181–198.
Keller, A., von Steiger, B., van der Zee, S. E. A. T. M. and Schulin, R. 2001. A stochastic empirical model for regional heavy-metal balances in agroecosystems. *J. Environ. Qual.* 30(6):1976–1989.
Kuo, S., Huang, B. and Bembenek, R. 2007. Release of cadmium from a triple superphosphate and a phosphate rock in soil. *Soil Sci.* 172(4):257–265.
Lambert, R., Grant, C. and Sauve, S. 2007. Cadmium and zinc in soil solution extracts following the application of phosphate fertilizers. *Sci. Total Environ.* 378(3):293–305.
Levi-Minzi, R. and Petruzzelli, G. 1984. The influence of phosphate fertilizers on Cd solubility in soil. *Water Air Soil Pollut.* 23:423–429.
Li, N., Li, Z., Fu, Q., Zhuang, P., Guo, B. and Li, H. 2013. Agricultural technologies for enhancing the phytoremediation of cadmium-contaminated soil by *Amaranthus hypochondriacus* L. *Water Air Soil Pollut.* 224(9).
Liu, Z. B., Ji, X. H., Peng, H., Tian, F. X., Wu, J. M. and Shi, L. H. 2012. Effects of phosphorous fertilizers on phytoavailability of cadmium in its contaminated soil and related mechanisms. *Chin. J. Appl. Ecol.* 23(6):1585–1590.
Loganathan, P., Hedley, M. J., Grace, N. D., Lee, J., Cronin, S. J., Bolan, N. S. and Zanders, J. M. 2003. Fertiliser contaminants in New Zealand grazed pasture with special reference to cadmium and fluorine: A review. *Aust. J. Soil Res.* 41(3):501–532.
Loganathan, P., Hedley, M. J., Gregg, P. E. H. and Currie, L. D. 1997. Effect of phosphate fertiliser type on the accumulation and plant availability of cadmium in grassland soils. *Nutr. Cycling Agroecosyst.* 46(3):169–178.
Loganathan, P., Mackay, A. D., Lee, J. and Hedley, M. J. 1995. Cadmium distribution in hill pastures as influenced by 20 years of phosphate fertilizer application and sheep grazing. *Aust. J. Soil Res.* 33(5):859–871.
Loganathan, P., Vigneswaran, S., Kandasamy, J. and Naidu, R. 2012. Cadmium sorption and desorption in soils: A review. *Crit. Rev. Environ. Sci. Technol.* 42(5):489–533.
Lorenz, S. E., Hamon, R. E., McGrath, S. P., Holm, P. E. and Christensen, T. H. 1994. Applications of fertilizer cations affect cadmium and zinc concentrations in soil solutions and uptake by plants. *Eur. J. Soil Sci.* 45:159–165.
Mahdy, A. M. 2012. Soil-solution speciation by MINEQL+4.6 model and plant uptake of Cd and Zn by barley (Hordeum vulgare L.) after application of different phosphate fertilizers. *J. Plant Nutr. Soil Sci.* 175(4):560–571.

Mäkelä-Kurtto, R., Erviö, R. and Sippola, J. 1991. Chemical changes in cultivated soils in 1974–1987. Report of Investigation Geological Survey of Finland, Geological Survey of Finland, Espoo, Finland, vol. 105, pp. 81–91. Available at http://tupa.gtk.fi/julkaisu/tutkimusraportti/tr_105.pdf.

McDowell, R. W., Taylor, M. D. and Stevenson, B. A. 2013. Natural background and anthropogenic contributions of cadmium to New Zealand soils. *Agr. Ecosyst. Environ.* 165:80–87.

McGowen, S. L., Basta, N. T. and Brown, G. O. 2001. Use of diammonium phosphate to reduce heavy metal solubility and transport in smelter contaminated soil. *J. Environ. Qual.* 30:493–500.

McGrath, D. and Tunney, H. 2010. Accumulation of cadmium, fluorine, magnesium, and zinc in soil after application of phosphate fertilizer for 31 years in a grazing trial. *J. Plant Nutr. Soil Sci.* 173(4):548–553.

McKenna, I. M., Chaney, R. L. and Williams, F. M. 1993. The effects of cadmium and zinc interactions on the accumulation and tissue distribution of zinc and cadmium in lettuce and spinach. *Environ. Pollut.* 79(2):113–120.

McLaughlin, M. J., Hamon, R. E., McLaren, R. G., Speir, T. W. and Rogers, S. L. 2000. Review: A bioavailability-based rationale for controlling metal and metalloid contamination of agricultural land in Australia and New Zealand. *Aust. J. Soil Res.* 38(6):1037–1086.

McLaughlin, M. J., Maier, N. A., Freeman, K., Tiller, K. G., Williams, C. M. J. and Smart, M. K. 1995. Effect of potassic and phosphatic fertilizer type, fertilizer Cd concentration and zinc rate on cadmium uptake by potatoes. *Fert. Res.* 40(1):63–70.

Mills, R. F., Peaston, K. A., Runions, J. and Williams, L. E. 2012. HvHMA2, a P 1B-ATPase from barley, is highly conserved among cereals and functions in Zn and Cd transport. *PLoS ONE* 7(8).

Mitchell, L. G., Grant, C. A. and Racz, G. J. 2000. Effect of nitrogen application on concentration of cadmium and nutrient ions in soil solution and in durum wheat. *Can. J. Soil Sci.* 80(1):107–115.

Molina, M., Aburto, F., Calderón, R., Cazanga, M. and Escudey, M. 2009. Trace element composition of selected fertilizers used in Chile: Phosphorus fertilizers as a source of long-term soil contamination. *Soil Sediment Contam.* 18(4):497–511.

Moraghan, J. T. 1984. Differential responses of five species to phosphorus and zinc fertilizers. *Commun. Soil Sci. Plant Anal.* 15:137–147.

Mortvedt, J. J. 1987. Cadmium levels in soils and plants from some long-term soil fertility experiments in the United States of America. *J. Environ. Qual.* 16(2):137–142.

Mulla, D. J., Page, A. L. and Ganje, T. J. 1980. Cd accumulation and bioavailability in soils from long-term fertilization. *J. Environ. Qual.* 9:408–412.

Munksgaard, N. and Lottermoser, B. 2011. Fertilizer amendment of mining-impacted soils from Broken Hill, Australia: Fixation or release of contaminants? *Water Air Soil Pollut.* 215(1–4):373–397.

Naidu, R., Bolan, N., Kookana, R. S. and Tiller, K. 1994a. Ionic-strength and pH effects on the sorption of cadmium and the surface charge of soils. *Eur. J. Soil Sci.* 45(4):419–429.

Naidu, R., Bolan, N. S., Kookana, R. S. and Tiller, K. G. 1994b. Ionic-strength and pH effects on the sorption of cadmium and the surface charge of soils. *Eur. J. Soil Sci.* 45(4):419–429.

Nicholson, F. A. and Jones, K. C. 1994. Effect of phosphate fertilizers and atmospheric deposition on long-term changes in the cadmium content of soils and crops. *Environ. Sci. Technol.* 28(12):2170–2175.

Nicholson, F. A., Smith, S. R., Alloway, B. J., Carlton-Smith, C. and Chambers, B. J. 2003. An inventory of heavy metaltrace metal inputs to agricultural soils in England and Wales. *Sci. Total Environ.* 311:205–219.

Nziguheba, G. and Smolders, E. 2008. Inputs of trace elements in agricultural soils via phosphate fertilizers in European countries. *Sci. Total Environ.* 390(1):53–57.

Oliver, D. P., Hamon, R., Tiller, K. G., Wilhelm, N. S., Merry, R. H. and Cozens, G. D. 1994. The effects of zinc fertilization on cadmium concentration in wheat grain. *J. Environ. Qual.* 23(4):705–711.

Pan, J., Plant, J. A., Voulvoulis, N., Oates, C. J. and Ihlenfeld, C. 2010. Cadmium levels in Europe: Implications for human health. *Environ. Geochem. Health* 32(1):1–12.

Pearse, S. J., Veneklaas, E. J., Cawthray, G. R., Bolland, M. D. A. and Lambers, H. 2006. Carboxylate release of wheat, canola and 11 grain legume species as affected by phosphorus status. *Plant Soil* 288(1–2):127–139.

Raynaud, X., Jaillard, B. and Leadley, P. W. 2008. Plants may alter competition by modifying nutrient bioavailability in rhizosphere: A modeling approach. *Am. Nat.* 171(1):44–58.

Richards, I. R., Clayton, C. J. and Reeve, A. J. K. 1998. Effects of long-term fertilizer phosphorus application on soil and crop phosphorus and cadmium contents. *J. Agric. Sci.* 131(2):187–195.

Richardson, A. E., Barea, J. M., McNeill, A. M. and Prigent-Combaret, C. 2009a. Acquisition of phosphorus and nitrogen in the rhizosphere and plant growth promotion by microorganisms. *Plant Soil* 321(1–2):305–339.

Richardson, A. E., Hocking, P. J., Simpson, R. J. and George, T. S. 2009b. Plant mechanisms to optimise access to soil phosphorus. *Crop Pasture Sci.* 60(2):124–143.

Rieuwerts, J. S., Ashmore, M. R., Farago, M. E. and Thornton, I. 2006. The influence of soil characteristics on the extractability of Cd, Pb and Zn in upland and moorland soils. *Sci. Total Environ.* 366(2–3):864–875.

Rivera-Becerril, F., Calantzis, C., Turnau, K., Caussanel, J. P., Belimov, A. A., Gianinazzi, S., Strasser, R. J. and Gianinazzi-Pearson, V. 2002. Cadmium accumulation and buffering of cadmium-induced stress by arbuscular mycorrhiza in three *Pisum sativum* L. genotypes. *J. Exp. Bot.* 53(371):1177–1185.

Roberts, A. H. C., Longhurst, R. D. and Brown, M. W. 1994. Cadmium status of soils, plants, and grazing animals in New Zealand. *N. Z. J. Agric. Res.* 37(1):119–129.

Sauvé, S., Hendershot, W. and Allen, H. E. 2000a. Solid-solution partitioning of metals in contaminated soils: Dependence on pH, total metal burden, and organic matter. *Crit. Rev. Environ. Sci. Technol.* 34:1125–1131.

Sauvé, S., Norvell, W. A., McBride, M. and Hendershot, W. 2000b. Speciation and complexation of cadmium in extracted soil solutions. *Environ. Sci. Technol.* 34:291–296.

Schipper, L. A., Sparling, G. P., Fisk, L. M., Dodd, M. B., Power, I. L. and Littler, R. A. 2011. Rates of accumulation of cadmium and uranium in a New Zealand hill farm soil as a result of long-term use of phosphate fertilizer. *Agr. Ecosyst. Environ.* 144(1):95–101.

Schjønning, P., Christensen, B. T. and Carstensen, B. 1994. Physical and chemical properties of a sandy loam receiving animal manure, mineral fertilizer or no fertilizer for 90 years. *Eur. J. Soi Sci.* 45:257–268.

Sheppard, M. I., Sheppard, S. C. and Grant, C. 2007. Solid/liquid partition coefficients to model trace element critical loads for agricultural soils in Canada. *Can. J. Soil Sci.* 87:189–201.

Sheppard, S. C., Grant, C. A. and Drury, C. F. 2009a. Trace elements in Ontario soils—Mobility, concentration profiles, and evidence of non-point-source pollution. *Can. J. Soil Sci.* 89(4):489–499.

Sheppard, S. C., Grant, C. A., Sheppard, M. I., De Jong, R. and Long, J. 2009b. Risk indicator for agricultural inputs of trace elements to Canadian soils. *J. Environ. Qual.* 38(3):919–932.

Siebers, N., Kruse, J. and Leinweber, P. 2013a. Speciation of phosphorus and cadmium in a contaminated soil amended with bone char: Sequential fractionations and XANES spectroscopy. *Water Air Soil Pollut.* 224(5).

Siebers, N. and Leinweber, P. 2013. Bone char: A clean and renewable phosphorus fertilizer with cadmium immobilization capability. *J. Environ. Qual.* 42(2):405–411.

Siebers, N., Siangliw, M. and Tongcumpou, C. 2013b. Cadmium uptake and subcellular distribution in rice plants as affected by phosphorus: Soil and hydroponic experiments. *J. Soil Sci. Plant Nutr.* 13(4):833–844.

Singh, B. R. 1990. Cadmium and fluoride uptake by oats and rape from phosphate fertilizers in two different soils. *Nor. J. Agric. Sci.*(4):239–249.

Singh, B. R. and Myhr, K. 1998. Cadmium uptake by barley from different Cd sources at two pH levels. *Geoderma* 84:135–194.

Smolders, E., Lambregts, R. M., McLaughlin, M. J. and Tiller, K. G. 1998. Effect of soil solution chloride on cadmium availability to Swiss chard. *J. Environ. Qual.* 27(2):426–431.

Smolders, E. and McLaughlin, M. J. 1996. Chloride increases cadmium uptake in Swiss chard in a resin-buffered nutrient solution. *Soil Sci. Soc. Am. J.* 60(5):1443–1447.

Sparrow, L. A., Salardini, A. A. and Bishop, A. C. 1993a. Field studies of cadmium in potatoes (*Solanum tuberosum* L.). I. Effects of lime and phosphorus on cv. Russet Burbank. *Aust. J. Agric. Res.* 44(4):845–853.

Sparrow, L. A., Salardini, A. A. and Bishop, A. C. 1993b. Field studies of cadmium in potatoes (*Solanum tuberosum* L.). II. Response of cvv. Russet Burbank and Kennebec to two double superphosphates of different cadmium concentration. *Aust. J. Agr. Res.* 44:855–861.

Strong, W. M. and Soper, R. J. 1974a. Phosphorus utilization by flax, wheat, rape, and buckwheat from a band or pellet-like application. I. Reaction zone proliferation. *Agron. J.* 66:597–601.

Strong, W. M. and Soper, R. J. 1974b. Phosphorus utilization by flax, wheat, rape, and buckwheat from a band or pellet-like application. II. Influence of reaction zone phosphorus concentration and soil phosphorus supply. *Agron. J.* 66:601–605.

Syers, J. K., Mackay, A. D., Brown, M. W. and Currie, L. D. 1986. Chemical and physical characteristics of phosphate rock materials of varying reactivity. *J. Sci. Food Agric.* 37:1057–1064.

Traina, S. J. 1999. The environmental chemistry of cadmium. *In* M. J. McLaughlin, B. R. Singh, eds. *Cadmium in Soils and Plants,* 11–38. Kluwer Academic Publishers, Dordrecht, Netherlands.

Willaert, G., Verloo, M. and Vulsteke, G. 1990. Influence of the cultivar on the mineral composition of self-blanching celery, celeriac and scorzonera. *Rev. Agri.* 43(4):599–614.

Williams, C. H. and David, D. J. 1973. The effect of superphosphate on the cadmium content of soils and plants. *Aust. J. Soil Res.* 11(1):43–56.

Williams, C. H. and David, D. J. 1976. The accumulation in soil of cadmium residues from phosphate fertilizers and their effect on the cadmium content of plants. *Soil Sci.* 121(2):86–93.

Zhang, M. and Zhang, H. 2010. Co-transport of dissolved organic matter and heavy metals in soils induced by excessive phosphorus applications. *J. Environ. Sci.* 22(4):598–606.

Zhao, Z. Q., Zhu, Y. G., Smith, S. E. and Smith, F. A. 2005. Cadmium uptake by winter wheat seedlings in response to interactions between phosphorus and zinc supply in soils. *J. Plant Nutr.* 28:1569–1580.

Zheng, R., Li, H., Jiang, R., Römheld, V., Zhang, F. and Zhao, F. J. 2011. The role of root hairs in cadmium acquisition by barley. *Environ. Pollut.* 159(2):408–415.

Zwonitzer, J. C., Pierzynski, G. M. and Hettiarachchi, G. M. 2003. Effects of phosphorus additions on lead, cadmium, and zinc bioavailabilities in a metal-contaminated soil. *Water Air Soil Pollut.* 143(1):193–209.

6

Multicomponent Modeling of Phosphate and Arsenic Reactions and Transport in Soils and Geological Media

Hua Zhang

CONTENTS

A tremendous amount of research has been devoted to fully elucidate the geochemical processes controlling the fate and behavior of phosphate (PO_4^{3-}) and arsenate (AsO_4^{3-}) because of their important roles in environmental processes. Mobility and bioavailability of PO_4^{3-} and AsO_4^{3-} in soil and water are largely dependent on adsorption–desorption processes on solid phases. It has been demonstrated that both PO_4^{3-} and AsO_4^{3-} oxyanions strongly adsorb soils and geological media and compete with each other for adsorption sites (Cui and Weng 2013; Hiemstra and Van Riemsdijk 1999; Hingston et al. 1971; Jain and Loeppert 2000; Manning and Goldberg 1996b; Neupane et al. 2014; Roy et al. 1986a; Smith et al. 2002; So et al. 2012; Violante and Pigna 2002; Zhang and Selim 2007; Zhao and Stanforth 2001). The existence of phosphorus (P) can substantially reduce arsenic (As) adsorption and increase As mobility in the vadose zone and aquifer (Bearup et al. 2012; Darland and Inskeep 1997; Melamed et al. 1995; Peryea 1991; Williams et al. 2003; Zhang and Selim 2007), which has caused environmental concern because of widespread As contamination around the world (Smedley and Kinniburgh 2002).

Geochemical characteristics including mineral properties, pH, As/P concentrations, residence time, and sequence of addition can affect competitive sorption between phosphate and arsenate. Both PO_4^{3-} and AsO_4^{3-} oxyanions specifically sorbed on mineral surfaces by forming similar types of inner-sphere surface complexes through ligand exchange; however, evidence has shown that some adsorption sites are only available for either PO_4^{3-} or AsO_4^{3-}. Based on competitive adsorption of anions on goethite and gibbsite, Hingston et al. (1971) proposed two types of adsorption sites on a mineral surface. The first type is available for both PO_4^{3-} or AsO_4^{3-} where competition takes place, while the second type of adsorption site is specifically available for either anion. Violante and Pigna (2002) demonstrated that minerals rich in aluminum (Al) have a greater affinity for PO_4^{3-} than AsO_4^{3-}, whereas metal oxides and phyllosilicates rich in Fe were more effective in adsorbing AsO_4^{3-} than PO_4^{3-}. Due to similar dissociation constants of PO_4^{-3} (pKa1 = 2.23, pKa2 = 7.2, pKa3 = 12.3) and AsO_4^{-3} (pKa1 = 2.20, pKa2 = 6.97, pKa3 = 11.53), adsorption of both anions decreases with increasing pH (Jain and Loeppert 2000; Manning and Goldberg 1996a,b). In addition, Jain and Loeppert (2000) reported that the effect of PO_4^{3-} on AsO_4^{3-} adsorption on ferrihydrite was greater at high pHs than at low pHs. Adsorption studies on soils often reveal that PO_4^{3-} is preferentially adsorbed rather than AsO_4^{3-} whether added separately or added simultaneously in equal molar ratios (Roy et al. 1986a,b; Smith et al. 2002), and sequence of addition might significantly affect As–P competition (Liu et al. 2001; Zhao and Stanforth 2001). Zhao and Stanforth (2001) confirmed that PO_4^{3-} and AsO_4^{3-} equally adsorbed on goethite when added simultaneously, and when added sequentially, desorption process was kinetically controlled, with a fraction of both PO_4^{3-} and AsO_4^{3-} remaining nonexchangeable.

The presence of P can greatly enhance the movement of arsenic in soils and geological media as a result of the competition between As and P. The miscible displace experiments of Melamed et al. (1995) showed that an increasing presence of PO_4^{3-} shifted AsO_4^{3-} breakthrough curves (BTCs) to the left, indicating reduced AsO_4^{3-} sorption and thus enhanced mobility of AsO_4^{3-} in soil. Similarly, Darland and Inskeep (1997) found that peak concentration and total recovery of As from a sand column containing free Fe oxides increased with increasing P addition. Moreover, they found that a significant fraction of As was retained even though the P loading exceeded the maximum adsorption capacity for P. Williams et al. (2003) compared the effects of pH, pore water velocity, and PO_4^{3-} additions on AsO_4^{3-} transport through columns of a subsurface soil. They concluded that among the three factors, the addition of PO_4^{3-} has the greatest impact on the mobility of AsO_4^{3-}.

A variety of numerical models have been employed with the objective of describing and predicting the competitive reaction and transport of As and P in soils and geological media. Surface complexation models have been used with varying degrees of success to simulate the competitive sorption under varying geochemical conditions (Cui and Weng 2013; Hiemstra and Van Riemsdijk 1999; Manning and Goldberg 1996a,b; So et al. 2012; Stachowicz et al. 2008). Empirical models of the Langmuir and Freundlich types have been extended into multicomponent models for describing the competition between anions (Hingston et al. 1971; Murali and Aylmore 1983; Roy et al. 1986a,b; Sheindorf et al. 1981). In addition, multicomponent models have been coupled with advective–dispersive transport equation (ADE) to simulate transport of As and P with presence of competing anions (Zhang and Selim 2007).

In this chapter, mathematical models of competitive As–P sorption and transport in geological porous media are reviewed. First, we examined surface complexation models commonly used to describe chemical mechanisms of competitive reaction between PO_4^{3-} and AsO_4^{3-}. Then equilibrium and kinetic multicomponent models based on the empirical Freundlich or Langmuir equations are presented. This is followed by a concluding remark on the challenges and directions of model development and applications.

Surface Complexation Models

Under various geochemical conditions and reaction times, surface complexes of PO_4^{3-} and AsO_4^{3-} can exist as inner-sphere surface complexes or outer-sphere surface complexes. The dominant process of AsO_4^{3-} sorption on metal oxides is inner-sphere surface complexation via ligand exchange mechanisms. Spectroscopic analysis revealed that inner-sphere surface complexes formed between As anions and metal oxides can be mononuclear monodentate, mononuclear bidentate, and binuclear bidentate. Macroscopic studies suggest the formation of nonprotonated, bidentate binuclear inner-sphere complexes of PO_4^{3-} (Arai and Sparks 2001).

Surface complexation models (SCMs) are widely used as an equilibrium-based modeling approach of sorption reactions on oxides, clay minerals, and other reactive phases. The most significant feature of the SCMs is the consideration of charges of both aqueous ion and reactive surface sites. Several variations of SCMs including the constant capacitance model (CCM), diffuse-layer model (DLM), and triple-layer models (TLM) have been proposed based on different assumptions of the distribution of surface electrostatic potential (Ψ_i). For a detailed discussion on the different approaches implemented in SCMs, the reader is referred to Goldberg (1992).

SCMs have been incorporated into geochemical modeling packages such as MINTEQA2 (Allison et al. 1991) and PHREEQC2 (Parkburst and Appelo 1999) to predict reaction and transport of As and P in soil and water environment. Databases of reaction constants have been compiled for surface reactions on hydrous ferric (Fe [III]) oxide (HFO), hydrous manganese (Mn) oxide (HMO), goethite, and crystalline Al oxide. Surface complexation constants of the generalized two-layer model are compiled for HFO (Dzombak and Morel 1990), gibbsite (Karamalidis and Dzombak 2010), and HMO (Tonkin et al. 2004).

SCMs are capable of describing and predicting the competitive reactions of PO_4^{3-} and AsO_4^{3-} anions on surface adsorption sites. For example, the set of reactions in Table 6.1 can be used to calculate competition between PO_4^{3-} and AsO_4^{3-} anions for monodentate-specific adsorption on goethite. Manning and Goldberg (1996a,b) applied CCM to As(V) and P adsorption data on oxides and clay minerals at varying pH and different As/P ratios with reasonable success. One-site (monodentate) and two-site (monodentate + bidentate) conceptualizations of the oxide surface gave comparable fits to experimental batch equilibrium data. Parameters optimized from single-ion adsorption data sets adequately described the competitive adsorption observed in binary adsorption envelopes.

Multicompetitive interaction of As and P oxyanions on goethite with the presence of calcium (Ca^{2+}), magnesium (Mg^{2+}), and carbonate (CO_3^{2-}) ions has been studied and the data have been modeled with the charge distribution (CD) model in combination with the extended Stern model option by Stachowicz et al. (2008). Independently calculated CD values and reaction constants are able to predict successfully the ion concentrations in the mixed two- and three-component systems as a function of pH and loading. Their results showed that Ca^{2+} and Mg^{2+} ions promote the adsorption of PO_4^{3-} and AsO_4^{3-} through electrostatic forces, and the CO_3^{2-} anion is a very weak competitor. Concentration of PO_4^{3-} is of critical importance in groundwater systems because of its strong interaction with As anions.

Recently, Sø et al. (2012) applied CCM and the CD multisite complexation (CD-MUSIC) models to study the competitive adsorption of PO_4^{3-} onto calcite. Their experimental results indicate that adsorption of AsO_4^{3-} onto calcite is of minor importance in most groundwater aquifers because of the strong competitive effects of the predominate phosphate. Fitting the CD-MUSIC model to single-component adsorption data resulted in several formulations with similar goodness-of-fit. The optimized parameters well-predicted competitive adsorption of arsenate and phosphate on the calcite surface in the binary sorbate system by the simple addition of single-component models. In contrast, CCM underpredicted the competition when individually fitted parameters were used.

In addition to mineral types, pH, and As/P ratios, competitive effects of natural organic matter (NOM) and synergistic effects of major cations (mainly Ca^{2+}) need to be included to accurately describe surface complexation of oxyanions in soil and sediments. Cui and Weng (2013) successfully applied the ligand and charge distribution (LCD) adsorption model to competitive adsorption of PO_4^{3-} and AsO_4^{3-} in five Dutch agricultural soil samples in a pH range of 4–8. Modeling results indicated that the synergistic effects

TABLE 6.1

Surface Complexation Reactions and CCM Intrinsic Surface Complexation Constants for Arsenate and Phosphate Adsorption on Goethite, Gibbsite, Kaolinite, Montmorillonite, and Illite

Reactions	Equilibrium Constants	Goethite	Gibbsite	KGa-1	SWy-1	IMt-2
$XOH + H^+ = XOH_2^+$	$K_{s+}(int) = \frac{[XOH_2^+]}{[XOH][H^+]} e^{\frac{F\Psi}{RT}}$	7.52	9.10	6.40	6.09	6.00
$XOH = XO^- + H^+$	$K_{s-}(int) = \frac{[XOH^-][H^+]}{[XOH]} e^{-\frac{F\Psi}{RT}}$	−10.6	−10.5	−10.4	−10.2	−10.5
$XOH + H_3AsO_4 = XH_2AsO_4 + H_2O$	$K_{sAs}^1(int) = \frac{[XH_2AsO_4]}{[XOH][H_3AsO_4]}$	8.40	8.60	10.0	8.52	9.24
$XOH + H_3AsO_4 = XHAsO_4^- + H_2O + H^+$	$K_{sAs}^2(int) = \frac{[XHAsO_4^-][H^+]}{[XOH][H_3AsO_4]} e^{-\frac{F\Psi}{RT}}$	3.40	2.20	2.96	3.65	3.66
$XOH + H_3AsO_4 = XAsO_4^{2-} + H_2O + 2H^+$	$K_{sAs}^3(int) = \frac{[XAsO_4^{2-}][H^+]^2}{[XOH][H_3AsO_4]} e^{-\frac{2F\Psi}{RT}}$	−2.30	−5.70	−4.55	−4.18	−4.35
$XOH + H_3PO_4 = XH_2PO_4 + H_2O$	$K_{sP}^1(int) = \frac{[XH_2PO_4]}{[XOH][H_3PO_4]}$	8.05	9.90	10.6		8.30
$XOH + H_3PO_4 = XHPO_4^- + H_2O + H^+$	$K_{sP}^2(int) = \frac{[XHPO_4^-][H^+]}{[XOH][H_3PO_4]} e^{-\frac{F\Psi}{RT}}$	3.40	3.20	3.66	4.12	3.38
$XOH + H_3PO_4 = XPO_4^{2-} + H_2O + 2H^+$	$K_{sP}^3(int) = \frac{[XPO_4^{2-}][H^+]^2}{[XOH][H_3PO_4]} e^{-\frac{2F\Psi}{RT}}$	−2.20	−4.70	−4.88	−4.03	−5.98

Source: Values of equilibrium constants compiled from Manning, B.A. and Goldberg, S., *Soil Sci Soc Am J* 60:121–131, 1996; Manning, B.A. and Goldberg, S., *Clays Clay Miner* 44:609–623, 1996.

Note: IMt-2 = illite, KGa-1 = kaolinite, SWy-1 = montmorillonite.

from calcium (Ca) adsorption on oxide surfaces are a main factor contributing to the maximum adsorption of PO_4^{3-} and AsO_4^{3-} around neutral pH in those soils. In addition, the one representative surface approach (e.g., using only goethite) may become insufficient for simulating competitive adsorption. Therefore, reaction surfaces should be treated as a combination of multiple mineral phases.

Empirical Equilibrium Models

Simple equilibrium-type empirical models—including linear, Freundlich, and Langmuir sorption models—are most commonly used for describing the distribution of chemicals between liquid and solid phases without explicitly accounting for retention mechanisms (Kinniburgh 1986). It has been demonstrated that a number of reactions including ion exchange, outer-sphere surface complexation, inner-sphere surface complexation, and/or surface precipitation can all be embraced in simple equilibrium models (Veith and Sposito 1977). Unlike SCMs, those equilibrium models can be applied to a wide range of conditions with minimal adjustable parameters. A major limitation of the empirical models is that the model parameters are not applicable outside the specific conditions under which they were obtained.

Selectivity Coefficient

The simplest and most widely used adsorption model is a linear function $S = K_D C$, where K_D(L^3 M^{-1}) is a linear distribution coefficient describing the partitioning of solute between dissolved concentration (C, M L^{-3}) and sorbed contents (S, M M^{-1}). The linear sorption equation assumes unlimited sorption capacity and uniform sorption affinity on the surface of the sorbent. Therefore, it lacks the ability to describe the competition between different solutes.

A widely used method to assess competitive adsorption is by assessing the relative preference of individual ions by estimating their selectivity coefficients in a multicomponent system. In the system, the selectivity coefficient (${}^{\mathrm{As}}_{\mathrm{P}}K_C$) of AsO_4^{3-} over PO_4^{3-} in soil was estimated using the following equation (De Brouwere et al. 2004):

$$ {}^{\mathrm{As}}_{\mathrm{P}}K_C = \frac{S_{\mathrm{As}}}{S_{\mathrm{P}}}\frac{C_{\mathrm{P}}}{C_{\mathrm{As}}} = \frac{K_{D\text{-}\mathrm{As}}}{K_{D\text{-}\mathrm{P}}} \tag{6.1} $$

where S_{As} and S_{P} are the amounts of As and P, respectively, adsorbed on soil, C_{As} and C_{P} are the aqueous concentrations of As and P, respectively, and $K_{D\text{-}\mathrm{As}}$ and $K_{D\text{-}\mathrm{P}}$ are distribution coefficients of As and P, respectively.

Experiments of De Brouwere et al. (2004) on a large set of soil samples showed that ${}^{As}_{P}K_C$ decreased from 3.1 to 0.2 with an increasing degree of P saturation of the amorphous oxides (Figure 6.1), which indicate that adsorption sites vary in their selectivity in respect of As and P (Figure 6.2). In addition, it was demonstrated by Zhang and Selim (2008) that estimated K_{As-P} values for three soils did not change with reaction time. However, increasing ${}^{As}_{P}K_C$ during the desorption process demonstrates that more P than As was desorbed from soils, which suggests that As bound more strongly with soil than P.

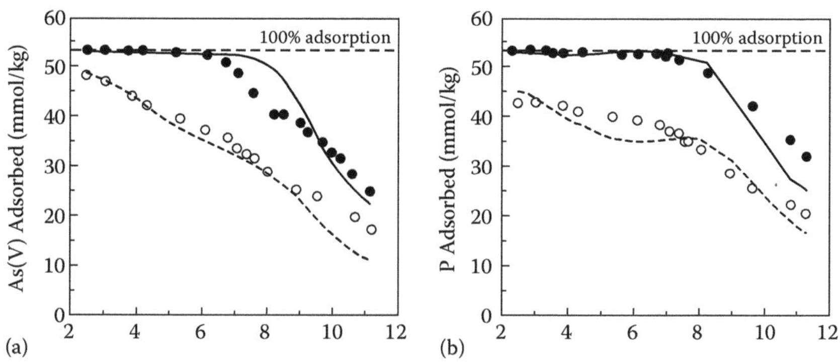

FIGURE 6.1
(a) Arsenate and (b) phosphate single-anion and binary anion adsorption envelopes on goethite (α-FeOOH) with constant capacitance model calculations using the one-site assumption. In all panels, symbols are experimental single-anion data (filled circles), binary anion data (open circles), single-anion model calculations (solid lines), and binary anion model calculations (dotted lines). Reaction conditions: 133 μM As(V) or P, 133 μM As(V) + 133 μM P, 0.1 M NaCl, 2.5 g L^{-1} α-FeOOH, reaction time = 4 h, T = 23°C. (Reproduced from Manning, B.A. and Goldberg, S., *Soil Sci Soc Am J* 60:121–131, 1996. With permission.)

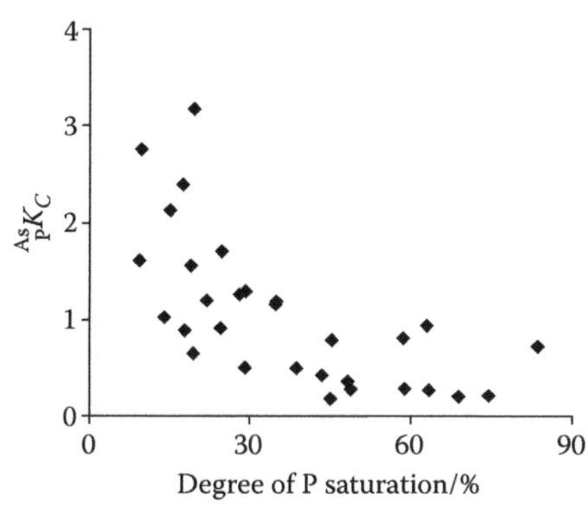

FIGURE 6.2
The As to P selectivity coefficient for adsorption in 30 grassland soils decreases with increasing degree of phosphorus saturation of Fe and Al oxides. (Reproduced from De Brouwere, K. et al., *Eur J Soil Sci* 55:165–173, 2004. With permission.)

Langmuir-Type Equilibrium Model

Experimentally observed adsorption isotherms (solute concentration vs. adsorbed content) are commonly fitted with the Langmuir and Freundlich equations (Kinniburgh 1986). The Langmuir equation has the advantage of providing a sorption maximum S_{max}(M M^{-1}) that can be correlated to soil sorption properties (Veith and Sposito 1977). It has the form

$$S = \frac{K_L C S_{max}}{1 + K_L C} \tag{6.2}$$

where K_L(L^3 M^{-1}) is a Langmuir coefficient related to the binding strength. To simulate the competitive sorption of heavy metals, the Langmuir equation can be extended to the multicomponent Langmuir equation in the form of (Murali and Aylmore 1983)

$$S_i = \frac{K_i C_i S_{max}}{1 + \sum_j K_j C_j} \tag{6.3}$$

where i and j indicate the ith and jth components in the mixture, respectively.

In a binary system of the competitive adsorption of AsO_4^{3-} and PO_4^{3-}, this equation set can be simplified to

$$S_{As} = \frac{K_{L-As} C_{As} S_{max}}{1 + K_{L-As} C_{As} + K_{L-P} C_P} \tag{6.4a}$$

$$S_P = \frac{K_{L-P} C_P S_{max}}{1 + K_{L-As} C_{As} + K_{L-P} C_P} \tag{6.4b}$$

with three parameters, K_{L-As}, and K_{L-P}. This multicomponent Langmuir model assumes that there is only one set of sorption sites (S_{max}) for all competing ions. Furthermore, the model also assumes that the presence of competing ions does not affect the sorption affinity (K_{L-As} and K_{L-P}) of others. However, studies have often shown that some adsorption sites are specifically available for either PO_4^{3-} or AsO_4^{3-}.

A Langmuir-type model with parameters of surface sites and reaction constants for each solute was developed by Hingston et al. (1971) to describe competitive adsorption of PO_4^{3-} or AsO_4^{3-}. Their model is essentially a very simplified (without electrostatic terms) version of the SCM model described in the "Surface Complexation Models" section. As a matter of fact, the multisite versions of Langmuir-type models could be mathematically viewed as a special case of SCMs under constant pH and ionic strength.

Freundlich-Type Equilibrium Model

The Freundlich equation is a widely used empirical adsorption model that can be expressed as (Kinniburgh 1986)

$$S = K_F C^N \tag{6.5}$$

where K_F is the Freundlich distribution or partition coefficient (M M^{-1} (M $L^{-3})^{-N}$), and N is the dimensionless reaction order commonly less than 1. It has been shown that the Freundlich equation can be derived from integrating a set of Langmuir isotherms with an exponential distribution of adsorption energies. The Freundlich reaction order N can be viewed as a measurement of the heterogeneity of surface reaction sites.

The Sheindorf–Rebhun–Sheintuch (SRS) equation has been developed to describe competitive or multicomponent sorption where it is assumed that the single-component sorption follows the Freundlich equation (Sheindorf et al. 1981). The derivation of the SRS equation was based on the assumption of an exponential distribution of adsorption energies for each component. Specifically, the SRS model was developed to describe competitive equilibrium sorption for multicomponent systems where the sorption isotherms of a single component follow the Freundlich equation. A general form of the SRS equation can be written as

$$S_i = K_F^i C_i \left[\sum_j (\alpha_j^i C_j) \right]^{N_i - 1} \tag{6.6}$$

where α_j^i is a dimensionless competition coefficient that describes the inhibition by component j to the adsorption of component i. It can be used to evaluate the relative selectivity of the sorbent to adsorbed components. By definition, α_j^i equals 1 when $i = j$. If there is no competition (i.e., $\alpha_j^i = 0$ for all components $j \neq i$), Equation 6.3 yields a single-species Freundlich equation (Equation 6.2) for component i. It should also be pointed out that if the single-species sorption isotherm is linear (i.e., $N_i = 1$), Equation 6.3 predicts the absence of competition. It should be noted that the SRS equation does not imply specific reaction mechanisms.

The SRS equation has been used by Roy et al. (1986a,b) for describing competitive adsorption isotherms of PO_4^{3-} or AsO_4^{3-} on three different soils (Figure 6.3). To apply the SRS equation, the single-component isotherms are first fitted to the Freundlich equation to obtain the distribution coefficient (K_F^{As} and K_F^{P}) and

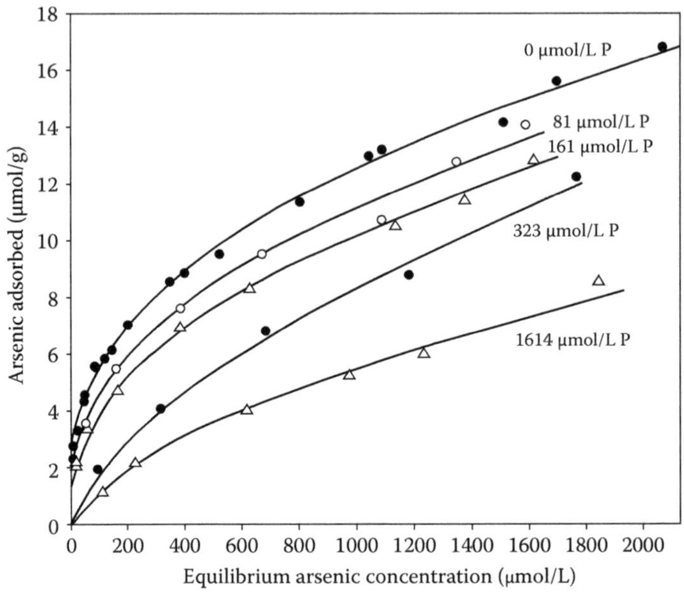

FIGURE 6.3
Adsorption isotherms of arsenate by EPA-14 at pH 4.5 and 298 K with phosphate added. The P concentrations shown are the initial concentrations added. (Reproduced with permission from Roy, W.R. et al., *Soil Sci Soc Am J* 50:1176–1182, 1986a.)

reaction exponent (N_{As} and N_P). Then competition coefficients (α_{As}^{P} and α_{P}^{As}) are estimated through fitting the experimental isotherms of binary mixtures to the SRS equation. Their results showed that adsorption of PO_4^{3-} or AsO_4^{3-} conforms to the SRS equation. The estimated competitive coefficient was 1.66 to 2.76 for PO_4^{3-} displacing AsO_4^{3-} (α_{P}^{As}) and 0.11–0.35 for AsO_4^{3-} displacing PO_4^{3-} (α_{P}^{As}), showing stronger competitions of PO_4^{3-} on AsO_4^{3-} adsorption in soils. They suggested that the coefficients should be regarded as empirical values describing the relative affinity of soils for a particular anion from anionic mixtures. Barrow et al. (2005) successfully used nonlinear curve fitting to determine the competitive coefficients between AsO_4^{-3} and PO_4^{-3} on an Andisol. They found that competition coefficients (α_{As}^{P} and α_{P}^{As}) were not symmetrical and suggested that in addition to competition for sorption sites, sorption of ions could produce a separate competition effect through change in electric potential of the surface.

Kinetic Multicomponent Models

Traditionally, competitive adsorption of PO_4^{3-} or AsO_4^{3-} has been studied with equilibrium batch experiments conducted within a relatively short time of reaction. However, several experiments have demonstrated that the adsorption of PO_4^{3-} or AsO_4^{3-} on minerals and soils is time-dependent (Zhang and Selim 2007, 2008a). Therefore, a number of formulations have been developed to describe the kinetics of competitive sorption in soils.

Langmuir-Type Kinetic Model

The kinetic equations of the Langmuir type have been developed into the reversible second-order model in the form of (Murali and Aylmore 1983)

$$\frac{\partial S}{\partial t} = k_{Lf}(\Gamma - S)\frac{\theta}{\rho}C - k_{Lb}S \tag{6.7}$$

where ρ is the soil bulk density (M L^{-3}), θ is the volumetric water content (L^3 L^{-3}), t is the reaction time (T), Γ is the amount of available adsorption sites (M M^{-1}), and k_{Lf} and k_{Lb} are the second-order forward and backward reaction rate coefficients (h^{-1}), respectively. Under equilibrium conditions (i.e., $\frac{\partial S}{\partial t} = 0$), Equation 6.7 yields equilibrium Langmuir equation (Equation 6.2) assuming $K_L = \frac{k_{Lf}}{k_{Lb}}\frac{\theta}{\rho}$ and $\Gamma = S_{\max}$. To simulate the time-dependent sorption of competing ions, the multicomponent second-order kinetic equation was proposed in the form of

$$\frac{\partial S_i}{\partial t} = k_{Lf}^i \frac{\theta}{\rho} C_i \left[\Gamma - \sum_j S_j\right] - k_{Lb}^i S_i \tag{6.8}$$

Under equilibrium condition, Equation 6.8 yields Equation 6.7 assuming that $K_L^i = \frac{k_{Lf}^i}{k_{Lb}^i}\frac{\theta}{\rho}$ and $\Gamma = S_{\max}$. On the other hand, Equation 6.8 yields a single species second-order kinetic sorption equation (Equation 6.3) without the presence of competing ions.

Freundlich-Type Kinetic Model

The kinetic (time-dependent) sorption models of the Freundlich type have been developed to simulate the sorption of solutes during their transport in soils and aquifers (Selim et al. 1992). The reversible nth-order (Freundlich-type) kinetic sorption equation is in the form of

$$\frac{\partial S}{\partial t} = k_{Ff}\frac{\theta}{\rho}C^n - k_{Fb}S \tag{6.9}$$

where k_{Ff} and k_{Fb} are the nth-order forward and backward reaction rate coefficients (h^{-1}), respectively, and n is a nonlinear parameter usually less than 1. Under equilibrium conditions (i.e., $\frac{\partial S}{\partial t} = 0$), Equation 6.9 yields the Freundlich equation (Equation 6.5) assuming $K_F = \frac{k_{Ff}}{k_{Fb}} \frac{\theta}{\rho}$ and $N = n$.

For time-dependent sorption of multiple components, the SRS equation can be extended such that for reversible nth-order multicomponent kinetic retention, the equation is of the form (Zhang and Selim 2007)

$$\frac{\partial S_i}{\partial t} = k_{Ff}^i \frac{\theta}{\rho} C_i \left[\sum_j (\alpha_j^i C_j) \right]^{n_i - 1} - k_{Fb}^i S_i \tag{6.10}$$

Under equilibrium condition, Equation 6.10 yields Equation 6.9 assuming that $K_F^i = \frac{k_{Ff}^i}{k_{Fb}^i} \frac{\theta}{\rho}$ and $N_i = n_i$. On the other hand, if there is no competition (i.e., $\alpha_j^i = 0$ for all $j \neq i$), Equation 6.10 yields a single species Nth-order kinetic sorption equation (Equation 6.6).

Multicomponent Multireaction Model

The multireaction model (MRM) has been applied successfully to simulate the kinetic adsorption and desorption of arsenate (Zhang and Selim 2008b) and phosphate (Wilson et al. 2004) in minerals and soils. To simulate the kinetics of competitive adsorption of PO_4^{3-} and AsO_4^{3-}, Zhang and Selim (2007) extended the MRM into a multicomponent multireaction model (MCMRM). Specifically, the equilibrium and kinetic adsorption equations were of the kinetic Freundlich type and can be described with the following equations:

$$S_e^i = K_e^i C_i \left[\sum_j (\alpha_j^i C_j) \right]^{n_i - 1} \tag{6.11}$$

$$\frac{\partial S_k^i}{\partial t} = k_1^i \frac{\theta}{\rho} C_i \left[\sum_j (\alpha_j^i C_j) \right]^{n_i - 1} - (k_2^i + k_3^i) S_k^i \tag{6.12}$$

$$\frac{\partial S_p^i}{\partial t} = k_3^i S_k^i \tag{6.13}$$

$$\frac{\partial S_s^i}{\partial t} = k_s^i \frac{\theta}{\rho} C_i \tag{6.14}$$

where S_e is the amount retained on equilibrium sites (M M^{-1}), S_k is the amount retained on kinetic type sites (M M^{-1}), S_p is the amount retained irreversibly by consecutive reaction (M M^{-1}), and S_s is the amount retained irreversibly by concurrent type of reaction (M M^{-1}). K_e is a dimensionless equilibrium constant, k_1 and k_2(T^{-1}) are the forward and backward reaction rates associated with kinetic sites, respectively, k_3(T^{-1}) is the irreversible rate coefficient associated with the kinetic sites, and k_s(T^{-1}) is the irreversible rate coefficient associated with solution. The total amount of adsorption (S_s, M M^{-1}) can be calculated as

$$S_T^i = S_e^i + S_k^i + S_p^i + S_s^i \tag{6.15}$$

The MCMRM model accounts for short- and long-term adsorption mechanisms of PO_4^{3-} and AsO_4^{3-} sorption in heterogeneous soils. The nonlinear competitive adsorption equations in the MCMRM model provide useful quantitative predictors of possible competitive retention interactions in soils.

The single component MRM model has been applied by Zhang and Selim (2008a) to simulate the time-dependent retention of AsO_4^{3-} and PO_4^{3-}. The estimated equilibrium reaction constant (K_e) and forward rate coefficients (k_1) for AsO_4^{3-} exhibited a marked decrease with increasing concentrations of applied PO_4^{3-}, which indicates the competition of PO_4^{3-} to AsO_4^{3-} adsorption. Furthermore, Zhang and Selim (2007) applied the MCMRM to simulate the competitive adsorption kinetics between AsO_4^{3-} and PO_4^{3-} on soils, where kinetic sorption parameters (K_e, k_1, k_2, k_3, and n) were obtained from single component MRM simulation, and the competitive coefficients (α_{As}^{P} and α_{P}^{As}) were based on the SRS equation (Equation 6.6). As shown in Figure 6.4, the MCMRM with the appropriate competitive coefficients has the capability of predicting competitive sorption kinetic between AsO_4^{3-} and PO_4^{3-}.

Multicomponent Transport Model

Competitive sorption of PO_4^{3-} and AsO_4^{3-} can result in complex spatial–temporal patterns during their transport in soils and geological media. Accurate simulation and prediction of PO_4^{3-} and AsO_4^{3-} transport require multicomponent models (MCMs) that sufficiently describe the competition processes. Therefore, the chemical reaction models described in the "Surface Complexation

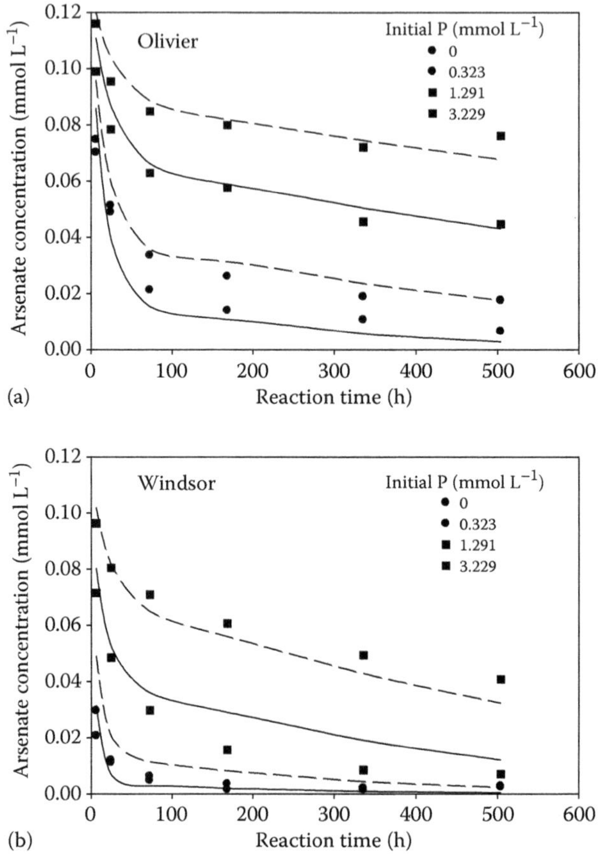

FIGURE 6.4
Arsenate concentrations versus reaction times with the presence of various concentrations of phosphate for (a) Olivier and (b) Windsor soils. The initial AsO_4^{3-} concentrations were 0.13 mmol L^{-1}. Symbols are for different initial PO_4^{3-} concentrations of 0, 0.32, 1.29, and 3.23 mmol L^{-1}. Solid and dashed curves are MCMRM simulations. (Reproduced with permission from Zhang, H. and Selim, H.M., *Soil Sci Soc Am J* 71:1267–1277, 2007.)

Models" and "Kinetic Multicomponent Models" sections need to be incorporated into the steady-state ADE in the form of (Zhang and Selim 2007)

$$\frac{\partial C_i}{\partial t} + \frac{\rho}{\theta}\frac{\partial S_i}{\partial t} = D\frac{\partial^2 C_i}{\partial x^2} - v\frac{\partial C_i}{\partial x} \tag{6.16}$$

where x is the distance (L), D is the dispersion coefficient ($L^2\ T^{-1}$), v (= q/θ) is the average pore water velocity ($L\ T^{-1}$), and q is Darcy's water flux density ($L\ T^{-1}$). The appropriate initial and boundary conditions for a finite soil column are

$$C_i(x) = C_o \quad t = 0 \tag{6.17a}$$

$$S_i(x) = S_o \quad t = 0 \tag{6.17b}$$

$$\left(-D\frac{\partial C_i}{\partial x} + vC_i\right)\bigg|_{x=0} = \begin{cases} vC_{p,i} & t \in T_{p,i} \\ 0 & t \notin T_{p,i} \end{cases} \tag{6.17c}$$

$$\frac{\partial C_i}{\partial x}\bigg|_{x=L} = 0 \quad t > 0 \tag{6.17d}$$

where C_o is the initial solution concentration (M L^{-3}), S_o is the initial amount of sorption (M M^{-1}), $C_{p,i}$ is the input solute concentration (M L^{-3}), T_p is the duration of applied solute pulses (T), and L is the length of the column (L).

Reaction on solid surface $\left(\frac{\partial S_i}{\partial t}\right)$ is the major mechanism controlling the reactive transport of PO_4^{3-} and AsO_4^{3-} in natural soil and water. As shown in the "Surface Complexation Models" section, SCMs, multicomponent Langmuir, and SRS equations describe the retention phenomena based on the local equilibrium assumption (LEA). It assumes that the reaction of an individual chemical species in the soil is sufficiently fast or instantaneous, and an apparent equilibrium condition is achieved in a timescale much smaller than advection and diffusion processes.

Although several experimental studies have clearly illustrated the importance of PO_4^{3-} competition on the transport of As in soils (Darland and Inskeep 1997; Melamed et al. 1995; Williams et al. 2003), little research has been conducted to model the competitive transport of PO_4^{3-} and AsO_4^{3-} under dynamic flow conditions, and there is still a lack of effective tools for modeling competitive transport because of the complexity of this process.

The MCMRM described in Equations 6.11 through 6.15 has been incorporated into the transport model of Equations 6.16 and 6.17 to describe the kinetics sorption and transport of competitive AsO_4^{3-} and PO_4^{3-} in soils (Zhang and Selim 2007). In a forward modeling mode, the MCMRM underestimated As and P retention by soil columns using retention parameters from adsorption experiments. Then inverse modeling was carried out by using a nonlinear least-squares optimization scheme to estimate necessary MCMRM parameters from observed BTCs. The use of fitted kinetic parameters provided adequate description of both As and P transport in soils (Figure 6.5).

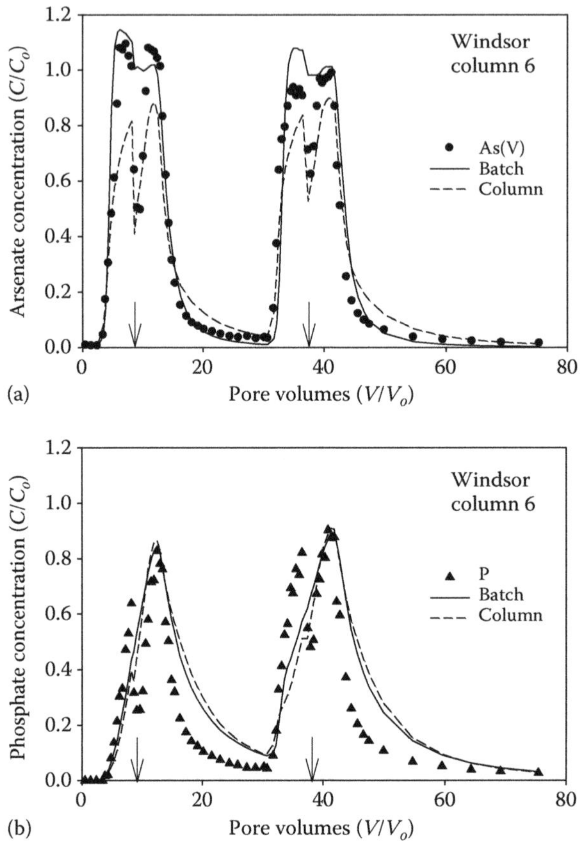

FIGURE 6.5
Experimental (a) arsenate and (b) phosphate BTCs in Windsor soil. Solid curves are MCMRM predictions using batch kinetic parameters. Dashed curves are MCMRM simulations using kinetic parameters obtained from single-component As(V) transport experiments. Arrows indicate pore volumes when flow interruptions occurred. (Reproduced from Zhang, H. and Selim, H.M., *Soil Sci Soc Am J* 71:1267–1277, 2007. With permission.)

Concluding Remarks

Competitive reactions between phosphate and arsenate have a major impact on the environmental fate and behavior of the two important anions. Mathematical models have been developed in the last three decades to simulate the competitive sorption of phosphate and arsenate on minerals and soils. In essence, the SCMs based on a detailed description of surface reaction mechanisms have the capability of describing competitive adsorption on minerals under the impact of geochemical factors including mineral properties, pH, and As/P concentrations. However, there is still much doubt on the practicality of SCMs because of the model requirement for a comprehensive database of thermodynamic reaction constants and surface properties of various types of minerals. The commonly used Langmuir and Freundlich equations have been extended to multicomponent models with the inclusion of competitive coefficients. The empirical types of models are rather easy to use but lack the transferability to a general purpose model.

The majority of the studies on phosphate and arsenate are of the equilibrium nature; however, kinetics studies often demonstrate the time-dependent behavior of the competition between these anions. Recently efforts have been made to extend the equilibrium types of MCMs to include the effects of reaction time and sequence of reactant addition. Heterogeneity of reaction sites on soil surfaces has also been considered with the extension to MCMRM. MCMRMs have been successfully used to simulate the reaction and transport of phosphate and arsenate in laboratory settings. Further research will be needed to model the physical and chemical mechanisms involved in the kinetic sorption and transport.

Acknowledgments

This research was financially supported by the National Natural Science Foundation of China (41271506 and 41230858) and Key Research Program of the Chinese Academy of Sciences (KZZD-EW-14). Dr. Hua Zhang was financially supported by the Recruitment Program of Global Young Experts (1000Plan).

References

Allison, J.D., Brown, D.S. and Novo-Gradac, K.J. (1991) MINTEQA2/PRODEFA2, *A Geochemical Assessment Model for Environmental Systems*, Version 3.0 User's Manual. EPA/600/3-91/021. U.S. Environmental Protection Agency, Washington, DC.

Arai, Y. and Sparks, D.L. (2001) ATR-FTIR spectroscopic investigation on phosphate adsorption mechanisms at the ferrihydrite-water interface. *J Colloid Interface Sci* 241:317–326. doi: 10.1006/jcis.2001.7773.

Barrow, N.J., Cartes, P. and Mora, M.L. (2005) Modifications to the Freundlich equation to describe anion sorption over a large range and to describe competition between pairs of ions. *Eur J Soil Sci* 56:601–606. doi: 10.1111/j.1365-2389.2005.00700.x.

Bearup, L.A., Navarre-Sitchler, A.K., Maxwell, R.M. and McCray, J.E. (2012) Kinetic metal release from competing processes in aquifers. *Environ Sci Technol* 46:6539–6547. doi: 10.1021/es203586y.

Cui, Y. and Weng, L. (2013) Arsenate and phosphate adsorption in relation to oxides composition in soils: LCD modeling. *Environ Sci Technol* 47:7269–7276. doi: 10.1021/es400526q.

Darland, J.E. and Inskeep, W.P. (1997) Effects of pH and phosphate competition on the transport of arsenate. *J Environ Qual* 26:1133–1139.
De Brouwere, K., Smolders, E. and Merckx, R. (2004) Soil properties affecting solid-liquid distribution of As(V) in soils. *Eur J Soil Sci* 55:165–173. doi: 10.1046/j.1365-2389.2003.00584.x.
Dzombak, D.A. and Morel, F.M.M. (1990) *Surface Complexation Modeling: Hydrous Ferric Oxide.* John Wiley & Sons, New York.
Goldberg, S. (1992) Use of surface complexation models in soil chemical systems. *Adv Agron* 47:233–329.
Hiemstra, T. and Van Riemsdijk, W.H. (1999) Surface structural ion adsorption modeling of competitive binding of oxyanions by metal (hydr)oxides. *J Colloid Interface Sci* 210:182–193. doi: 10.1006/jcis.1998.5904.
Hingston, F.J., Posner, A.M. and Quirk, J.P. (1971) Competitive adsorption of negatively charged ligands on oxide surfaces. *Discuss Faraday Soc* 52:334–342.
Jain, A. and Loeppert, R.H. (2000) Effect of competing anions on the adsorption of arsenate and arsenite by ferrihydrite. *J Environ Qual* 29:1422–1430.
Karamalidis, A.K. and Dzombak, D.A. (2010) Surface Complexation Modeling: Gibbsite. John Wiley & Sons, New York.
Kinniburgh, D.G. (1986) General-purpose adsorption-isotherms. *Environ Sci Technol* 20:895–904. doi: 10.1021/es00151a008.
Liu, F., De Cristofaro, A. and Violante, A. (2001) Effect of pH, phosphate and oxalate on the adsorption/desorption of arsenate on/from goethite. *Soil Sci* 166:197–208. doi: 10.1097/00010694-200103000-00005.
Manning, B.A. and Goldberg, S. (1996a) Modeling arsenate competitive adsorption on kaolinite, montmorillonite and illite. *Clays Clay Miner* 44:609–623. doi: 10.1346/ccmn.1996.0440504.
Manning, B.A. and Goldberg, S. (1996b) Modeling competitive adsorption of arsenate with phosphate and molybdate on oxide minerals. *Soil Sci Soc Am J* 60:121–131.
Melamed, R., Jurinak, J.J. and Dudley, L.M. (1995) Effect of adsorbed phosphate on transport of arsenate through an oxisol. *Soil Sci Soc Am J* 59:1289–1294.
Murali, V. and Aylmore, L.A.G. (1983) Competitive adsorption during solute transport in soils. 1. Mathematical-models. *Soil Sci* 135:143–150. doi: 10.1097/00010694-198303000-00002.
Neupane, G., Donahoe, R.J. and Arai, Y. (2014) Kinetics of competitive adsorption/desorption of arsenate and phosphate at the ferrihydrite-water interface. *Chem Geol* 368:31–38. doi: 10.1016/j.chemgeo.2013.12.020.
Parkhurst, D.L. and Appelo, C.A.J. (1999) *User's Guide to PHREEQC Version 2—A Computer Program for Speciation, Batch Reaction, One-Dimensional Transport and Inverse Geochemical Calculations.* Water Resources Investigations, Report 99-4259, U.S. Geological Survey, Lakewood, CO.
Peryea, F.J. (1991) Phosphate-induced release of arsenic from soils contaminated with lead arsenate. *Soil Sci Soc Am J* 55:1301–1306.
Roy, W.R., Hassett, J.J. and Griffin, R.A. (1986a) Competitive coefficients for the adsorption of arsenate, molybdate, and phosphate mixtures by soils. *Soil Sci Soc Am J* 50:1176–1182.
Roy, W.R., Hassett, J.J. and Griffin, R.A. (1986b) Competitive interactions of phosphate and molybdate on arsenate adsorption. *Soil Sci* 142:203–210. doi: 10.1097/00010694-198610000-00003.

Selim, H.M., Buchter, B., Hinz, C. and Ma, L. (1992) Modeling the transport and retention of cadmium in soils—Multireaction and multicomponent approaches. *Soil Sci Soc Am J* 56:1004–1015.

Sheindorf, C., Rebhun, M. and Sheintuch, M. (1981) A Freundlich-type multicomponent isotherm. *J Colloid Interface Sci* 79:136–142. doi: 10.1016/0021-9797(81)90056-4.

Smedley, P.L. and Kinniburgh, D.G. (2002) A review of the source, behaviour and distribution of arsenic in natural waters. *Appl Geochem* 17:517–568. doi: 10.1016/s0883-2927(02)00018-5.

Smith, E., Naidu, R. and Alston, A.M. (2002) Chemistry of inorganic arsenic in soils: II. Effect of phosphorus, sodium, and calcium on arsenic sorption. *J Environ Qual* 31:557–563.

Sø, H.U., Postma, D., Jakobsen, R. and Larsen, F. (2012) Competitive adsorption of arsenate and phosphate onto calcite; experimental results and modeling with CCM and CD-MUSIC. *Geochim Cosmochim Acta* 93:1–13. doi: 10.1016/j.gca.2012.06.021.

Stachowicz, M., Hiemstra, T. and van Riemsdijk, W.H. (2008) Multi-competitive interaction of As(III) and As(V) oxyanions with Ca^{2+}, Mg^{2+}, PO^{43-}, and CO^{32-} ions on goethite. *J Colloid Interface Sci* 320:400–414. doi: 10.1016/j.jcis.2008.01.007.

Tonkin, J.W., Balistrieri, L.S. and Murray, J.W. (2004) Modeling sorption of divalent metal cations on hydrous manganese oxide using the diffuse double layer model. *Appl Geochem* 19:29–53.

Veith, J.A. and Sposito, G. (1977) Use of Langmuir equation in interpretation of adsorption phenomena. *Soil Sci Soc Am J* 41:697–702.

Violante, A. and Pigna, M. (2002) Competitive sorption of arsenate and phosphate on different clay minerals and soils. *Soil Sci Soc Am J* 66:1788–1796.

Williams, L.E., Barnett, M.O., Kramer, T.A. and Melville, J.G. (2003) Adsorption and transport of arsenic(V) in experimental subsurface systems. *J Environ Qual* 32:841–850.

Wilson, G.V., Rhoton, F.E. and Selim, H.M. (2004) Modeling the impact of ferrihydrite on adsorption-desorption of soil phosphorus. *Soil Sci* 169:271–281. doi: 10.1097/01.ss.0000126841.03965.3a.

Zhang, H. and Selim, H.M. (2007) Modeling competitive arsenate-phosphate retention and transport in soils: A multi-component multi-reaction approach. *Soil Sci Soc Am J* 71:1267–1277. doi: 10.2136/sssaj2006.0422.

Zhang, H. and Selim, H.M. (2008a) Competitive sorption-desorption kinetics of arsenate and phosphate in soils. *Soil Sci* 173:3–12. doi: 10.1097/ss.06013e31815ce750.

Zhang, H. and Selim, H.M. (2008b) Reaction and transport of arsenic in soils: Equilibrium and kinetic modeling. *Adv Agron* 98:45–115.

Zhao, H.S. and Stanforth, R. (2001) Competitive adsorption of phosphate and arsenate on goethite. *Environ Sci Technol* 35:4753–4757, doi: 10.1021/es010890y.

7

Influence of Phosphates on Fractionation, Mobility, and Bioavailability of Soil Metal(loid)s

Sabry M. Shaheen and Christos D. Tsadilas

CONTENTS

Contamination of elevated heavy metal(loid)s (HMs) in soils can adversely affect soil ecology, agricultural productivity, quality of agricultural products and water resources, as well as human and animal health (Raicevic et al. 2005; Thawornchaisit and Polprasert 2009; Kabata-Pendias 2011; Shaheen et al. 2013, 2014a,b). The improper management of metallurgical industrial wastes and mining wastes usually causes soil contamination with toxic HMs (e.g., arsenic [As], cadmium [Cd], chromium [Cr], cobalt [Co], copper [Cu], mercury [Hg], nickel [Ni], lead [Pb], selenium [Se], and zinc [Zn]) (Hooda 2010). Increasing awareness of the environmental and public health hazard of toxic HMs pressurizes society to develop management strategies to remediate or restore the contaminated areas. As land treatment is considered an important waste management practice, soil becomes a major source of HMs entering the food chain mainly through plant uptake and animal transfer (Rogival et al. 2007; Zhuang et al. 2009; Shaheen et al. 2014b,c).

Mobility, bioavailability, and entry of soil-borne HMs into the food chain depend on the source and speciation of metal(loid)s input (Rinklebe and Shaheen 2014; Shaheen and Rinklebe 2014), the properties of the soil (Shaheen 2009; Shaheen et al. 2013), bioavailability of HMs, the rate and magnitude of uptake by plants, and the extent of absorption by birds and animals used for egg, milk, and meat production (Adriano 2001). Physicochemical interaction of HMs with soil is the primary factor that controls their bioavailability. Metal(loid)s in the solid or sorbed phase are usually unavailable, and the uptake by organisms, including higher plants, usually occurs when these metal(loid)s become soluble and reach the solution phase or pore water. Thus, processes influencing the equilibrium between soluble and solid or bound phases of HMs (i.e., sorption reactions) impact their overall bioavailability.

Applying fertilizers and amendments to multimetal contaminated soils may have contradictory effects on the speciation, mobility, and bioavailability of specific HMs depending on the amendment. Application of phosphates either as fertilizers and/or soil amendments is very common. Thus, application of phosphates to soils might affect the forms and mobility of soil HMs and their downward movement and uptake by the grown plants. A large number of studies have provided conclusive evidence for the potential value of both water-soluble (e.g., diammonium phosphate [DAP]) and water-insoluble (e.g., hydroxyapatite [HA] and phosphate rocks [PR]) phosphate (P) compounds to immobilize metals in soils, thereby reducing their bioavailability for plant uptake (e.g., Ma et al. 1997; Boisson et al. 1999; Cao et al. 2004; Chen et al. 2007; Marchat et al. 2007; Thawornchaisit and Polprasert 2009; Mignardi et al. 2012; Bolan et al. 2013a,b; Shaheen and Tsadilas 2013). It is, however, important to recognize that depending on the nature of P compounds and the HM species, application of these materials can cause either mobilization or immobilization of the metals.

Furthermore, some of these materials contain high levels of metals and can act as agents of metal introduction to soils. Accordingly, these materials should be scrutinized before their large-scale use as immobilizing agents in contaminated sites. Although mobilization by certain P compounds enhances the bioavailability of metals, immobilization inhibits their plant uptake and reduces their transport in soils and subsequent groundwater contamination. Whenever phytoremediation of contaminated sites is practicable, appropriate P compounds can be used to enhance the bioavailability of metals for plant uptake. Removal of metals through phytoremediation techniques and the subsequent recovery of the metals or their safe disposal are attracting research and commercial interests. Phosphate compounds can be used to enhance the solubilization of metals, leading to their increased uptake by plants. However, when it is not possible to remove the metals from the contaminated sites by phytoremediation, other viable options such as *in situ* immobilization should be considered as an integral part of risk management. One way to facilitate such immobilization is by altering the physicochemical properties of the metal–soil complex by introducing an anion, such as phosphate, that enhances metal adsorption via anion-induced negative charge (i.e., cation exchange capacity [CEC]) and metal precipitation. It is important to recognize that large-scale use of P compounds can lead to surface and groundwater contamination of this element. Therefore, this chapter focuses on reviewing and reporting the research results of the influence of P compounds on the speciation, mobility, and bioavailability of HMs in soils as well as the role of phosphates on *in situ* and phytoremediation of HM-contaminated soils.

Phosphates and Geochemical Fractions of Soil Metal(loid)s

Soil contamination by HMs represents a worldwide environmental concern mainly because these HMs can be transferred to the hydrosphere and biosphere, thereby posing a hazard to the environment and human health. Consequently, the mobility and bioavailability of HMs in soil play an important role in the dynamics of these HMs in the environment as well as their uptake by vegetation and animals. Total metal content of soils is useful for many geochemical applications; however, it is not sufficient to assess the environmental impact of metal contaminants (Shaheen and Rinklebe 2014). The toxicity and mobility of HMs in soils depend on their specific chemical form, binding state, metal properties, environmental factors, and soil properties (Rinklebe and Shaheen 2014). The chemical fractionation of HMs must be taken into account in environmental pollution studies since total content in soils provides, in most cases, limited information on the mobility and bioavailability of these metals (Ma and Rao 1997; Kabala and Singh 2001; Nannoni et al. 2011; Rinklebe and Shaheen 2014) and can be misleading when assessing environmental effects due to a potential overestimation of exposure risk (Ahmad et al. 2014; Shaheen and Rinklebe 2014). Speciation of soil HMs is more of an interest agriculturally in terms of what is biologically extractable. Additionally, the assessment of metal distribution in the various organic and mineral phases found in soils may be more useful for the prediction of metal behavior, including solubility, mobility, bioavailability, and thus toxicity (Kabata-Pendias 2011). The chemical fractionation, that is, the association of HMs with specific soil fractions, such as exchangeable (EX), carbonate bound (CB), organic bound (OC), amorphous Fe and Al oxides-bound (OX), and residual (RES) fraction, determines how strongly they are retained in soil and how easily they may be mobilized to the hydrosphere and biosphere. Water-soluble + EX are considered to be bioavailable and mobile; oxide- and CB, OC, and OX fractions may be potentially mobile; while the RES fraction is mainly not available to either plants or microorganisms (Ma and Rao 1997; He et al. 2013; Shaheen and Rinklebe 2014).

Assessment of the impact of soil amendments on HMs speciation and mobility in soils is a key issue in many environmental studies because such information aids our understanding of HM behavior in the environment and is therefore of widespread interest. Assuming that the non-RES metal (sum of the EX, OC, CB, and OX fractions) is more bioavailable than the RES metal, the effectiveness of soil amendments, that is, phosphates on metal mobility, can be assessed using a fractionation scheme (Ma and Rao 1997; Basta and

Gradwohl 2000; Basta and McGowan 2004). The more effective treatments convert greater amounts of metal from the non-RES to the RES fraction. Application of phosphates to soils is very common. *In situ* transformation of soil metals by phosphates shows great promise in minimizing the risk from HMs at contaminated sites. Thus, knowledge about the influence of these compounds on the fractionation and mobility of HMs in soils will be required to answer both scientific and practical questions regarding protection of groundwater and plants and sustainable management of soils, or to explain the fate of environmentally harmful substances.

Amendments with different P sources were evaluated for their effectiveness in reducing the mobility of HMs in soils. A large number of studies have provided conclusive evidence for the potential impact of phosphate compounds on the availability and fractionation of HMs. A pot trial was conducted by Chen et al. (2007) to assess the efficiency of P-induced metal fractions and mobility in soils. Natural HA, PR, triple superphosphate (TSP), and DAP were evaluated for their ability to reduce Cd, Pb, and Zn bioavailability in an artificially metal-contaminated soil with three addition levels: T0 (without metals added), T1 (Cd/Pb/Zn:0.6:100:66 mg kg^{-1}), and T2 (Cd/Pb/Zn: 1.5:300:200 mg kg^{-1}). Phosphate compounds were applied at 2500 mg P_2O_5 kg^{-1} soil for each metal level with five treatments: control, TSP, DAP, PR, and HA, respectively (Figure 7.1). Sequential extraction results indicated that the phosphate amendments converted significant amounts of the soil Pb, Zn, and Cd from the non-RES fractions to RES fraction. Among the P treatments evaluated, PR and HA were the most effective in converting Pb from the non-RES to the RES fraction, while the TSP treatment was the least effective. This can probably be attributed to the fact that more Pb phosphate or Pb sorption was formed from HA and PR than from TSP treatment. Amendments with P sources also modified Zn and Cd distribution in soil, with an increase in RES Zn and Cd. The main transformation for Cd was the reduction of EX and OX fractions, and for Zn was the reduction in EX, OX, and CB fractions (Figure 7.1).

Solubility products of Zn and Cd phosphate are much greater than that of Pb phosphate (Lindsay 1979). Therefore, Pb phosphate was probably formed prior to Cd and Zn. However, hopeite ($Zn_3(PO_4)_2 \cdot 4H_2O$) is much more soluble than pyromorphite, and therefore, Zn phosphate may not control Zn solubility in this case. The reduction in extractable Zn by phosphate can be due to sorption onto the phosphate mineral rather than to precipitation of hopeite (Xu and Schwartz 1994). Hettiarachchi et al. (2001) reported that PR was more effective than TSP or phosphoric acid in reducing Pb bioavailability for Zn–Pb contaminated soils.

Spuller et al. (2007) studied the impact of DAP ($(NH_4)_2HPO_4$) and calcium phosphate monobasic (CPM) ($Ca(H_2PO_4)_2$) on the mobility and fractions of antimony (Sb), Cu, and Pb. They mixed the soil with DAP and CPM by 0.5%, 1%, and 2%. The soil amendments were added to aliquots of the soil and thoroughly mixed. Batch experiments were done with 50 g of treated/untreated soil and 500 mL of deionized water. The effects of phosphatic amendments

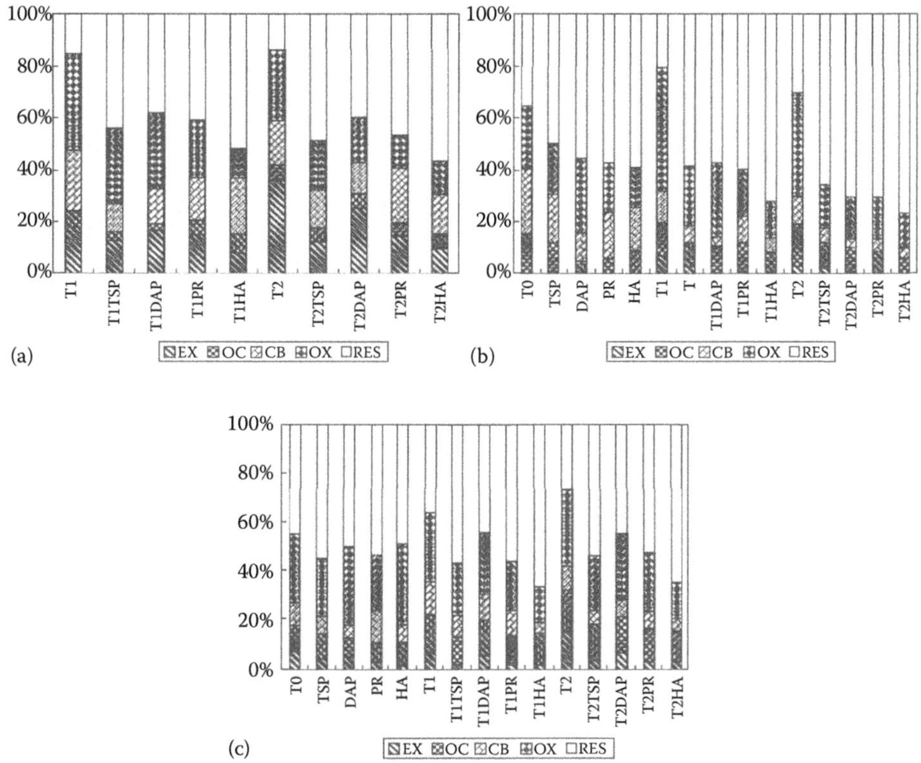

FIGURE 7.1
Relative percentages of Cd (a), Pb (b), and Zn (c) in each fraction of the soils from the CK and P-treated pots (mean, $n = 4$) (the concentration of Cd in T0 was below the limit, and therefore the data were omitted). T0 = uncontaminated (control), T1 = 0.6:100:66 mg kg^{-1} Cd/Pb/Zn, T2 = 1.5:300:200 $mgkg^{-1}$ Cd/Pb/Zn, TSP = uncontaminated + TSP amendment, T1TSP = T1 + TSP amendment, T2TSP = T2 + TSP amendment, DAP = uncontaminated + DAP amendment, T1DAP = T1 + DAP amendment, T2DAP = T2 + DAP amendment, PR = uncontaminated + PR amendment, T1PR = T1 + PR amendment, T2PR = T2 + PR amendment, HA = uncontaminated + HA amendment, T1HA = T1 + HA amendment, T2HA = T2 + HA amendment. (Reprinted from *Ecotoxicol. Environ. Saf.* 67, Chen, S., Xu, M., Ma, Y., and Yang, J., Evaluation of different phosphate amendments on availability of metals in contaminated soil, 278–285, Copyright 2007, with permission from Elsevier.)

on bulletborne metals showed that although DAP and CPM significantly decreased the mobility of Pb, both amendments significantly increased Cu and Sb mobility with obvious increase in DAP compared to CPM (Figure 7.2).

Average Pb concentrations were reduced by 45% (DAP) and 99% (CPM). In the phosphate-amended soil, Pb was shifted from the mobile/exchangeable fractions to organic matter/amorphous iron oxide pools. The formation of Pb phosphate minerals could not be proved since they are not driven by the extractants employed. For Cu and Sb, an increase in the organic/amorphous

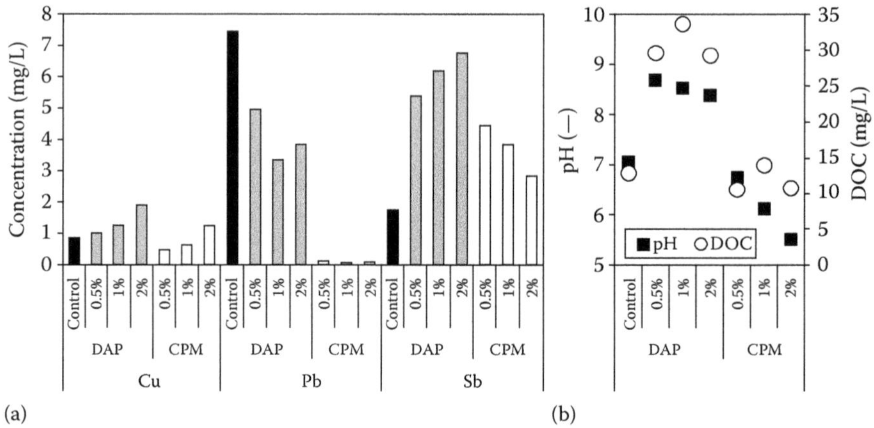

FIGURE 7.2
Contaminant (a) and DOC concentrations/pH values (b) in batch extracts of control and phosphate-amended soil. DAP = diammonium phosphate, CPM = calcium phosphate monobasic. (Reprinted from *J. Hazard. Mater.* 141, Spuller, C., Weigand, H. and Marb, C., Trace metal stabilisation in a shooting range soil: Mobility and phytotoxicity, 378–387, Copyright 2007, with permission from Elsevier.)

iron fractions resulted. This conflicts to some extent with the observed leaching behavior. Mass balances showed that 1.8% and 16.4% of the initial Cu and Sb contents were leached from the DAP/CPM treated column compared to 0.2% (Cu) and 4.0% (Sb) in the control column. Therefore, it can be hypothesized that cotransport and competitive sorption may not only cause the mobilization of contaminants but also affect their redistribution among different constituents of the solid phase (Spuller et al. 2007).

Lead stabilization may have been due to the formation of sparingly soluble Pb phosphate mineral phases such as lead(II) phosphate [$Pb_3(PO_4)_2$] or lead oxide phosphate [$Pb_4O(PO_4)_2$] following dissolution of the phosphate source (Badawy et al. 2002). Further important lead sequestering minerals of very low solubility Pb hydroxypyromorphite (log K_{sp} = −76.8) and Pb chloropyromorphite (log K_{sp} = −84.4) have been identified in phosphate-amended soils (McGowen et al. 2001; Cao et al. 2002). The lower stabilization by DAP compared to CPM is consistent with (1) the limited availability of phosphate in this compound and (2) the concomitant increase in dissolved organic carbon (DOC) concentrations in the DAP treatment. Release of DOC may also be responsible for constant or even increased Cu concentrations in the P-amended soil relative to the control soil due to formation of DOC–Cu complexes (Han and Thompson 2003). Copper is known to bind stronger to organic matter than any other divalent metal (McBride 1994). Consistently, the mobilization of organic matter as highlighted by the elevated DOC concentrations favors the leaching of mobile copper-DOC complexes.

In the study of Spuller et al. (2007), a pronounced mobilization of Sb was observed in the phosphate-amended samples compared to the control. This agrees with a competitive displacement of antimonate by phosphate at the surfaces of pedogenic iron (hydr)oxides similar to the behavior of arsenic (Pereya 1991). Antimony in soil environments predominantly exists in the oxidation states III and V. The corresponding species in soil water are $Sb(OH)_3$ and $Sb(OH)_6^-$, both known to undergo inner sphere complexation with iron (hydr)oxides (Johnson et al. 2005). Thus, a direct competition for sorption sites can be expected in the presence of high phosphate concentrations, which may explain Sb mobilization in the DAP/CPM-amended column. The different levels of Sb release are in accordance with Johnson et al. (2005), who observed a strong increase in aqueous Sb concentrations at pH values above 8.0. Thus, the release of phosphoric acid by CPM is considerably higher resulting in lower pH values due to the limited acid neutralizing capacity of the soil. The pH dependence of Sb release may thus explain the different Sb concentrations observed in the DAP and CPM treatments (Spuller et al. 2007).

Scheckel et al. (2005) determined speciation of Pb in phosphate-amended soils and reported that the amount of Pb in the residual fraction increased as a function of P concentration and amendment solubility (Figure 7.3).

Reduction of Pb mobility in the (Pb)-contaminated soils via phosphate amendments to sequester the Pb as pyromorphite ($Pb_5(PO_4)_3X$, where X = Cl, OH,

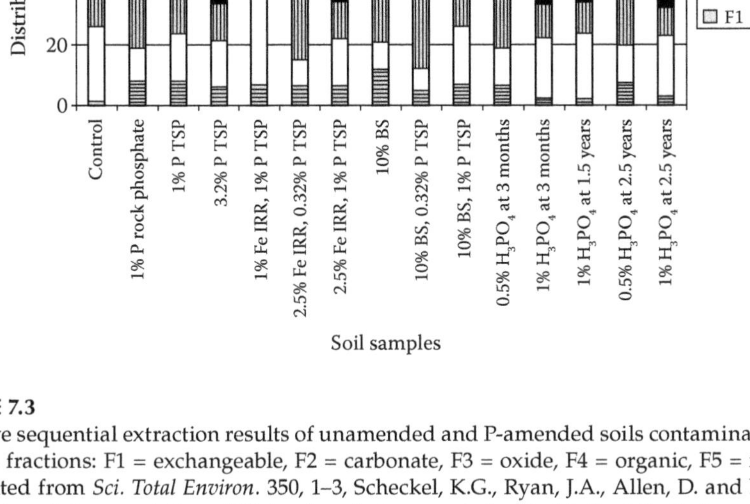

FIGURE 7.3
Selective sequential extraction results of unamended and P-amended soils contaminated with Pb. SSE fractions: F1 = exchangeable, F2 = carbonate, F3 = oxide, F4 = organic, F5 = residual. (Reprinted from *Sci. Total Environ.* 350, 1–3, Scheckel, K.G., Ryan, J.A., Allen, D. and Lescano, N.V., Determining speciation of Pb in phosphate-amended soils: Method limitations, 261–272, Copyright 2005, with permission from Elsevier.)

or F) has been extensively studied in the literature (Ruby et al. 1994; Cotter-Howells 1996; Laperche et al. 1997; Traina and Laperche 1999; Hettiarachchi et al. 2000; Ryan et al. 2001; Stanforth and Qiu 2001; Cao et al. 2003; Melamed et al. 2003). Selective sequential extraction (SSE) procedures have shown that a substantial portion of the soil-Pb was transformed to pyromorphite in phosphate-amended soils as evidenced by an increase in Pb concentration in the residual fraction relative to a control (Ma and Rao 1997; Ryan et al. 2001; Agbenin 2002; Cao et al. 2002, 2003; Chen et al. 2003); however, these studies cannot rule out or disprove the probability that pyromorphite formed as a result of the extraction procedure due to the likely nonequilibrium of the amended soil and to rapid kinetic formation of pyromorphite from soluble Pb and P (Scheckel and Ryan 2002). This phenomenon was illustrated by Scheckel et al. (2003) in SSE experiments with pure Pb components within a sand matrix (Pb-acetate, cerussite [$PbCO_3$], anglesite [$PbSO_4$], galena [PbS], and chloropyromorphite) in the presence and absence of solid calcium phosphate. Employing x-ray diffraction (XRD) and x-ray absorption fine structure (XAFS) spectroscopies to examine the solid residue after extraction steps, it was evident that pyromorphite easily formed when initially independent sources of Pb and P were reacted during extraction steps. In fact, pyromorphite was identified after the first (least aggressive) extraction step via XRD and XAFS spectroscopies (Scheckel et al. 2003). These results illustrate that pyromorphite forms during the extraction steps of SSE of a pure experimental system and may be a speculated reason why researchers that employ SSE methods to P-amended, Pb-contaminated soils achieve residual fraction values greater than values recorded for spectroscopic studies (Scheckel and Ryan 2004). Addition of K_2HPO_4, $Ca_3(PO_4)_2$, $Ca_5(PO_4)_3OH$, and $Ca(H_2PO_4)_2{\cdot}2H_2O$ in the contaminated soil at the P application rate of 2.283 mg P/kg-soil caused the transformation of easily available HMs species to more stable forms (Wu et al. 2013).

Hashimoto et al. (2011) studied the enhanced transformation of Pb speciation in rhizosphere soils using P amendments and phytostabilization. Lead speciation and solubility in rhizosphere soils of five different plant species were investigated using extended x-ray absorption fine structure (EXAFS) spectroscopy and chemical extraction. The EXAFS analysis indicated that Pb occurred as $PbCO_3$ (37%), Pb sorbed to organic matter (Pb-org: 15%), and Pb sorbed to pedogenic birnessite and/or ferrihydrite (Pb-ox: 36%) in the bulk soil. Comparison of the EXAFS spectra between bulk and rhizosphere soils demonstrated notable differences in fine structure, indicating that Pb species had been modified by rhizosphere processes. The estimated proportion of $PbCO_3$ (25%) in the buckwheat soil was smaller than the other rhizosphere soils (35%–39%). The addition of P significantly reduced Pb solubility in the bulk and rhizosphere soil except in the rhizosphere of buckwheat, for which the Pb solubility was 10-fold greater than in the other P-amended soils. This larger solubility in the buckwheat rhizosphere could not be explained by the total Pb speciation in the soil but was presumably related to the acidifying effect of buckwheat, resulting in a decrease in the soil pH by 0.4 units. The reduced Pb solubility by P amendment resulted from the transformation of preexisting $PbCO_3$ (37%) into Pb $_5(PO_4)_3Cl$ (26%–32%) in the bulk and rhizosphere soils. In the P-amended rhizosphere soils, Pb-org species were no longer detected, and the Pb-ox pool increased by 51%–57%. The study demonstrated that rhizosphere processes modify Pb solubility and speciation in P-amended soils and that some plant species, like buckwheat, may impair the efficiency of Pb immobilization by P amendments.

Scheckel et al. (2003) investigated the potential formation of pyromorphite ($Pb_5(PO_4)_3Cl$) during the sequential extraction steps of Pb-spiked samples with and without calcium phosphate amendments and examined the differences in the operationally defined distribution of Pb in samples with and without the presence of P. The systems that were examined in the absence of phosphate behaved, for the most part, adequately according to the operational definitions of the extraction procedure. However, when the samples were amended with phosphate, the results were drastically changed with a significant shift of extractable Pb to the residual phase. This redistribution was attributed to pyromorphite formation during the extraction procedure as confirmed by x-ray diffraction and x-ray absorption spectroscopies (XASs).

The efficiency of three phosphate fertilizers including TSP, DAP, and PR on Cd fractions was studied by Thawornchaisit and Polpraser (2009). Results from the assessment of Cd speciation via sequential extraction procedures revealed that the soluble–exchangeable fraction and the surface adsorption fraction of Cd in the soils treated with phosphate fertilizers, especially with TSP, have been reduced considerably (Figure 7.4). In addition, it was found that the reduction was correspondingly related with the increase in more stable forms of cadmium (i.e., the metal bound to manganese oxides and the metal bound to crystalline iron oxides). Treatment efficiency increased as the phosphate dose (based on the molar ratio of PO_4/Cd) increased.

Melamed et al. (2003) found that phosphate treatments effectively transformed up to 60% of total soil Pb from the nonresidual fraction (sum of water soluble and exchangeable, carbonate, Fe-Mn oxide, and organic fractions) to the residual fraction relative to the control. These results suggest that the P compounds could significantly reduce the mobility and increase the geochemical stability of HMs in contaminated soils.

Greenhouse experiments were conducted by Huang et al. (2013) to examine the effect of different P sources on Zn phytoextraction by *Sedum alfredii* (*S. alfredii*) in aged Zn-contaminated paddy soil. In these experiments, Zn accumulation, soil pH, microbial biomass and enzyme activity, available Zn changes, and Zn phytoremediation efficiency in soil after plant harvest were determined. It was found that upon addition of P, Zn uptake of *S. alfredii* increased significantly. Mehlich-3 extractable or the fractions of exchangeable and carbonate-bound soil Zn were significantly increased at higher P applications. Soil pH significantly decreased with increasing P application

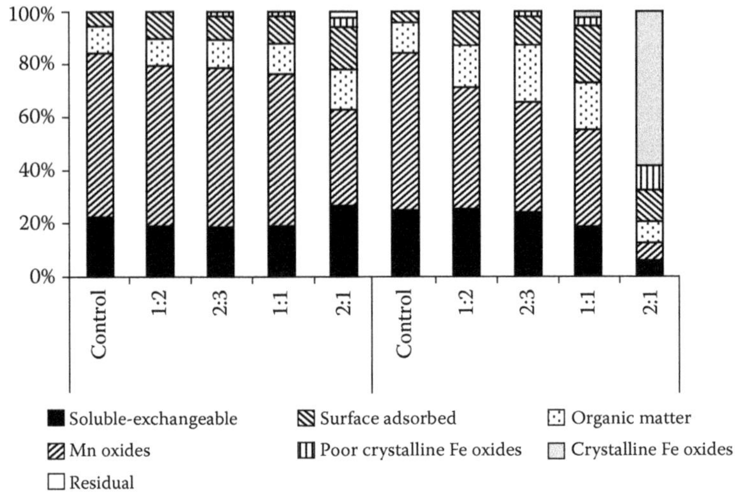

FIGURE 7.4
Relative Cd distribution in the control and treated soils at various phosphate doses after 28-day stabilization. (Reprinted from *J. Hazard. Mater.* 165, Thawornchaisit, U. and Polprasert, C., Evaluation of phosphate fertilizers for the stabilization of cadmium in highly contaminated soils, 1109–1113, Copyright 2009, with permission from Elsevier.)

rates. These results indicate that application of P fertilizers has the potential to enhance Zn mobility and uptake by hyperaccumulating plant *S. alfredii*, and thus to increase phytoremediation efficiency of Zn-contaminated soils.

The influence of P on Zn speciation in soils has not been extensively investigated. Baker et al. (2012) examined Zn speciation in proximity to phosphate application points in a Pb/Zn smelter-contaminated soil. They hypothesized that additions of P to a Pb/Zn-contaminated soil would induce Zn phosphate mineral formation, and fluid P sources would be more effective than granular P amendments. A combination of different synchrotron-based techniques, namely, spatially resolved micro-x-ray fluorescence (μ-XRF), microextended x-ray absorption fine structure spectroscopy (μ-EXAFS), and micro-x-ray diffraction (μ-XRD), were used to speciate Zn at two incubation times in the proximity of application points (0 to 4 mm) for fluid and granular P amendments in a Pb/Zn smelter-contaminated soil. Phosphate rock, TSP, monoammonium phosphate (MAP), and fluid ammonium polyphosphate induced Zn phosphate formation. Ammonium polyphosphate was more effective at greater distances (up to 3.7 mm) from the point of P application. Phosphoric acid increased the presence of soluble Zn species because of increased acidity. Soluble Zn has implications with respect to Zn bioavailability, which may negatively impact vegetation and other sensitive organisms. Although additions of P may immobilize Pb, this practice needs close monitoring due to potential increases in Zn solubility in a Pb/Zn smelter-contaminated soil.

Phosphates and (Im)mobilization and Bioavailability of Soil Metal(loid)s

Unlike organic contaminants, HMs do not undergo microbial or chemical degradation, and the total concentration in soils persists for a long time after their introduction (Adriano et al. 2004). With greater public awareness of the implications of contaminated soils on human and animal health, there has been increasing interest throughout the scientific community in the development of technologies to remediate HM-contaminated sites. For diffuse distribution of metal(loid)s such as fertilizer-derived Cd input in soils, remediation options generally include amelioration of soils to minimize their bioavailability. Mobilizing amendments such as desorbing agents increase the bioavailability and mobility of metal(loid)s, while immobilizing amendments such as precipitating agents and sorbent materials decrease the bioavailabilty and mobility of metal(loid)s. Mobilizing agents can be used to enhance the removal of HMs through plant uptake and soil washing. Immobilizing agents can be used to reduce the transfer of HMs to the food chain via plant uptake and leaching to groundwater. One of the major limitations of the mobilizing technique is susceptibility to leaching of the mobilized HMs in the absence of active plant uptake. Similarly, in the case of the immobilization technique, the long-term stability of the immobilized HMs needs to be monitored (Bolan et al. 2014).

Bioavailability of HMs plays a vital role in the remediation of contaminated soils. Bioavailability can be minimized through chemical and biological immobilization of HMs using a range of amendments including P compounds (Kumpiene et al. 2008; Park et al. 2011). Phosphate compounds have often been used to increase the immobilization of HMs such as Pb and Zn, thereby reducing their mobility and bioavailability in soil. However, the P compounds might increase the metal mobility based on the amendment and particular metal (Bolan et al. 2003, 2013a,b, 2014; Song et al. 2009).

Several studies have demonstrated that *in situ* HMs immobilization using phosphate is a cost-effective and environmentally positive remediation technology (Ma et al. 1995; Cao et al. 2002, 2008; Bolan et al. 2014). Phosphate treatment transforms reactive and bioavailable heavy metal fractions in more stable forms with reduced mobility and bioavailability (Cao et al. 2002, 2004). Phosphate minerals (i.e., HA and PR) can effectively immobilize HMs in water and soil (e.g., Boisson et al. 1999; Cao et al. 2004, 2009; Chen et al. 2007; Corami et al. 2007, 2008a,b; Marchat et al. 2007; Islam et al. 2010). Although PR has lower removal capacity than synthetic HA (e.g., Cao et al. 2004; Chen et al. 2007), natural low-grade (<30% P_2O_5) PR, which is presently considered as waste for fertilizer industries, is relatively inexpensive and needs to be explored as a low-cost adsorbent for processing contaminated media.

Significant efforts have been made to evaluate the effectiveness of phosphates on *in situ* immobilization of contaminated water and soils (Ma et al. 1997; Ma and Rao 1997; Basta et al. 2001; Bolan et al. 2013a). Depending on the characteristics of soils (acidic or calcareous) and on the source of P, various mechanisms have been attributed to the effects of P amendments on Cd, Pb, and Zn immobilization in soils. These include precipitation of metals after dissolution of natural phosphate, surface complexation, adsorption, dissolution of hydroxyapatite followed by coprecipitation, or an ion exchange process at the surface of phosphate minerals. These mechanisms depend on the solubility of phosphates (Eighmy et al. 1997; Bolan et al. 2003b; Miretzky and Fernandez Cirelli 2008) and on the characteristics of HMs (Xu et al. 1994; Chen et al. 1997a,b; Hamon et al. 2002; Mavropoulos et al. 2002; Bolan et al. 2003a; Cao et al. 2004; Sneddon et al. 2006; Chen et al. 2007).

The reaction mechanisms for metal immobilization by phosphate minerals include (1) ion exchange process, (2) surface complexation, (3) dissolution of the original phosphate minerals and precipitation of new metal phosphates, and (4) substitution of Ca in phosphate by other metals during recrystallization (coprecipitation) (Jeanjean et al. 1994; Ma et al. 1994; Xu et al. 1994). Precipitates generally isolated after the reactions between water-soluble phosphate compounds and metals were cadmium orthophosphate [$Cd_3(PO_4)_3$], lead(IV) phosphate [$Pb_3(PO_4)_3$], zinc phosphate [$Zn_3(PO_4)_3$], and lead chlorapatite [$Pb_{10}(PO_4)_6(Cl)_2$] (McGowen et al. 2001; Singh et al. 2001), while with water-insoluble phosphates, the cadmium hydroxide [$Cd(OH)_2$], $Pb_3(PO_4)_3$, $Zn_3(PO_4)_3$, $Zn_3(PO_4)_2{\cdot}4H_2O$, hydroxyl lead apatite [$Pb_{10}(PO_4)_6(OH)_2$], lead fluorapatite [$Pb_{10}(PO_4)_6F_2$], or hydrocerussite [$Pb_3(CO_3)_2(OH)_2$] compounds were identified (Xu et al. 1994; Chen et al. 1997a,b; Zhang et al. 1998; Basta et al. 2001). Lead immobilization by water-insoluble phosphate compounds may be explained by the formation of pyromorphite, which is highly stable even at low pH (Ruby et al. 1994). A more stable Pb phosphate such as $Pb_xCa(10-x)(PO_4)_6(OH)_2$ is based on an ion exchange mechanism where aqueous Pb replaces Ca on the hydroxyapatite lattice formed (Takeuchi and Arai 1990; Cotter-Howells 1996). A second mechanism involves hydroxyapatite dissolution followed by precipitation of Pb with phosphate to give pure pyromorphite (Ma et al. 1993). Also for PR, solubilization with subsequent precipitation of a pyromorphite-like mineral ($Pb_{10}(PO_4)_6X_2$, X = OH, Cl, F) was proposed (Ma et al. 1995). For this amendment, complexation on the PR surface sites was also proposed (Cao et al. 2004), similar to the sorption of Cd and Zn onto HA (Xu et al. 1994). Most previous studies concerned Pb immobilization using various P sources. The main amendments were bone meal, PR, HA (natural or synthetic), apatite (natural, synthetic, or biological), calcium phosphate [$Ca_3(PO_4)_2$], ammonium hydrogen phosphate [$(NH_4)_2HPO_4$], sodium hydrogen phosphate (Na_2HPO_4), potassium hydrogen phosphate (K_2HPO_4), calcium hydrogen phosphate ($CaHPO_4$), sodium dihydrogen phosphate (NaH_2PO_4), potassium dihydrogen phosphate (KH_2PO_4), calcium dihydrogen phosphate [$Ca(H_2PO_4)_2$], and phosphoric acid (H_3PO_4) (Bolan et al. 2003a,b; Miretzky and Fernandez Cirelli 2008). For soluble phosphates, the choice of amendment was widely debated because cations (K^+, NH_4^+, Na^+) associated with the phosphate ions could increase or decrease the mobility of HMs (Pierzynski and Schwab 1993; McGowen et al. 2001;

Bolan et al. 2003a). Although use of phosphate amendments has benefits, an excess of available P in the amended soil may be a potential source of eutrophication of surface and groundwaters (Melamed et al. 2003; Brown et al. 2004; Koopmans 2004). To avoid these risks, combined use of soluble phosphates and biochemical actions was proposed (Soares and Siqueira 2008). An alternative solution is to mix two different types of P sources, the first one coming from soluble phosphates and the second one from insoluble phosphates (Ma and Rao 1997; Ma et al. 1997). Thus, recent studies showed that it was possible to reduce the mobility of pollutants (Basta et al. 2001; Cao et al. 2003) and their bioavailability (Hettiarachchi and Pierzynski 2002; Zhu et al. 2004) with this type of mixture. Furthermore, less leaching of soluble P and lower impacts on soil pH were observed (McGowen et al. 2001; Cao et al. 2003a,b; Zhu et al. 2004; Chen et al. 2007; Tang et al. 2009).

Sources of phosphate for stabilization of HMs include PO_4-containing minerals (e.g., natural or synthetic apatites and hydroxyapatites) and soluble phosphates such as phosphate-based salts and phosphoric acid. Upon the dissolution of phosphates, leachable metals in contaminated soil will react with phosphates and form metal phosphates with less solubility and with enhanced geochemical stability in a wide range of environmental conditions (Cao et al. 2003). This mechanism has been widely observed in the case of Pb-contaminated soils where formation of Pb phosphate minerals such as pyromorphite upon addition of phosphate has been confirmed using x-ray adsorption spectroscopy (Ryan et al. 2001). Stabilization efficiency can vary according to the kind of contaminant. For example, addition of TSP along with a high Fe by-product reduced the *in vitro* extractable Pb in the soil; however, plant Cd concentration in the same treatment was increased over the untreated soils (Brown et al. 2004). In case of mineral apatite, the material has shown to reduce the solubility and bioavailability of Pb, Cd, and Zn in contaminated soils (Chen et al. 1997). In contrast, McGowen (2000) found that apatite treatments are mostly ineffective for reducing the release and transport of Cd and Zn from contaminated soils. The different findings, particularly on Cd stabilization, suggest that the effectiveness of using phosphate to immobilize other metals such as Cd is not well understood. Further research on Cd stabilization with various phosphate sources is also needed. In addition to types of contaminants, different phosphate sources with different solubility may also impact the effectiveness of metal stabilization (Cao et al. 2003). The use of readily soluble phosphoric acid and phosphates with moderate or less solubility, like $Ca(H_2PO_4)_2$ or phosphate rock, generally facilitates the effectiveness of Pb stabilization (Cao et al. 2003). Wang et al. (2001) have shown that phosphate-based salts such as brushite ($CaHPO_4 \cdot 2H_2O$), sodium dihydrogen phosphate dodecahydrate ($Na_2HPO_4 \cdot 12H_2O$), and magnesium hydrogen orthophosphate trihydrate ($MgHPO_4 \cdot 3H_2O$) reduced the leachable Cd, Cu, Ni, Pb, and Zn; however, both Na and Mg phosphate-based salts were not as effective as $CaHPO_4 \cdot 2H_2O$. While phosphates containing minerals and phosphate-based salts have been widely tested on metal stabilization, the efficiency of phosphate fertilizers that may cost less than phosphate minerals and phosphate-based salts was inadequately studied.

The application of phosphate amendments to HM-contaminated soils has been identified as a potentially efficient *in situ* remediation method (Laperche et al. 1997; Hettiarachchi et al. 2001; Cao et al. 2004). A number of natural or synthetic materials such as phosphorus-containing amendments (Ma et al. 1995) have been tested for their ability to stabilize HMs in soils. The ability of phosphate-rich material to reduce HMs uptake by plants has been well documented (e.g., Ma et al. 1994; Cao et al. 2004; Zhu et al. 2004; Bolan et al. 2013).

Pot trials were conducted by Chen et al. (2007) to assess the efficiency of HA, PR, TSP, and DAP in reducing Cd, Pb, and Zn bioavailability and uptaking by Chinese rape (*Brassica campestris* L.) grown in an artificially metal-contaminated soil. Phosphate compounds were applied at 2.500 mg P_2O_5 kg^{-1} soil for each metal level with five treatments: CK (control), TSP, DAP, PR, and HA, respectively. The immobilization and bioavailability of Cd, Pb, and Zn were determined by plant uptake, scanning electron microscope (SEM), and sequential extraction. The results showed that phosphate application improved plant growth. The positive effect of phosphate amendments on plant growth may be attributed to either the effect of P on decreasing metal toxicity, as estimated with tissue metal analyses, or improved P nutrition, or a combination of both processes. The results also showed that the HA was superior to all other materials for immobilizing Pb, Zn, and Cd and reducing their bioavailability and uptake in shoot with reductions of 34.6%–53.3% for Pb, 31.2%–47.3% for Zn, and 39.1%–42.4% for Cd, respectively, compared to control treatment. The mechanisms of Pb immobilization in the soil and the decreased Pb translocation from the plant root to shoot induced by added phosphate were identified with SEM equipped with energy dispersive x-ray elemental spectrometry (EDXES). Sequential extraction results indicated that the phosphate amendments converted significant amounts of the soil Pb, Zn, and Cd from EX, OC, CB, amorphous Fe, and Al OX (nonresidual [RES] fractions) to RES fraction. In general, the effect of different phosphates on plant uptake of Pb, Zn, and Cd followed the order HA > PR > DAP > TSP. Since pH is one of the most important parameters affecting metal availability to plants, the increase in soil pH in the HA treatment was an additional mechanism of P-induced reduction in metal bioavailability. Among the phosphate amendments, application of DAP in the control (T0) treatment decreased soil pH significantly (5.56–5.23) compared to T0 treatment with P. Other researchers have investigated the effects of chemical treatments on soil pH and found for example that ammonium-based fertilizers can result in soil acidification and that DAP amendment decreased soil pH from 7.1 to 6.5 after 2 months and then remained constant for 2 more months (Basta and McGowan 2004). The phosphate amendments investigated generally decreased the uptake of Cd, Pb, and Zn by plants in the contaminated soils. Results of Chen et al.

(2007) concluded that HA and PR amendments could significantly immobilize soil Pb, Zn, and Cd in contaminated soils and reduce their bioavailability and uptake by the grown plants.

Ma et al. (1994) and Ma and Rao (1997) reported that PR and hydroxyapatite efficiently decreased the extractability of HMs, including Pb, Cd, Cu, and Zn, from contaminated soils. The formation of pyromorphite in the soil and association of P with Pb in the roots appear to be responsible for the reduction of Pb contents in plant shoots. The decrease in Zn availability may be due to Zn adsorption on or coprecipitation with pyromorphite (Ma et al. 1994). Another point worth mentioning is that P concentrations in the plants were generally high among the amendments. Maintaining adequate plant P nutrition during remediation of metal-contaminated soils may be problematic due to the formation of insoluble metal phosphate.

Basta and Gradwohl (1998) investigated the remediation of HM-contaminated soils using PR and reported that treatment of smelter-contaminated soil with RP reduced Zn phytotoxicity and reduced plant-available Cd, Zn, and Pb. Oral bioavailability of Pb, Cd, and Zn was also reduced by RP treatment, but the extent of reduction was uncertain. Small reductions in oral bioavailability were measured by the physiologically based extraction test (PBET) gastric phase, but large reductions were obtained by the PBET intestinal phase.

Lead immobilized by RP is very stable to soil acidification. In fact, acidification increases the amount of Pb immobilized. Increased available Cd and Zn suggest that some of the fraction immobilized by RP may not be stable to soil acidification. However, available Cd and Zn in acidified RP-treated soil were less than those in the acidified control soil. Rock phosphate may serve as an inexpensive alternative method for remediation of smelter-contaminated soil.

Batch experiments were conducted by Seaman et al. (2001) to evaluate the ability of HA to reduce the solubility of metals, including the primary contaminants of concern, nickel (Ni) and uranium (U), from contaminated sediments located on the Department of Energy's Savannah River Site. Hydroxyapatite was added to the sediments at application rates of 0, 5, 15.8, and 50 g kg^{-1}. After equilibrating in either 0.02 molar potassium chloride (M KCl) or 0.01 molar calcium chloride (M $CaCl_2$), the samples were centrifuged and the supernatants filtered prior to metal, dissolved organic carbon (C), and phosphate (PO_4) analyses. The treated soils were then air-dried, and changes in solid-phase metal distribution were evaluated using sequential extractions and electron-based microanalysis techniques. Hydroxyapatite was effective at reducing the solubility of U and, to a lesser degree, Ni. Hydroxyapatite was also effective in reducing the solubility of Al, barium (Ba), Cd, Co, manganese (Mn), and Pb. Sequential extractions indicated that HA transfers such metals from chemically more labile forms, such as the water-soluble and exchangeable fractions, by altering solid-phase speciation in favor of secondary phosphate precipitates. Hydroxyapatite effectiveness was somewhat reduced in the presence of soluble organics that likely increased contaminant metal solubility through complexation. Arsenic and Cr solubility increased with HA addition, suggesting that the increase in pH and competition from PO_4 reduced sorption of oxyanion contaminants. Energy dispersive x-ray (EDX) analysis conducted in the transmission electron microscope (TEM) confirmed that HA amendment sequesters U, Ni, Pb, and possibly other contaminant metals in association with secondary Al-phosphates.

Waterlot et al. (2011) studied the effects of a phosphorus amendment and the pH of water used for watering on the mobility and phytoavailability of Cd, Pb, and Zn in three highly contaminated kitchen garden soils. The three soils were amended with a mixture of HA and DAP. At 6 months, extracting procedures were carried out to evaluate the effects of the amendment on the mobility of Cd, Pb, and Zn. This step was then supplemented by an evaluation of the impact of the amendment on the phytoavailability of pollutants, which was determined in plant uptake studies with ryegrass (*Lolium perenne* L.) by considering only the pollutant concentrations in their shoots. Two experiments were carried out for that. In the first one, unamended and amended soils and ryegrass were supplied with distilled water (pH = 7). In the second one, osmosed water (pH = 5.5) was used to evaluate the effects of the acid water–phosphate amendment system on the mobility and phytoavailability of Cd, Pb, and Zn. Six months after the start of the experiments, the selective extractions showed that the effectiveness of the amendment studied depended on the element, the soil, and the water's pH. Reductions of metal eluted from the contaminated soils were 1.5%–37.9% for Cd and 9.1%–80.9% for Pb. Application of P amendment to the combination of osmosed water was generally the most effective for immobilizing Cd and Pb elution. In contrast, the mixture of HA and DAP was ineffective for reducing Zn elution. The plant-fresh biomass yield was significantly ($p < 0.05$) increased by the combination of P amendment and distilled water, whereas a reduction of biomass was recorded with the combined amendment and osmosed water. Addition of P amendment generally reduced Pb uptake in ryegrass shoots (1%–47%), while both Cd and Zn were increased by 17.9%–79% and 0.45%–100%, respectively.

The efficiency of three phosphate fertilizers including TSP, DAP, and PR as stabilizing agents of Cd-contaminated soils has been assessed by Thawornchaisit and Polpraser (2009). Results of the study showed that after the 60-day stabilization, the leachable concentrations of Cd in PR-, DAP-, and TSP- treated soils reduced from 306 mg/kg (the control) to 140.34 and 12 mg/kg with the stabilization efficiency as TSP > DAP > PR. In addition, it was observed that stabilization was most effective when using the molar ratio of PO_4/Cd at 2:1 and at least 21- and 28-day stabilization time for TSP and DAP, respectively.

Mignardi et al. (2012) evaluated the effectiveness of phosphate treatment for the remediation of mine waste soils contaminated with Cd, Cu, Pb, and Zn. This study investigated the effectiveness of HA and natural PR in immobilizing Cd, Cu, Pb, and Zn in mine waste soils. The phosphate treatment of the polluted soils significantly reduced the water solubility of the metals with reduction percentage ranging generally from about 84% to 99%. The proposed immobilization mechanism involves both surface complexation of the heavy metals on the phosphate grains and partial dissolution of the phosphate amendments and precipitation of heavy-metal-containing phosphates. Between the two amendments evaluated, HA was slightly more effective with respect to PR in immobilizing heavy metals. Although it has a lower effectiveness, the application of phosphate rock to reduce metal solubility in contaminated soils may minimize the potential risk of eutrophication due to the use of highly soluble phosphate amendments.

Cao et al. (2009) examined the effectiveness of PA and/or PR in immobilizing Pb, Cu, and Zn in two contaminated soils. The effectiveness was evaluated using water extraction, plant uptake, and a simple bioaccessibility extraction test (SBET) mimicking metal uptake in the acidic environment of the human stomach. The possible mechanisms for metal immobilization were elucidated using x-ray diffraction, scanning electron microscopy, and the chemical speciation program Visual MINTEQ. Compared to the control, all P amendments significantly reduced Pb water solubility, phytoavailability, and bioaccessibility by 72%–100%, 15%–86%, and 28%–92%, respectively. The Pb immobilization was probably attributed to the formation of insoluble Pb phosphate minerals. Phosphorus significantly reduced Cu and Zn water solubility by 31%–80% and 40%–69%, respectively, presumably due to their sorption on minerals (e.g., calcite and phosphate phases) following calcium oxide (CaO) addition. However, P had little effect on the Cu and Zn phytoavailability, while the acid extractability of Cu and Zn induced by SBET (pH 2) was even elevated by up to 48% and 40%, respectively, in the H_3PO_4 treatments (PA and PR + PA). The results of Cao et al. (2009) indicated that all amendments reduced Pb availability in this experiment, showing a significant reduction of water-soluble, phytoavailable, and bioaccessible Pb, presumably due to formation of insoluble Pb phosphate (e.g., pyromorphite-like mineral). Phosphate amendments also lowered water solubility of Cu and Zn, probably due to their sorption onto soil minerals, such as calcite, and phosphate minerals following CaO addition. However, P had little effect on the phytoavailability of Cu and Zn, while the acid extractability of Cu and Zn was even increased in the acidic soluble H_3PO_4 treatments (PA and PR + PA). This study again demonstrated that one single amendment approach is not suitable for immobilizing all metals of concern, and the effectiveness of metal immobilization also depends on the methods of assessment.

Shahid et al. (2014) studied the influence of plant species and phosphorus amendments on metal speciation and bioavailability in a smelter-impacted soil as a case study of a food chain.

Their study aimed at assessing the influence of two P amendments on metal speciation in rhizosphere soil and their transfer to the soil–plant system. Complementary experiments were performed—field experiments on a contaminated cultivated soil and laboratory experiments on an uncultivated contaminated soil—to highlight the mechanisms involved in metal–P interactions. In the laboratory experiments, P amendments were added at 120 mg P/kg of soluble KH_2PO_4 amendment and 9.000 mg P/kg of solid $Ca_5(PO_4)_3OH$ amendment. Field-culture results showed the possible food-chain contamination due to Pb, Cd, Cu, and Zn phytoaccumulation by pea and mustard plants from a cultivated agricultural soil. Moreover, P–metal complexes were observed by microscopy in the rhizosphere soil. In the laboratory experiments, the application of P amendments significantly increased the Pb and Zn levels in rhizosphere soil compared to control. Phosphate amendments significantly increased metal–P fraction and decreased oxides and organic matter fractions of Pb and Zn. Soluble P amendment was more effective than solid P amendment in changing Pb and Zn speciation. The changes in metal speciation are higher in the rhizosphere soil of pea than tomato. Application of P amendments increased Pb and Zn TF root/soil but decreased TF shoot/root. The study concluded that the effectiveness of the *in situ* metal immobilization technique varies with the type and quantity of applied P amendment as well as plant and metal type.

Wu et al. (2013) concluded that the addition of K_2HPO_4, $Ca_3(PO_4)_2$, $Ca_5(PO_4)_3OH$, and $Ca(H_2PO_4)_2 \cdot 2H_2O$ was effective in reducing water-soluble and exchangeable Pb, Zn, Cu, and Cd, and minimizing TCLP-extractable Pb, Zn, and Cu. The sequential extraction test and the TCLP indicated that $Ca_5(PO_4)_3OH$ treatment had a higher potential in immobilizing Pb, Zn, Cu, and Cd but a slight enhancement of As mobility compared with other phosphate treatments.

In situ immobilization field experiments were carried out by Liang et al. (2011) in a Cd- and Pb-contaminated agricultural field by adding natural clays combined with phosphate fertilizer. *Lactuca sativa* L., *Brassica campestris* L., and *Raphanus sativus* were selected as model vegetables. Phosphate fertilizer, natural clays such as sepiolite and bentonite, and composites of the above were employed as immobilization materials for heavy metal pollution. The effects of immobilization materials on the biomass of vegetables, the concentrations of Cd and Pb in the edible parts, and speciation transformation of Cd and Pb were investigated. It was found that the phosphate fertilizer and its composite with sepiolite or bentonite could increase the biomass of vegetables. All the immobilization materials decreased the Cd and Pb contents in the vegetables to different degrees. The composite of sepiolite and phosphate fertilizer had the best performance, and the contents of Cd in *Lactuca sativa* L. and *Brassica campestris* L. were below the level of the national standard for food safety. All the immobilization treatments inhibited the bioavailability of Cd and Pb by reducing the exchangeable fraction contents and increasing the residual fraction contents of Cd and Pb. Therefore, using natural clays combined with phosphate fertilizer seems to be efficient for the *in situ* remediation of Cd- and Pb-contaminated agricultural fields.

Impellitteri (2005) studied the effects of pH and phosphate on metal distribution and mobilization in soils from a Pb smelting site and found that the presence of phosphate causes an increase in solution As due to an increase in As(V). At pH 4, phosphate causes a 10-fold reduction in solution Pb and a fivefold reduction of solution Cd, and reduces solution concentrations of Cu, Fe, and Al. Zinc solution concentrations are slightly reduced. Solution Pb is also reduced at pH 6.

From the above-mentioned literature-based studies, we can conclude that phosphates can immobilize and mobilize soil metal(loid)s. However, phosphates were more effective in immobilizing soil Pb and reducing its mobility and bioavailability but were more effective in mobilizing soil As and increasing its mobility and bioavailability. Thus, we will present in more detail the influence of phosphates on both these elements.

Phosphates and Mobility of Soil Pb

Pb is a toxic element occurring in waters and soils mainly as a result of human activities. Soil metal contamination such as Pb is a threat to the ecosystem and human health. The bioavailability of Pb ions can be decreased by complexation with various substances in order to decrease their toxicity. *In situ* stabilization of Pb-contaminated soils can be achieved by adding phosphorus. *In situ* phosphate treatment is being evaluated as a cost-effective remedial strategy for reducing the human health and ecological risks associated with Pb-contaminated soils. Phosphate treatments have been shown to effectively stabilize toxic metals particularly for Pb^{2+} from contaminated soils. It is well documented that Pb phosphates are the most insoluble and stable forms of Pb in soils, and they can be formed rapidly in the presence of adequate Pb and P (Cao et al. 2003; Scheckel et al. 2003, 2013; Chrysochoou et al. 2007; Kumpiene et al. 2008).

A long-term, integrated risk assessment was performed by Tang et al. (2009) to examine the stability of P-immobilized soil Pb and verify the environmental safety of phosphate treatments. Urban soil in the Jasper County Superfund Site, Missouri, containing ~4.000 mg Pb kg^{-1}, was treated with phosphoric acid (H_3PO_4) at a rate of 10 g kg^{-1} by three application methods: rototilling (RT), surface application (SA), and pressure injection (PI). Soils and plant samples were collected 8–10 years posttreatment and analyzed for Pb bioaccessibility, Pb leachability, plant uptake of Pb, microbial toxicity, chemical fractionation of Pb, and solid-phase Pb speciation. Results showed that in all cases, P treatment significantly reduced soil bioaccessible and leachable Pb, and the reductions were maintained throughout the sampling period. Rototilling appeared to be the most effective for overall risk reductions. Lead uptake in plant tissues was positively correlated with soil Pb bioaccessibility, and no significant negative impacts of P treatment on soil microbial toxicity were observed. Chemical fractionation analyses indicated that risk reduction by P treatment was achieved through the reactions that transformed labile Pb species to relatively insoluble forms, presumably through the formation of pyromorphite-like minerals. This study demonstrates that soil treatments using soluble P would result in long-term health and ecological risk reductions of Pb-contaminated soil without an adverse effect on the environment.

Miretzky and Fernandez-Cirelli (2008) reviewed the efficiency of phosphates for Pb immobilization in soils and reported that Pb chemical immobilization using phosphate addition is a widely accepted technique to immobilize Pb from aqueous solution and contaminated soils. The application of different P amendments may cause a shift of soil Pb from forms with high availability to the most strongly bound Pb fractions. The increase in Pb in the residual or insoluble fraction results from formation of pyromorphite $Pb_5(PO_4)_3X$ where X = F, Cl, Br, OH, the most stable environmental Pb compounds under a wide range of pH and Eh (redox potential) natural conditions. Accidental pyromorphite ingestion does not yield bioavailable Pb because pyromorphite is insoluble in the intestinal tract. Numerous natural and synthetic phosphate materials have been used to immobilize Pb such as apatite and HA, biological apatite, RP, soluble phosphate fertilizers as MAP, DAP, PA, biosolids rich in P, and phosphatic clay and mixtures. The identification of pyromorphite in phosphate-amended soils has been carried out by different nondestructive techniques such as x-ray diffraction, scanning electron microscopy coupled with energy dispersive x-ray spectroscopy, XAFS, TEM, and electron microprobe analysis. The effectiveness of *in situ* Pb immobilization has also been evaluated by selective sequential extraction, by the toxicity leaching procedure, and by a physiologically based extraction procedure simulating metal ingestion and gastrointestinal bioavailability to humans. Efficient Pb immobilization using P amendments requires an increase in the phosphate phase solubility and Pb species phase by inducing acid conditions.

Basta et al. (2001) evaluated the influence of P and other soil amendments on five metal-contaminated soils and mine wastes. In their study, seven treatments were used: unamended control; 2.500 mg of P/kg as TSP, PR, acetic acid followed by TSP, and phosphoric acid (PA); and 5.000 mg of P/kg as TSP or PR. A significant reduction in bioavailable Pb, as determined by PBET compared with the control upon addition of P, was observed in all materials tested. Increasing the amount of P added from 2.500 to 5.000 mg/kg also resulted in a significantly greater reduction in bioavailable Pb. Phosphate rock was equally or more effective than TSP or PA in reducing bioavailable Pb in four out of five soils tested. Preacidification produced significantly lower bioavailable Pb compared with the same amount of P from TSP or PR in only one material. Reductions in Pb bioavailability as measured by PBET were evident 3 days after treatment, and it may indicate that the reactions between soil Pb and P occurred *in situ* or during PBET. No further reductions were noted over 365 days. X-ray diffraction data suggested the formation of pyromorphite-like minerals induced by P additions. This study showed that P addition reduced bioavailable Pb by PBET and has potential for *in situ* remediation of Pb-contaminated soils.

Lead distribution in 11 contaminated soils among five physical size fractions was studied by Ma et al. (1997). Lead was present in elevated concentrations in these soils, ranging from 198 to 12.523 mg kg^{-1}. Although Pb in these soils was randomly distributed among the five size fractions studied, it was slightly concentrated in the silt and clay (<53 μm) fractions with some exceptions. Several single chemical extractants were used to evaluate Pb availability in the contaminated soils. The effects of incubation and phosphate rock on Pb availability were also investigated. Significant amounts of Pb were extracted with all the extractants. The extractability of the extractants increased in the order H_2O < $Ca(NO_3)_2$ < HOAc < EDTA-NH_4OAc. The percentage of Pb extracted generally decreased with the increase in the incubation time as the soils gradually lost moisture during incubation. Net extractable Pb reduction with incubation time was not correlated with soil pH, organic matter, or total Fe and Mn contents in soils. Reduced Pb extractability with incubation may be attributed to precipitation/adsorption of Pb, rendering Pb less available. Addition of phosphate rock reduced $Ca(NO_3)_2$ and HOAc extractable Pb from Pb-contaminated soils, but had little effect on EDTA-NH_4OAc extractable Pb. Net extractable Pb reduction by phosphate rock was not correlated with soil pH, organic matter, or other metals such as Zn, Cu, Ni, and Cd in soils, but it was highly correlated with total Pb. These results indicated that physical–chemical speciation of Pb-contaminated soils can be a useful tool in evaluating Pb availability.

A pilot-scale field demonstration was conducted by Melamed et al. (2003) at a Pb-contaminated site to assess the effectiveness of Pb immobilization using P amendments. The test site was contaminated by past battery recycling activities, with average soil Pb concentration of 1.16%. Phosphate amendments were applied at a molar ratio of P/Pb 4.0 with three treatments (i.e., T1, H_3PO_4 + 5% phosphate rock). Soil samples were collected and characterized 220 days after P application. Phosphate treatments effectively reduced toxicity characteristic leaching procedure (TCLP) Pb from 82 mg L^{-1} to below the Environmental Protection Agency's (EPA) regulatory level of 5 mg L^{-1} in the surface soil. Scanning electron microscopy–energy dispersive x-ray elemental analysis and x-ray diffraction analysis indicated formation of insoluble chloropyromorphite [$Pb_5(PO_4)_3Cl$] mineral in the P-treated soils. Although H_3PO_4 is necessary to dissolve metastable Pb in soil for further lead immobilization, it should be used with caution due to its potential secondary contamination. A mixture of H_3PO_4 and $Ca(H_2PO_4)_2$ or phosphate rock was effective in immobilizing Pb with minimum adverse impacts associated with pH reduction.

Wu et al. (2013) found that the addition of P compounds, (K_2HPO_4, $Ca(H_2PO_4)_2 \cdot 2H_2O$, $Ca_3(PO_4)_2$, and $Ca_5(PO_4)_3OH$) reduced Pb concentrations in the standard EPA toxicity characteristic leaching procedure (TCLP) extraction solution from 90.47 to 56.05, 83.80, 67.78, and 86.32 mg/kg (38.0%, 7.36%, 25.1%, and 4.59% reduction), respectively.

Phosphates and Mobility of Soil As

Mobilization of contaminants can be achieved through solubilization, desorption, chelation, and complexation reactions that result in the redistribution of contaminants from solid phase to solution phase, thereby increasing their bioavailability (Bolan et al. 2014). Although P addition seems to be highly effective, excess P in soil and its potential effect on eutrophication of surface water and the possibility of As enhanced leaching remain a concern. The use of mixed treatments may be a useful strategy to improve their effectiveness in reducing lead phyto- and bioavailability. The role of P in influencing the mobility and bioavailability of As has been reported by a number of workers. Therefore, there is a growing interest in using P fertilizer to enhance plant uptake of As in contaminated sites, thereby facilitating phytoremediation. Large quantities of phosphate fertilizers are used as a nutrient source in agricultural and pasture soils. Phosphorus influences As uptake by plants by two processes: First, it is well known that arsenate [As(V)] acts as a P analog and is taken up by plants via a P transporter system (Meharg and Macnair 1992). While P deficiency can enhance As uptake, a high concentration of P in soil solution can inhibit the uptake of As (Pigna et al. 2009; Lei et al. 2012). Second, P is more strongly adsorbed in soil than As(V) and hence competes for sorption sites, thereby facilitating the desorption of As to soil solution and its subsequent uptake by plants (Ravenscroft et al. 2001). Thus, the influence of P on As mobilization depends on the charge characteristics of soils.

Although a number of studies have examined the potential value of P compounds in the immobilization of Pb and Zn in contaminated soils (Kumpiene et al. 2008; Cao et al. 2009), there has been limited work on the value of P compounds in the mobilization of As in variable charge soils, thereby increasing its bioavailability.

The effect of phosphate-induced mobility of As on its bioavailability was examined using a naturally contaminated sheep dip soil by Bolan et al. (2013b). The results indicated that phosphate addition increased the mobility and bio availability of As by 4.34- and 3.35-fold, respectively, in the sheep dip soil (Figure 7.5). However, the net effect of P on As phytoavailability depends on the extent of P-induced As mobilization in soils and P-induced competition for As uptake by roots.

Bolan et al. (2013a) examined P–As interaction in variable-charge allophanic soils in relation to P-induced As mobilization and bioavailability. In this work, the effect of P on arsenate [As(V)] adsorption and desorption was examined using a number of allophanic and nonallophanic soils that vary in their anion adsorption capacity. The effect of P on As uptake by Indian mustard (*Brassica juncea* L.) plants was examined using a solution culture, a soil plant growth experiment involving two As-spiked allophanic and nonallophanic soils that vary in their anion adsorption capacity, and a field

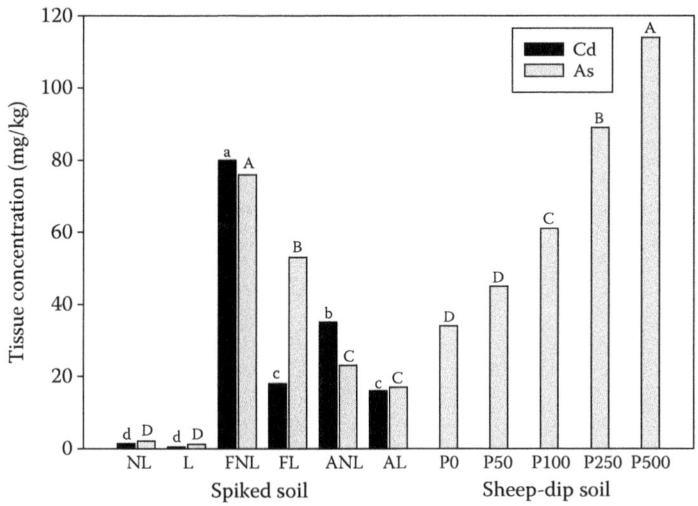

FIGURE 7.5
Effect of liming, aging, and phosphate application on Cd and As concentration in mustard plants. NL = no lime control, L = lime control, FNL = fresh no lime, FL = fresh lime, ANL = aged no lime, AL = aged lime, P0–P500 = 500 mg P/kg soil. Within each soil and metal(loid) treatment, means followed by the same letter are not significantly different (Ducman's multiple range test, $p < 0.05$). (Reprinted from *J. Hazard. Mater.* 2261, Bolan, N., Mahimairaja, S., Kunhkrishnan, A. and Naidu, R., Sorption-bioavailability nexus of arsenic and cadmium in variable-charge soils, 725–732, Copyright 2013, with permission from Elsevier.)

As-contaminated sheep dip soil (Figure 7.6). The addition of P resulted in an increase in As desorption, and the effect was more pronounced in the case of allophanic soil. In the case of both As-spiked soils and field-contaminated sheep dip soil, application of P increased the desorption of As, thereby increasing its bioavailability. Thus, the P-induced mobilization of As could be employed in the phytoremediation of As-contaminated sites.

Phosphate addition to soil influences As uptake by plants through its effects on sorption, root absorption, and translocation from root to shoot. The concentration of As in the leachate was approximately 10 times higher when $Ca(H_2PO_4)_2$ was used to leach the soil column as compared to the $CaSO_4$ solution. In the presence of the $Ca(H_2PO_4)_2$ solution, a maximum As concentration of 800 mg L^{-1} was found in the leachate, much higher than the World Health Organization (WHO) standard of 0.05 mg L^{-1} for drinking water. The P application increased both the transfer and translocation coefficients indicating that P enhanced the bioavailability of As in soils and its subsequent translocation from root to shoot.

Wu et al. (2013) evaluated the feasibility and efficiency of different forms of phosphates on As mobility. Four pure chemical P reagents—K_2HPO_4, Ca $(H_2PO_4)_2 \cdot 2H_2O$, $Ca_3(PO_4)_2$, and $Ca_5(PO_4)_3OH$—were added to the soil in solution form as the metal-stabilizing additives at a P application rate equivalent

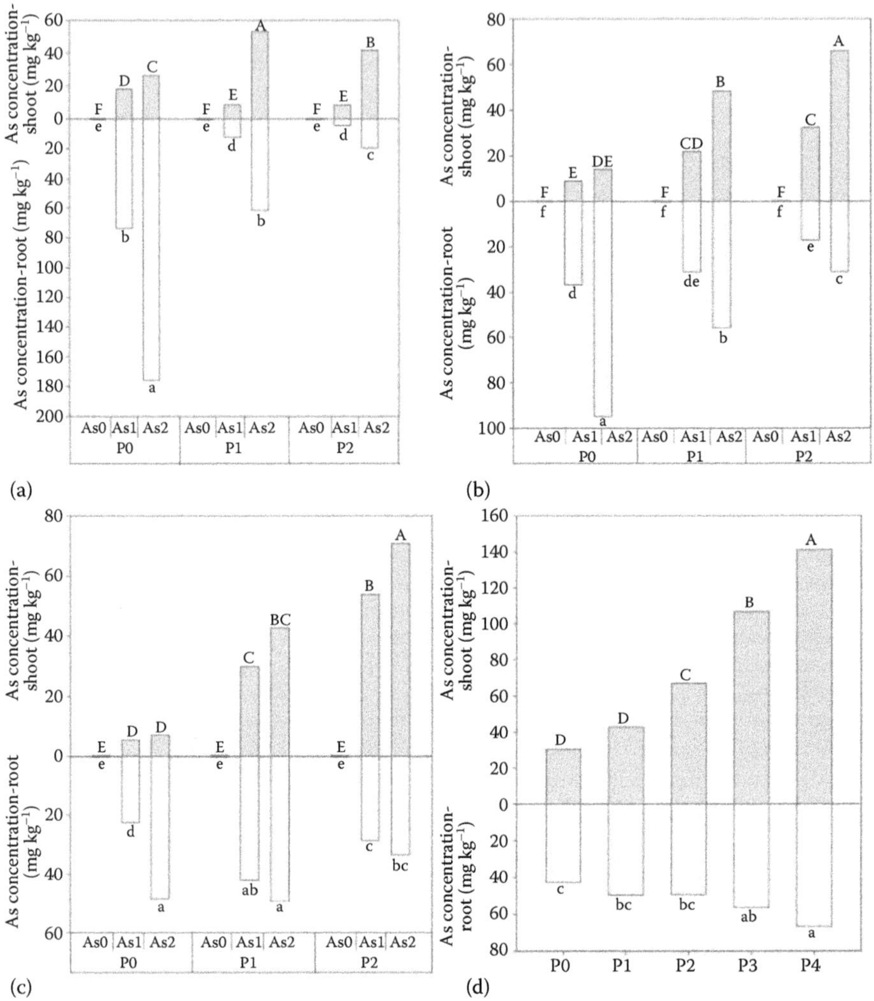

FIGURE 7.6
Effect of P and As on dry matter yield in (a) solution culture, (b) nonallophanic (Manawatu) soil, (c) allophanic (Patua) soil, and (d) sheep-dip (Hamilton) soil (solution culture: P0 = 0, P1 = 0.05, P2 = 0.25, As0 = 0, As1 = 0.05, As2 = 0.25 mM; allophanic and nonallophanic soils: P0 = 0, P1 = 3.38, P2 = 6.76 mmol kg^{-1}, As0 = 0, As1 = 3.38, As2 = 6.76 mmol kg^{-1}; sheep-dip soil: P0 = 0, P1 = 1.69, P2 = 3.38, P3 = 5.07, P4 = 6.76 mmol kg^{-1}). Means followed by the same letter (uppercase for shoot and lowercase for root) are not significantly different (Duncan's multiple-range test, $p < 0.05$). (Reprinted from *Sci. Total Environ.* 463–464, Bolan, N., Mahimairaja, S., Kunhikrishnan, A. and Choppala, G., Phosphorus–arsenic interactions in variable-charge soils in relation to arsenic mobility and bioavailability, 1154–1162, Copyright 2013, with permission from Elsevier.)

to 2.28 mg P/kg-soil. Shifts in trace metal speciation were determined using sequential extraction procedures, and relative toxicities were evaluated using the standard EPA TCLP. The TCLP for As in the soil increased from its initial value of 0.23 to 2.1, 0.70, 0.67, and 0.77 mg/kg, respectively, for the four treatments. The TCLP leachable As concentration of the K_2HPO_4-treated soil was about nine times that from the untreated soil. Thus, the sequential extraction test and the TCLP indicate that $Ca_5(PO_4)_3OH$ treatment has a higher potential in enhancement of As mobility compared with other phosphate treatments.

Qafoku et al. (1999) noticed that the leaching of As in a column containing mineral soil incorporated with As-rich poultry manure increased with the addition of a P compound. Peryea (1991) reported that desorption of As was dependent on the soil type, and only large additions of P (>400 mg kg^{-1}) would affect the As solubility in volcanic soils (Chen et al. 2002; Smith et al. 2002). In a hydroponic experiment, Wang et al. (2002) noticed that Chinese brake fern (*Pteris vittata* L.) plants accumulated As in the fronds up to 27.000 mg/kg dry weight, and the frond As to root As concentration ratio varied between 1.3 and 6.7. Increasing the P supply decreased the As uptake markedly, with the effect being greater on root As concentration than on shoot concentration. Tu and Ma (2003) suggested that P application may be an important strategy for the efficient use of *P. vittata* to phytoremediate As-contaminated soils. Similarly, Quaghebeur and Rengel (2004) reported that P-adequate canola (*Brassica napus* L.) plants contained a significantly higher proportion of As in the shoot than do P-deficient plants. The translocation of As from root to shoot was 8%–14% of the total As uptake in P-adequate plants but only 2% in P-deficient plants.

Jackson and Miller (2000) studied the effectiveness of phosphate and hydroxide for desorption of arsenic and selenium species from iron oxides. In this study, the relative effectiveness of PO_4 at 0.1 and 0.5 M and pH values of 3 and 6.7 and 0.1 M OH^- to extract As(III), As(V), dimethylarsinic acid (DMA), monomethylarsonic acid (MMA), p-arsanilic acid (p-ASA), roxarsone (ROX), Se(IV), and Se(VI) sorbed to geothite and an amorphous Fe oxide were compared, and the speciation in the resulting extract was determined. The extent to which 0.1 M PO_4 added to 0.25 M $NH_2OH \cdot HCl$ or 0.175 M Na oxalate/0.1 M oxalic acid prevents readsorption of As(V) or Se(IV) to geothite during the dissolution of an amorphous Fe oxide was also assessed. Hydroxide was the most effective extractant for desorption of all species except As(III) from both oxide surfaces. Arsenite was extracted most efficiently by 0.5 M PO_4 at low pH; however, amorphous Fe oxide exhibited a strong affinity for As(III) with a maximum of 18% of As(III) extracted by 0.5 M PO_4 at pH 2.8. Partial oxidation of As(III) to As(V) occurred in all extractions where an Fe oxide solid phase was present, but only in the hydroxide extract in the absence of an Fe solid phase. Addition of 0.1 M PO_4 to extractants used for the dissolution of the amorphous Fe oxide prevented the readsorption of As(V) and Se(IV) to geothite.

Phosphate may effectively compete with arsenic (especially arsenate) for surface binding sites on solids. Thus, there may be concerns with regard to As desorption caused by the addition of phosphates for *in situ* immobilization of metals. Phosphate has been shown to affect arsenite and arsenate sorption on iron oxides and soils (Qafoku et al. 1999; Jackson and Miller 2000; Smith et al. 2002).

The specificity of a solid for arsenate versus phosphate may simply be related to the order of sorption. Results from experiments by Liu et al. (2001) showed that arsenate displaces previously sorbed phosphate to a slightly greater degree than when arsenate is the initially sorbed species and phosphate had been added. They concluded that arsenate and phosphate compete for similar surface sites on the goethite surface. Thus, the ability of P to displace As (or vice versa) from solid materials will depend on a variety of factors including pH, mineralogy, solution speciation and concentration, chemical/physical kinetics, steric factors, and sorption reaction activation energy barriers. Impellitteri (2005) studied the effects of pH and phosphate on As speciation and mobilization in soils from a Pb smelting site and found that when phosphate is not present, solution As is mainly in the form of As(III). The presence of phosphate causes an increase in solution As due to an increase in As(V).

Summary

Applying phosphates to multimetal-contaminated soils can have contradictory effects on the fractionation of specific HMs, depending on the amendment. Thus, application of phosphates to soils might affect the mobility and bioavailability of soil HMs and their downward movement and uptake by the grown plants.

The effectiveness of amendments with different P sources was evaluated for their effectiveness in reducing the bioavailability of HMs in soils. The results indicated that several P amendments are efficient in transforming HMs (i.e., Pb, Zn, and Cd) from nonresidual into residual forms. However, P amendments are more efficient in transforming bioavailable Pb (nonresidual) into a less bioavailable form (residual) than for Zn and Cd. The effective immobilization of Pb was most likely due to the *in situ* formation of insoluble pyromorphite-like minerals after the application of P amendment.

There are some contraries in the influence of phosphates on the mobility of Cu and Zn. Although some studies indicated that the phosphates significantly decrease the mobility of Cu and Zn (Xu and Schwartz 1994; Chen et al. 2007), other studies reported that some phosphates increased Cu and Zn mobility (Spuller et al. 2007; Baker et al. 2012).

Also, the results of many studies indicated that the phosphates increase the mobility of soil As and Sb. These results indicate that application of phosphates has the potential to enhance the mobility of some elements (i.e., As and Sb and sometimes Cu and Zn). Therefore, there is a growing interest in using P fertilizer to enhance plant uptake of these elements in contaminated sites, thereby facilitating phytoremediation.

In conclusion, phosphate compounds might increase the metal (im)mobility based on the amendment and particular metal.

Acknowledgments

The authors express warm thanks to the contributors and the publishers for their permission to reuse data from publications mentioned in the text.

References

Adriano, D.C., 2001. *Trace Elements in Terrestrial Environments: Biogeochemistry, Bioavailability and Risks of Metals*, Second Edition, Springer, New York.

Adriano, D.C., Wenzel, W.W., Vangronsveld, J. and Bolan, N.S., 2004. Role of assisted natural remediation in environmental cleanup. *Geoderma* 122, 121–142.

Agbenin J.O., 2002. Lead in a Nigerian savanna soil under long-term cultivation. *Sci. Total Environ.* 286, 1–14.

Ahmad, M., Lee, S.S., Lim, J.E., Lee, S.E., Cho, J.S., Moon, D.H., Hashimoto, Y. and Ok, Y.S., 2014. Speciation and phytoavailability of lead and antimony in a small arms range soil amended with mussel shell, cow bone and biochar: EXAFS spectroscopy and chemical extractions. *Chemosphere* 95, 433–441.

Badawy, S.H., Helal, M.I.D., Chaudri, A.M., Lawlor, K. and McGrath, S.P., 2002. Soil solid-phase controls lead activity in soil solution. *J. Environ. Qual.* 31, 162–167.

Baker, L.R., Pierzynski, G.M., Hettiarachchi, G.M., Scheckel, K.G. and Newville, M., 2012. Zinc speciation in proximity to phosphate application points in a lead/zinc smelter-contaminated soil. *J. Environ. Qual.* 41, 1865–1873

Basta, N.T. and Gradwohl, R., 1998. Remediation of heavy metal-contaminated soil using rock phosphate. *Better Crops Plant Food* 82, 29–31.

Basta, N.T. and Gradwohl, R., 2000. Estimation of Cd, Pb, and Zn bioavailability in smelter-contaminated soils by a sequential extraction procedure. *J. Soil Contam.* 9, 149–164.

Basta, N.T., Gradwohl, R., Snethen, K.L. and Schroder, J.L., 2001. Chemical immobilization of lead, zinc, and cadmium in smelter-contaminated soils using biosolids and rock phosphate. *J. Environ. Qual.* 30, 1222–1230.

Basta, N.T. and McGowan, S.L., 2004. Evaluation of chemical immobilization treatments for reducing heavy metal transport in a smelter-contaminated soil. *Environ. Pollut*. 127, 73–82.

Boisson, J., Ruttens, A., Mench, M. and Vangronsveld, J., 1999. Evaluation of hydroxyapatite as a metal immobilizing soil additive for the remediation of polluted soils. Part 1. Influence of hydroxyapatite on metal exchangeability in soil, plant growth and plant metal accumulation. *Environ. Pollut*. 104, 225–233.

Bolan, N., Kunhikrishnan, A., Thangarajan, R., Kumpiene, J., Park, J., Makino, T., Kirkham, M.B. and Scheckel, K., 2014. Remediation of heavy metal(loid)s contaminated soils—To mobilize or to immobilize? (Review). *J. Hazard. Mater*. 266, 141–166.

Bolan, N., Mahimairaja, S., Kunhikrishnan, A. and Choppala, G., 2013a. Phosphorus–arsenic interactions in variable-charge soils in relation to arsenic mobility and bioavailability. *Sci. Total Environ*. 463–464, 1154–1162.

Bolan, N., Mahimairaja, S., Kunhikrishnan, A. and Naidu, R., 2013b. Sorption–bioavailability nexus of arsenic and cadmium in variable-charge soils. *J. Hazard. Mater*. 261, 725–732.

Bolan, N.S., Adriano, D.C., Mani, P.A. and Duraisamy, A., 2003a. Immobilization and phytoavailability of cadmium in variable charge soils. II. Effect of lime addition. *Plant Soil* 251, 187–198.

Bolan, N.S., Adriano, D.C. and Naidu, R., 2003b. Role of phosphorus in (im)mobilization and bioavailability of heavy metals in the soil-plant system. *Rev. Environ. Contam. Toxicol*. 177, 1–44.

Brown, S., Chaney, R., Hallfrisch, J., Ryan, J.A. and Berti, W.R., 2004. In situ soil treatments to reduce the phyto- and bioavailability of lead, zinc and cadmium. *J. Environ. Qual*. 33, 522–531.

Cao, R.X., Ma, L.Q., Chen, M., Singh, S.P. and Harris, W.G., 2002. Impacts of phosphate amendments on lead biogeochemistry at a contaminated site. *Environ. Sci. Technol*. 36, 5296–5304.

Cao, R.X., Ma, L.Q., Chen, M., Singh, S.P. and Harris, W.G., 2003. Phosphate-induced metal immobilization in a contaminated site. *Environ. Pollut*. 122, 19–28.

Cao, X., Dermatas, D., Xu, X. and Shen, G., 2008. Immobilization of lead in shooting range soils by means of cement, quicklime, and phosphate amendments. *Environ. Sci. Pollut. Res*. 15, 120–127.

Cao, X., Wahbi, A., Ma, L., Li, B. and Yang, Y., 2009. Immobilization of Zn, Cu, and Pb in contaminated soils using phosphate rock and phosphoric acid. *J. Hazard. Mater*. 164, 555–564.

Cao, X.D., Ma, L.Q., Rhue, D.R. and Appel, C.S., 2004. Mechanisms of lead, copper and zinc retention by phosphate rock. *Environ. Pollut*. 131, 435–444.

Chen, M., Ma, L.Q., Singh, S.P., Cao, R.X. and Melamed, R., 2003. Field demonstration of in situ immobilization of soil Pb using P amendments. *Adv. Environ. Res*. 8, 93–102.

Chen, S., Xu, M., Ma, Y. and Yang, J., 2007. Evaluation of different phosphate amendments on availability of metals in contaminated soil. *Ecotoxicol. Environ. Saf*. 67, 278–285.

Chen, T.B., Fan, Z.L., Lei, M., Huang, Z.C. and Wei, C.Y., 2002. Effect of phosphorus on arsenic accumulation in As-hyperaccumulator *Pteris vittata* L. and its implication. *Chin. Sci. Bull*. 47, 1876–1879.

Chen, X., Wright, J.V., Conca, J.L. and Peurrung, L.M., 1997a. Effect of pH on heavy metal sorption on mineral apatite. *Environ. Sci. Technol.* 31, 624–631.
Chen, X., Wright, J.V., Conca, J.L. and Peurrung, L.M., 1997b. Evaluation of heavy metal remediation using mineral apatite. *Water Air Soil Pollut.* 98, 57–78.
Corami, A., Mignardi, S. and Ferrini, V., 2007. Copper and zinc decontamination from single- and binary-metal solutions using hydroxyapatite. *J. Hazard. Mater.* 146, 164–170.
Corami, A., Mignardi, S. and Ferrini, V., 2008a. Cadmium removal from single- and multi-metal (Cd + Pb + Zn + Cu) solutions by sorption on hydroxyapatite. *J. Colloid Interface Sci.* 317, 402–408.
Corami, A., Mignardi, S. and Ferrini, V., 2008b. Removal of lead, copper, zinc and cadmium from water using phosphate rock. *Acta Geol. Sin.–Engl.* 82, 1223–1228.
Cotter-Howells, J., 1996. Lead phosphate formation in soils. *Environ. Pollut.* 93, 9–16.
Eighmy, T.T., Crannell, B.S., Butler, L.G., Cartledge, F.K., Emery, E.F., Oblas, D., Krzanowski, J.E., Eusden, J.D., Shaw, E.L. and Francis, C.A., 1997. Heavy metal stabilization in municipal solid waste combustion dry scrubber residue using soluble phosphate. *Environ. Sci. Technol.* 37, 3330–3338.
Hamon, R.E., McLaughlin, M.J. and Cozens, G., 2002. Mechanisms of attenuation of metal availability in in situ remediation treatments. *Environ. Sci. Technol.* 36, 3991–3996.
Han, N. and Thompson, M.L., 2003. Impact of dissolved organic matter on copper mobility in aquifer material, *J. Environ. Qual.* 32, 1829–1836.
Hashimoto, Y., Takaoka, M. and Shiota, K., 2011. Enhanced transformation of lead speciation in rhizosphere soils using phosphorus amendments and phytostabilization: An x-ray absorption fine structure spectroscopy investigation. *J. Environ. Qual.* 40, 696–703.
He, Q., Ren,Y., Mohamed, I., Ali, M., Hassan, W. and Zeng, F., 2013. Assessment of trace and heavy metal distribution by four sequential extraction procedures in a contaminated soil. *Soil Water Res.* 8, 71–76.
Hettiarachchi, G.M., Pierzynski, G.M. and Ransom, M.D., 2000. In situ stabilization of soil lead using phosphorus and manganese oxide. *Environ. Sci. Technol.* 34, 4614–4619.
Hettiarachchi, G.M., Pierzynski, G.M. and Ransom, M.D., 2001. In situ stabilization of soil lead using phosphorus. *J. Environ. Qual.* 30, 1214–1221.
Hooda, P.S. (ed.), 2010. *Trace Elements in Soils*. John Wiley & Sons Ltd., Chichester, United Kingdom.
Huang, H., Wang, K., Zhu, Z., Li, T., He, Z., Yang, X. and Gupta, D.K., 2013. Moderate phosphorus application enhances Zn mobility and uptake in hyperaccumulator *Sedum alfredii*. *Environ. Sci. Pollut. Res.* 20, 2844–2853.
Impellitteri, C.A., 2005. Effects of pH and phosphate on metal distribution with emphasis on As speciation and mobilization in soils from a lead smelting site. *Sci. Total Environ.* 345, 175–190.
Islam, M., Mishra, P.C. and Patel, R., 2010. Physicochemical characterization of hydroxyapatite and its application towards removal of nitrate from water. *J. Environ. Manage.* 91, 1883–1891.
Jackson, B.P. and Miller, W.P., 2000. Effectiveness of phosphate and hydroxide for desorption of arsenic and selenium species from iron oxides. *Soil Sci. Soc. Am. J.* 64, 1616–1622.

Jeanjean, J., Vincent, U. and Fedoroff, M., 1994. Structural modification of calcium hydroxyapatite induced by sorption of cadmium ions. *J. Solid State Chem.* 108, 68–72.

Johnson, C.A., Moench, H., Wersin, P., Kugler, P. and Wenger, C., 2005. Solubility of antimony and other elements in samples taken from shooting ranges, *J. Environ. Qual.* 34, 248–254.

Kabala, C. and Singh, B.R., 2001. Fractionation and mobility of copper lead and zinc in soil profiles in the vicinity of a copper smelter. *J. Environ. Qual.* 30, 485–492.

Kabata-Pendias, A., 2011. *Trace Elements in Soils and Plants*, Fourth Edition. CRC Press, Boca Raton, FL.

Koopmans, G.F., 2004. *Characterization, Desorption, and Mining of Phosphorus in Non-calcareous Sandy Soils*. Wageningen Universiteit, The Netherlands.

Kumpiene, J., Lagerkvist, A. and Maurice, C., 2008. Stabilization of As, Cr, Cu, Pb, and Zn in soil using amendments—A review. *Waste Manage.* 28, 215–225.

Laperche, V., Logan, T.J., Gaddam, P. and Traina, S., 1997. Effect of apatite amendments on plant uptake of lead from contaminated soil. *Environ. Sci. Technol.* 31, 2745–2753.

Lei, M., Wan, X., Huang, Z., Chen, T., Li, X. and Li, Y., 2012. First evidence on different transportation modes of arsenic and phosphorus in arsenic hyperaccumulator *Pteris vittata*. *Environ. Pollut.* 161, 1–7.

Lindsay, W.L., 1979. *Chemical Equilibria in Soils*. Wiley, New York.

Ma, L.Q., Choate, A.L. and Rao, G.N., 1997. Effects of incubation and phosphate rock on lead extractability and speciation in contaminated soils. *J. Environ. Qual.* 26, 801–807.

Ma, L.Q. and Rao, G.N., 1997. Chemical fractionation of cadmium, copper, nickel and zinc in contaminated soils. *J. Environ. Qual.* 26, 259–264.

Ma, Q.Y., Logan, T.J. and Traina, S.J., 1995. Lead immobilization from aqueous solutions and contaminated soils using phosphate rocks. *Environ. Sci. Technol.* 29, 1118–1126.

Ma, Q.Y., Traina, S.J., Logan, T.J. and Ryan, J.A., 1994. Effects of aqueous Al, Cd, Cu, Fe(II), Ni, and Zn on Pb immobilization by hydroxyapatite. *Environ. Sci. Technol.* 28, 1219–1228.

Marchat, D., Bernache-Assollant, D. and Champion, E., 2007. Cadmium fixation by synthetic hydroxyapatite in aqueous solution—Thermal behaviour. *J. Hazard. Mater.* 139, 453–460.

Mavropoulos, E., Rossi, A.M., Costa, A.M., Perez, C.A.C., Moreira, J.C. and Saldanha, M., 2002. Studies on the mechanisms of lead immobilization by hydroxyapatite. *Environ. Sci. Technol.* 36, 1625–1629.

McBride, M., 1994. *Environmental Chemistry of Soils*. Oxford University Press, Oxford, United Kingdom.

McGowen, S.L., 2000. In situ chemical treatments for reducing heavy metal solubility and transport in smelter contaminated soil. PhD dissertation, Oklahoma State University, Stillwater, OK.

McGowen, S.L., Basta, N.T. and Brown, G.O., 2001. Use of diammonium phosphate to reduce heavy metal solubility and transport in smelter-contaminated soil. *J. Environ. Qual.* 30, 493–500.

Meharg, A.A. and Macnair, M.R., 1992. Suppression of the high affinity phosphate uptake system: A mechanism of arsenate tolerance in *Holcus lanatus* L. *J. Exp. Bot.* 43, 519–524.

Melamed, R., Cao, X., Chen, M. and Ma, L.Q., 2003. Field assessment of lead immobilization in a contaminated soil after phosphate application. *Sci. Total Environ.* 305, 117–127.

Mignardi, S., Corami, A. and Ferrini, V., 2012. Evaluation of the effectiveness of phosphate treatment for the remediation of mine waste soils contaminated with Cd, Cu, Pb, and Zn. *Chemosphere* 86, 354–360.

Miretzky, P. and Fernandez Cirelli, A., 2008. Phosphates for Pb immobilization in soils: A review. *Environ. Chem. Lett.* 6, 121–133.

Nannoni, F., Protano, G. and Riccobono, F., 2011. Fractionation and geochemical mobility of heavy elements in soils of a mining area in northern Kosovo. *Geoderma* 161, 63–73.

Park, J.H., Lamb, D.T., Paneerselvam, P., Choppala, G., Bolan, N.S. and Chung, J.W., 2011. Role of organic amendments on enhanced bioremediation of heavy metal(loid) contaminated soils. *J. Hazard. Mater.* 185, 549–574.

Pereya, F.J., 1991. Phosphate-induced release of arsenic from soils contaminated with lead arsenate. *Soil Sci. Soc. Am. J.* 55, 1301–1306.

Pierzynski, G.M. and Schwab, A.P., 1993. Heavy metals in the environment: Bioavailability of zinc, cadmium, and lead in a metal contaminated alluvial soil. *J. Environ. Qual.* 22, 247–254.

Pigna, M., Cozzolino, V., Violante, A. and Meharg, A.A., 2009. Influence of phosphate on the arsenic uptake by wheat (*Triticum durum* L.) irrigated with arsenic solution at three different concentrations. *Water Air Soil Pollut.* 197, 371–380.

Qafoku, N.P., Kulier, U., Summer, M.E. and Radcliffe, D.E., 1999. Arsenate displacement from fly ash in amended soils. *Water Air Soil Pollut.* 114, 185–198.

Quaghebeur, M. and Rengel, Z., 2004. Phosphate and arsenate interactions in the rhizosphere of canola (*Brassica napus*). *Funct. Plant Biol.* 31, 1085–1094.

Raicevic, S., Kaludjerovic-Radoicic, T. and Zouboulis, A.I., 2005. In situ stabilization of toxic metals in polluted soils using phosphates: Theoretical prediction and experiment verification. *J. Hazard. Mater.* B117, 41–53.

Ravenscroft, P., McArthur, J. and Hoque, B., 2001. Geochemical and palaeohydrological controls on pollution of groundwater by arsenic. In: W.R. Chappell, C.O. Abernathy, R. Calderon (eds.), *Arsenic Exposure and Health Effects*, IV, Elsevier Science Ltd., Oxford, United Kingdom, pp. 53–77.

Rinklebe, J. and Shaheen, S.M., 2014. Assessing the mobilization of cadmium, lead, and nickel using a seven-step sequential extraction technique in contaminated floodplain soil profiles along the central Elbe River, Germany. *Water Air Soil Pollut.* 225, 1–20. doi: 10.1007/s11270-014-2039-1.

Rogival, D., Scheirs, J. and Blust, R., 2007. Transfer and accumulation of metals in a soil-diet-wood mouse food chain along a metal pollution gradient. *Environ. Pollut.* 145, 516–528.

Ruby, M.V., Davis, A. and Nicholson, A., 1994. In-situ formation of lead phosphates in soils as a method to immobilize lead. *Environ. Sci. Technol.* 28, 646– 654.

Ryan, J.A., Zhang, P.C., Hesterberg, D., Chou, J. and Sayers, D.E., 2001. Formation of chloropyromorphite in a lead-contaminated soil amended with hydroxyapatite. *Environ. Sci. Technol.* 35, 3798–3803.

Scheckel, K.G., Diamond, G.L., Burgess, M.F., Klotzbach, J.M., Maddaloni, M., Miller, B.W., Partridge, C.R. and Serda, S.M., 2013. Amending soils with phosphate as means to mitigate soil lead hazard: A critical review of the state of the science. *J. Toxicol. Environ. Health B* 16, 337–380.

Scheckel, K.G., Impellitteri, C.A., Ryan, J.A. and McEvoy, T., 2003. Assessment of a sequential extraction procedure for perturbed lead-contaminated samples with and without phosphorus amendments. *Environ. Sci. Technol.* 37, 1892–1898.
Scheckel, K.G. and Ryan, J.A., 2002. Effects of aging and pH on dissolution kinetics and stability of chloropyromorphite. *Environ. Sci. Technol.* 36, 2198–2204.
Scheckel, K.G. and Ryan, J.A., 2004. Spectroscopic speciation and quantification of Pb in phosphate amended soils. *J. Environ. Qual.* 33, 1288–1295.
Scheckel, K.G., Ryan, J.A., Allen, D. and Lescano, N.V., 2005. Determining speciation of Pb in phosphate-amended soils: Method limitations. *Sci. Total Environ.* 350, 261–272.
Seaman, J.C., Arey, J.S. and Bertsch, P.M., 2001. Immobilization of nickel and other metals in contaminated sediments by hydroxyapatite addition. *J. Environ. Qual.* 30, 460–469.
Shaheen, S.M., 2009. Sorption and lability of cadmium and lead in different soils from Egypt and Greece. *Geoderma* 153, 61–68.
Shaheen, S.M. and Rinklebe, J., 2014. Geochemical fractions of chromium, copper, and zinc and their vertical distribution in soil profiles along the Central Elbe River, Germany. *Geoderma* 228–229, 142–159.
Shaheen S.M., Rinklebe, J., Frohne, T., White, J. and DeLaune, R., 2014a. Biogeochemical factors governing Co, Ni, Se, and V dynamics in periodically flooded Egyptian north Nile delta rice soils. *Soil Sci. Soc. Am. J.* 78, 1065–1078.
Shaheen, S.M., Rinklebe, J., Rupp, H. and Meissner, R., 2014b. Lysimeter trials to assess the impact of different flood-dry-cycles on the dynamics of pore water concentrations of As, Cr, Mo, and V in a contaminated floodplain soil. *Geoderma* 228–229, 5–13.
Shaheen, S.M., Rinklebe, J., Rupp, H. and Meissner, R., 2014c. Temporal dynamics of soluble Cd, Co, Cu, Ni, and Zn and their controlling factor in a contaminated floodplain soil using undisturbed groundwater lysimeter. *Environ. Pollut.* 191, 223–231.
Shaheen, S.M. and Tsadilas, C.D., 2013. Phosphorus sorption and availability to canola grown on an Alfisol amended with various soil amendments. *Commun. Soil Sci. Plant Anal.* 44, 89–103.
Shaheen, S.M., Tsadilas, C.D. and Rinklebe, J., 2013. A review of the distribution coefficient of trace elements in soils: Influence of sorption system, element characteristics, and soil colloidal properties. *Adv. Colloid Interface Sci.* 201–202, 43–56.
Singh, S.P., Ma, L.Q. and Harris, W.G., 2001. Heavy metal interactions with phosphatic clay: Sorption and desorption behavior. *J. Environ. Qual.* 30, 1961–1968.
Smith, E., Naidu, R. and Alston, A.M., 2002. Chemistry of inorganic arsenic in soils. II. Effect of phosphorus, sodium, and calcium on arsenic sorption. *J. Environ. Qual.* 31, 557–563.
Sneddon, I.R., Orueetxebarria, M., Hodson, M.E., Schofield, P.F. and Valsami-Jones, E., 2006. Use of bone meal amendments to immobilise Pb, Zn and Cd in soil: A leaching column study. *Environ. Pollut.* 144, 816–825.
Soares, C.R.F.S. and Siqueira, J.O., 2008. Mycorrhiza and phosphate protection of tropical grass species against heavy metal toxicity in multi-contaminated soil. *Biol. Fert. Soils* 44, 833–841.
Song, Y.C., Sivakumar, S., Nguyen, T.T., Kim, S.H. and Kim, B.G., 2009. The immobilization of heavy metals in biosolids using phosphate amendments—Comparison of EPA (6010 and 3051) and selective sequential extraction methods. *J. Hazard Mater.* 167, 1033–1037.

Spuller, C., Weigand, H. and Marb, C., 2007. Trace metal stabilisation in a shooting range soil: Mobility and phytotoxicity. *J. Hazard. Mater.* 141, 378–387.
Stanforth, R. and Qiu, J., 2001. Effect of phosphate treatment on the solubility of lead in contaminated soil. *Environ. Geol.* 41, 1–10.
Takeuchi, Y. and Arai, H.J., 1990. Removal of coexisting Pb^{2+}, Cu^{2+}, and Cd^{2+} ions from water by addition of hydroxyapatite powder. *J. Chem. Eng. Jpn.* 23, 75–80.
Thawornchaisit, U. and Polprasert, C., 2009. Evaluation of phosphate fertilizers for the stabilization of cadmium in highly contaminated soils. *J. Hazard. Mater.* 165, 1109–1113.
Traina, S.J. and Laperche, V., 1999. Contaminant bioavailability in soils, sediments, and aquatic environments. *Proc. Natl. Acad. Sci. U. S. A.* 196, 3365–3371.
Tu, C. and Ma, L.Q., 2003. Effects of arsenate and phosphate on their accumulation by an arsenic-hyperaccumulator *Pteris vittata* L. *Plant Soil* 249, 373–382.
Wang, J.R., Zhao, F.J., Meharg, A.A., Raab, A., Feldmann, J. and McGrath, S.P., 2002. Mechanisms of arsenic hyperaccumulation in Pteris vittata: Uptake kinetics, interactions with phosphate, and arsenic speciation. *Plant Physiol.* 130, 1552–1561.
Wang, Y.M., Chen, T.C., Yeh, K.J. and Shue, M.F., 2001. Stabilization of an elevated heavy metal contaminated site. *J. Hazard. Mater.* B88, 63–74.
Waterlot, C., Pruvot, C., Ciesielski, H. and Douay, F., 2011. Effects of a phosphorus amendment and the pH of water used for watering on the mobility and phytoavailability of Cd, Pb, and Zn in highly contaminated kitchen garden soils. *Eco. Eng.* 37, 1081–1093.
Wu, W.H., Xie, Z.M., Xu, J.M., Wang, F., Shi, J.-C., Zhou, R.-B., and Jin, Z.-F., 2013. Immobilization of trace metals by phosphates in contaminated soil near lead/zinc mine tailings evaluated by sequential extraction and TCLP. *J. Soils Sediments* 13, 1386–1395.
Xu, Y. and Schwartz, F.W., 1994. Lead immobilization by hydroxyapatite in aqueous solutions. *J. Contam. Hydrol.* 15, 187–206.
Xu, Y., Schwartz, F.W. and Traina, S.J., 1994. Sorption of Zn^{2+} and Cd^{2+} on hydroxyapatite surfaces. *Environ. Sci. Technol.* 28, 1472–1480.
Yoon, J.K., Cao, X. and Ma, L.Q., 2007. Application methods affect phosphorus-induced lead immobilization from a contaminated soil. *J. Environ. Qual.* 36, 373–378.
Zhang, P., Ryan, J.A. and Yang, J., 1998. In vitro soil Pb solubility in the presence of hydroxyapatite. *Environ. Sci. Technol.* 32, 2763–2768.
Zhu, Y.-G., Chen, S.-B. and Yang, J.-C., 2004. Effects of soil amendments on lead uptake by two vegetable crops from a lead-contaminated soil from Anhui. *China Environ. Int.* 30, 351–356.
Zhuang, P., McBride, M.B., Xia, H., Li, N. and Li, Z., 2009. Health risk from heavy metals via consumption of food crops in the vicinity of Dabaoshan mine, South China. *Sci. Total Environ.* 407, 1551–1561.

8

Effect of Phosphate Addition on Mobility and Phytoavailability of Heavy Metals in Soils

Shiwei Zhou, Zhengguo Song, Minggang Xu, and Shibao Chen

CONTENTS

The pollution sources of heavy metals in soils are mainly derived from anthropogenic activities associated with atmospheric pollution from motor vehicles, combustion of fossil fuels, agricultural fertilizers and pesticides, organic manures, disposal of urban and industrial wastes, metallurgical industries, and mining and smelting of nonferrous metals (Alloway 1990). During the last three decades, heavy metal pollution in China's agricultural soils has become a serious situation due to highly intensive cultivation and rapid industrialization and urbanization. As a result, about 20 million ha of cultivated land (20% of the country's arable land) has been polluted (Chen 1996), bringing serious threats to the ecosystem and public health. Recently, an occurrence of Cd toxicity in rice caused widespread public concern and anxiety, highlighting China's dependence on food derived from contaminated soils (Bi et al. 2013; Liu et al. 2013).

Disposal of metal-contaminated soils is very time-consuming and costly. Over the past two decades, *in situ* chemical immobilization has been put forth for consideration as an alternative to the remediation of contaminated soils because of its low cost and high efficiency (Mulligan et al. 2001; Guo et al. 2006; Kumpiene et al. 2008; Udeigwe et al. 2011). Since Gauglitz et al. (1992) showed that hydroxyapatite (HA) has the capacity to immobilize heavy metals, the use of phosphate materials as amendments has been a promising way to remediate heavy-metal-contaminated soils (Bolan et al. 2003a; Guo et al. 2006; Kumpiene et al. 2008; Miretzky and Fernandez-Cirelli 2008). Because high-quality agricultural soil resources are insufficient to supply food for a very large population, many crops are still planted in contaminated soils in many countries, especially in China, giving rise to two issues that need to be considered: (1) the ability of the soils to retain lower metal solubility and thus mobility for reducing the risk of heavy metal transport to waters and plants, and (2) the ability to provide adequate nutrients for plant growth. Phosphate materials that can perform as both fertilizers and amendments address each of these issues.

Phosphate addition does not generally change the content but does affect the existing species of heavy metals in soils, accordingly impacting their solubility/mobility and availability. Therefore, we provide here a review that details the immobilization of heavy metals in soils by phosphate natural and synthetic materials with special focus on the chemical speciation and the ionic interaction between phosphate and heavy metal.

Evaluation Methods and Techniques for the Immobilization of Heavy Metals in Phosphate-Amended Soils

Application of phosphate amendments caused heavy metals in soils to shift from forms with high availability/mobility to the most strongly bound fractions such as residual forms. Thus, the methods and techniques commonly used for the evaluation of immobilization effectiveness of heavy metals were often based on their speciation besides availability/toxicity and solubility/mobility.

Heavy Metal Speciation Methods

Extraction Procedures of Chemical Speciation of Heavy Metals in Soils

Metal phytoavailability has generally been estimated by various chemical extractants such as neutral salts, mild acids, and organic chelators. Alloway (1990) suggested that the most generally useful extractants were 0.005 M DTPA and 0.05 M EDTA because of their good predictors of plant uptake. However, Menzies et al. (2007) found that neutral salt extractants provided the most useful indication of the phytoavailability of cadmium (Cd), zinc (Zn), nickel (Ni), copper (Cu), and lead (Pb). It may be that many of these single extractions are only relevant to specific crops or particular soil types, and more importantly, that they are unlikely to provide useful information on metal speciation in soils. Thus, only a few studies reported the single extraction fractionation of heavy metals in phosphate-amended soils (Ma et al. 1997; Lin et al. 2005; Fang et al. 2012). In contrast, a series of sequential extraction procedures (SEPs), although more time-consuming, were widely used to evaluate the distribution, transport, and transformation of heavy metal species in soils amended with different phosphate materials (Ma and Rao 1997; Basta et al. 2001; Scheckel et al. 2003; Cajuste et al. 2006; Chen et al. 2007; Padmavathiamma and Li 2010; Wu et al. 2013; Al-Hwaiti et al. 2014). However, there are two main problems with SEP: (1) the nonspecific selectivity of extractants that results in metal species intersecting, and (2) the redistribution (adsorption or precipitation) of metals dissolving in extractants among phases during extraction (Ramos et al. 1994). Obviously, in the presence of anion additives such as phosphate, the redistribution by precipitation with phosphate would no doubt underestimate the available species and environmental risk of heavy metals in soils.

Chemical Modeling of Metal Speciation in Soil Solutions

Some chemical speciation programs such as MINTEQ, GEOCHEM, and WHAM have been used to calculate the concentrations of free-metal ions and complexes of metal with inorganic and organic ligands in soil solutions (Dudal and Gérard 2004). Activity-ratio diagrams were then constructed on this basis, where the precipitates or minerals of metal–phosphate could be derived (McGowen et al. 2001; Cao et al. 2002, 2009; Bolan et al. 2003b).

Chemical speciation models have played important roles in soil chemistry due to successfully explaining the chemical reactions in soils and predicting the speciation concentration of elements (Huang 1998). However, these models were built on the assumption that the equilibrium conditions in soils were the same as those in pure water, implying their limitation and uncertainty. In fact, there were remarkable differences between the two systems;

for example, the pH values for surface precipitation of hydroxides of metals were lower than that for bulk precipitation (Anderson and Rubin 1981). Therefore, the most important challenge for the future is to update continuously the data of equilibrium constants for metal dissolution and adsorption available on these chemical speciation models (Huang 1998).

Spectroscopic and Microscopic Techniques

The identification of minerals or precipitates of metal–phosphate has been widely carried out by different nondestructive techniques such as x-ray absorption fine structure (XAFS) (Cotter-Howells et al. 1999; Scheckel and Ryan 2004; Chrysochoou et al. 2007; Hashimoto et al. 2009a, 2011; Siebers et al. 2013), x-ray diffraction (XRD) (Cotter-Howells and Caporn 1996; Hettiarachchi et al. 2001; Scheckel et al. 2003; Cao et al. 2003, 2009; Abbaspour and Golchin 2011), scanning electron microscopy coupled with energy dispersive x-ray spectroscopy (SEM-EDX) (Cotter-Howells and Caporn 1996; Singh et al. 2001; Cao et al. 2003, 2009; Chen et al. 2007; Kassir et al. 2012), transmission electron microscopy (TEM) (Cotter-Howells et al. 1999; Seaman et al. 2001; Mavropoulos et al. 2004; Kamiishi and Utsunomiya 2013), and electron microprobe analysis (EMPA) (Yang et al. 2001; Strawn et al. 2007; Munksgaard and Lottermoser 2011). The application of these advanced, molecular-level spectroscopic and microscopic methods can provide more direct, conclusive, and accurate results than the use of SEP analysis or chemical speciation models.

Of particular interest is the fact that XAFS is leading the development of soil science and has resulted in a new discipline, molecular environmental soil science (Xu and Sparks 2013), because it is a powerful structural detection technique that can probe directly the electronic and structural nature of elements (interatomic distances, structural disorder, and number and kind of neighboring atoms at a given distance). However, these spectroscopic and microscopic techniques also have some limitations; for example, they commonly depend on relatively high concentrations of target elements and are usually limited to qualitative analysis.

Methods for Assessing Metal Phytoavailability/Toxicity and Mobility

Application of phosphate materials to contaminated soils could markedly change metal phytoavailability/toxicity and mobility, and therefore some methods were developed for assessing the phytoavailability/toxicity and mobility of heavy metals in phosphate-amended soils, in which the plant concentrations were assumed to represent the phytoavailable metal content of soils. Some heavy-metal-susceptible plants were tested, such as lettuce (Basta et al. 2001; Cao et al. 2009; Trevizam et al. 2010; Sanderson et al. 2014), Chinese cabbage (Chen et al. 2007; Wang et al. 2008; Fang et al. 2012), soybean (Cao et al. 2009), tomato (Khan and Jones 2008), barley (Mahdy 2012), and corn (Cajuste et al. 2006). In addition, a physiologically based extraction test

(PBET) simulating metal ingestion and gastrointestinal bioavailability to humans has been applied in some studies (Hettiarachchi et al. 2000; Basta et al. 2001; Liu and Zhao 2007; Cao et al. 2009).

Currently, there are many specialized methods for estimating the toxicity of heavy metals in soils to organisms such as earthworms, plant roots, and microbes. However, these toxicity tests, especially the root elongation and microbial test, were rarely used to evaluate the immobilization effectiveness of heavy metals in phosphate-amended soils (Conder et al. 2001; Maenpaa et al. 2002; Tang et al. 2009; Park et al. 2012; Sanderson et al. 2014). In contrast, relative toxicities using the standard Environmental Protection Agency (EPA) toxicity characteristic leaching procedure (TCLP) were widely evaluated (Chen et al. 1997; Singh et al. 2001; Theodoratos et al. 2002; Cao et al. 2004, 2013; Liu and Zhao 2007; Wang et al. 2008; Majs 2011; Wu et al. 2013).

Column leaching experiments using various leaching agents have been also used for assessing the mobility and thus the immobilization efficiency of heavy metals in phosphate-amended soils (Theodoratos et al. 2002; Chen et al. 2006a; Adhikari and Singh 2008; Jalali and Moharami 2010; Fang et al. 2012; Cao et al. 2013).

The last two decades of research have produced immense progress in the methods available to understand the species, mobility, bioavailability, and fate of heavy metals in soils, but no method is perfect and will work with all environmental conditions. Geebelen et al. (2003) has emphasized the necessity of using a diverse suite of bioavailability methodology when *in situ* metal immobilization is assessed because different amendments do not behave consistently across different soils with different sources of contamination.

Immobilization Mechanisms of Heavy Metals in Phosphate-Amended Soils

Numerous natural and synthetic phosphate materials have been used to immobilize the heavy metals in soils, including insoluble apatite, HA, and phosphate rock (PR), and soluble phosphate fertilizers such as diammonium phosphate (DAP), triple superphosphate (TSP), monocalcium phosphate (MCP), potassium dihydrogen phosphate (PDP), and phosphoric acid (PA). There were different effects of different phosphate sources due to different solubility and complexing capacity towards different heavy metals, and the interaction of phosphate with Pb was usually much stronger than that with other heavy metals, most probably because of different mechanisms of metal immobilization that were very complex and were suggested to be attributed to the following three main processes: phosphate-induced adsorption of metal ions, the formation of metal phosphate precipitates or minerals, and surface adsorption or complexation of metals.

Phosphate-Induced Adsorption of Metal Ions

Phosphate-induced adsorption in variable charge soils, occurring widely throughout the tropics and subtropics, has been widely observed, where specific adsorption of phosphate on hydrous oxides of iron (Fe), aluminum (Al), manganese (Mn), or kaolinite can occur by ligand exchange with surface hydroxyls, increasing negative charge and/or pH, accordingly increasing metal retention by soils (Yu 1997; Agbenin 1998; Bolan et al. 2003b; Pardo 2004).

As described in Equations 8.1 and 8.2 (Parfitt et al. 1975; Bolland et al. 1977), phosphate was adsorbed by ligand exchange with two A-type hydroxyl (OH) groups to give a binuclear complex, then metal ions were adsorbed on the bridging phosphate hydrous oxides through polyvalent bonds. Some studies have reported the formation of stable mineral–phosphate–metal ternary surface complexes (Bolland et al. 1977; Agbenin 1998; Bolan et al. 1999; Pérez-Novo et al. 2011; Elzinga and Kretzschmar 2013; Tiberg et al. 2013), and the formation of ternary complexes was further confirmed by means of attenuated total reflectance Fourier transform infrared spectroscopy (ATR-FTIR) (Elzinga and Kretzschmar 2013) and extended x-ray absorption fine structure (EXAFS) spectroscopy (Tiberg et al. 2013). However, some researchers reported evidence against this mechanism; for example, Collins et al. (1999) found by EXAFS measurement that Cd and P were retained on different sites on goethite surfaces, and the enhancement of Cd^{2+} adsorption in the presence of phosphate was solely by electrostatic interaction where no ternary complexes were observed.

$$\begin{matrix}\equiv Fe-OH \\ \equiv Fe-OH\end{matrix} + HPO_4^{2-} \rightarrow \begin{matrix}\equiv Fe-O \\ \equiv Fe-O\end{matrix}\!\!>\!P\!\begin{matrix}{=}O \\ \diagdown OH\end{matrix} + 2OH^- \quad (8.1)$$

$$\begin{matrix}\equiv Fe-O \\ \equiv Fe-O\end{matrix}\!\!>\!P\!\begin{matrix}{=}O \\ \diagdown OH\end{matrix} + M^{2+} \rightarrow \begin{matrix}\equiv Fe-O \\ \equiv Fe-O\end{matrix}\!\!>\!P\!\begin{matrix}{=}O \\ \diagdown OM^+\end{matrix} + H^+ \quad (8.2)$$

Metal Phosphate Precipitates or Minerals

The formation of precipitates and/or minerals is the main mechanism responsible for the immobilization of heavy metal in soils, especially in mining/smelting soils, using phosphate addition. It should be noted that according to the solubility of metal phosphates, which is presented in Table 8.1, Pb phosphates, especially pyromorphites ($Pb_5(PO_4)_3X$, where X = fluorine (F), chlorine (Cl), bromine (Br), OH, representing fluoro-, chloro-, bromo-, and hydroxypyromorphite, respectively), are the most stable compounds, and hence, the interaction of Pb and P through the formation of pyromorphites is an important immobilization mechanism controlling the mobility and availability of heavy metals in soils. Moreover, chloropyromorphite is several orders of magnitude less soluble than hydroxyl-, bromo-, and fluoropyromorphite (Table 8.1), and due to the ubiquity of chloride in nature, chloropyromorphite was expected to be the dominant form in the immobilization of Pb in

TABLE 8.1

Theoretical Solubility of Metal Phosphate Minerals/Precipitates at 25°C

Mineral/Precipitate	Equilibrium Reaction	Log K_{sp}
Plumbogummite	$PbAl_3(PO_4)_2(OH)_5H_2O \rightleftharpoons Pb^{2+} + 3Al^{3+} + 2PO_4^{3-} + 5OH^- + H_2O$	–99.3[a]
Chloropyromorphite	$Pb_5(PO_4)_3Cl\ (c) \rightleftharpoons 5Pb^{2+} + 3PO_4^{3-} + Cl^-$	–83.7
Bromopyromorphite	$Pb_5(PO_4)_3Br\ (c) \rightleftharpoons 5Pb^{2+} + 3PO_4^{3-} + Br^-$	–78.14
Hydroxypyromorphite	$Pb_5(PO_4)_3OH\ (c) \rightleftharpoons 5Pb^{2+} + 3PO_4^{3-} + OH^-$	–76.79
Fluoropyromorphite	$Pb_5(PO_4)_3F\ (c) \rightleftharpoons 5Pb^{2+} + 3PO_4^{3-} + F^-$	–71.63
$Pb_3(PO_4)_2$	$Pb_3(PO_4)_2\ (c) \rightleftharpoons 3Pb^{2+} + 2PO_4^{3-}$	–44.36
$PbHPO_4$	$PbHPO_4\ (c) \rightleftharpoons Pb^{2+} + HPO_4^{2-}$	–11.45
Zinc pyromorphite	$Zn_5(PO_4)_3OH \rightleftharpoons 5Zn^{2+} + 3PO_4^{3-} + OH^-$	–63.1[a]
Hopeite	$Zn_3(PO_4)_24H_2O \rightleftharpoons 3Zn^{2+} + 2PO_4^{3-} + 4H_2O$	–35.3
Cornetite	$Cu_3PO_4(OH_3) \rightleftharpoons 3Cu^{2+} + PO_4^{3-} + 3OH^-$	–48.0[a]
$Cu_3(PO_4)_2$	$Cu_3(PO_4)_2\ (c) \rightleftharpoons 3Cu^{2+} + 2PO_4^{3-}$	–36.86
$Cu_3(PO_4)_2{\cdot}2H_2O$	$Cu_3(PO_4)_22H_2O\ (c) \rightleftharpoons 3Cu^{2+} + 2PO_4^{3-} + 2H_2O$	–38.76
$Cd_3(PO_4)_2$	$Cd_3(PO_4)_2\ (c) \rightleftharpoons 3Cd^{2+} + 2PO_4^{3-}$	–38.1

Source: Lindsay, W. L.: *Chemical Equilibria in Soils.* 1979. Copyright Wiley-VCH Verlag GmbH & Co. KGaA. Reproduced with permission.

[a] With kind permission from Springer Science+Business Media: *Phosphate Minerals,* Nriagu, J. O., Moore, P. B. (eds.), Formation and stability of base metal phosphates in soils and sediments, 1984, pp. 318–329, Nriagu, J. O.

phosphate-amended soils (Ryan et al. 2001). Numerous studies have proved that the residual fraction of Pb increased after application of different phosphates resulting from the formation of pyromorphites (Ma and Rao 1997; Cao et al. 2002; Scheckel et al. 2003; Zhu et al. 2004; Chen et al. 2006b; Abbaspour and Golchin 2011; Wu et al. 2013), and the identification of pyromorphites formed in phosphate-amended soils has also been carried out by different techniques, especially XRD, XAFS, and SEM-EDX (Cotter-Howells and Caporn 1996; Hettiarachchi et al. 2001; Cao et al. 2003, 2009; Scheckel and Ryan 2004; Hashimoto et al. 2009a, 2011; Kamiishi and Utsunomiya 2013).

It is clear from Table 8.1 that the affinity sequence for metal–phosphate binding is as follows: Pb ≫ Zn > Cu > Cd; however, it is unclear whether other metal phosphate minerals except pyromorphites form in general soil environments. For example, some studies suggested the formation of zinc phosphate minerals (Cotter-Howells and Caporn 1996; McGowen et al. 2001), while others thought that only zinc phosphate precipitates instead of minerals were generated (Cao et al. 2003, 2004). For Cu or Cd, it is more difficult to form phosphate minerals in most soils, which has been verified (McGowen et al. 2001; Cao et al. 2003, 2004; Raicevic et al. 2005).

Adsorption or Complexation of Metals on Phosphate Surfaces

The surface of insoluble phosphate materials such as apatite, HA, and PR can directly adsorb or complex with heavy metals, effectively removing contaminants from soil solution, but there are different contributions from the surface adsorption or complexation for different heavy metals. Cao et al. (2004) showed that the formation of fluoropyromorphite was primarily responsible for Pb immobilization by PR (up to 78.3%) with less contribution from the surface adsorption or complexation (21.7%) compared to 74.5% for Cu and 95.7% for Zn. Xu et al. (1994) also revealed that surface complexation and coprecipitation were the most important mechanisms for the immobilization of Zn and Cd by HA, and Cd coprecipitation appeared to be more significant than Zn coprecipitation.

Adsorption or complexation occurs on the HA (or PR) surface, partially displacing H^+, as shown in the following equations (Xu et al. 1994; Cao et al. 2004):

$$\equiv P - OH + M^{2+} \rightarrow \equiv P - OM^{+} + H^{+} \tag{8.3}$$

$$\equiv P - O^{-} + M^{2+} \rightarrow \equiv P - OM^{+} \tag{8.4}$$

$$\equiv Ca - OH + M^{2+} \rightarrow \equiv Ca - OM^{+} + H^{+} \tag{8.5}$$

In addition, metal retention by HA or PR is also controlled by HA/PR dissolution, followed by precipitation. For Pb, dissolved phosphate ions react with Pb^{2+} and pure hydroxypyromorphite or fluoropyromorphite formed (Equations 8.6 through 8.9) (Cao et al. 2004). But for Cd or Zn, coprecipitation of Cd^{2+} with Ca^{2+} occurred, resulting in the substitution of Ca by other cations in HA crystal structure (Equations 8.10 and 8.11) (Xu et al. 1994). Because cations whose ionic radii were smaller than Ca^{2+} (0.099 nm) may be incorporated in the apatite structure to a much lesser extent than those cations of larger ionic radii, Zn^{2+} (0.069 nm) may become much more difficult to coprecipitate with Ca^{2+} into apatite structure than Cd^{2+} (0.097 nm) (Xu et al. 1994). As such it should be more obvious for Pb^{2+} (0.119 nm) coprecipitation; that is, a significant proportion of Pb–Ca HA should be found. However, even at low Pb^{2+} concentration, the reaction product was pyromorphite and not Pb–Ca HA. The principal mechanism involved was still dissolution of apatite followed by precipitation of pyromorphite, as mentioned previously (Equations 8.6 through 8.9). In fact, Mavropoulos et al. (2002, 2004) demonstrated that the dissolution of HA was followed by the formation of $(Pb_xCa_{10-x})(PO_4)_6(OH)_2$, with Pb ions mostly occupying Ca (II) sites, and the Ca/Pb molar ratio of Pb–Ca HA decreased continuously until it reached the structure of a pure hydroxypyromorphite. Thus, this reinforced the above hypothesis that Pb has a higher probability of being incorporated into the HA structure by isomorphic substitution (coprecipitaiton), but Pb–Ca HA was evolved continuously during coprecipitation, where HA crystals were continuously dissolved and recrystallized in order to form more stable structures (hydroxypyromorphite) with higher Pb content.

$$\text{HA dissolution: } Ca_{10}(PO_4)_6(OH)_2\ (s) + 14H^{+} \rightarrow 10Ca^{2+} + 6H_2PO_4^{-} + 2H_2O \tag{8.6}$$

$$\text{PR dissolution: } Ca_{10}(PO_4)_6F_2\ (s) + 12H^+ \rightarrow 10Ca^{2+} + 6H_2PO_4^- + 2F^- \quad (8.7)$$

Hydroxypyromorphite precipitation:

$$10Pb^{2+} + 6H_2PO_4^- + 2H_2O \rightarrow Pb_{10}(PO_4)_6(OH)_2\ (s) + 14H^+ \quad (8.8)$$

Fluoropyromorphite precipitation:

$$10Pb^{2+} + 6H_2PO_4^- + 2F^- \rightarrow Pb_{10}(PO_4)_6F_2\ (s) + 12H^+ \quad (8.9)$$

Coprecipitation:

$$xCd^{2+} + (10-x)Ca^{2+} + 6H_2PO_4^- + 2H_2O \rightarrow (Cd_xCa_{10-x})(PO_4)_6(OH)_2\ (s) + 14H^+ \quad (8.10)$$

Coprecipitation:

$$xCd^{2+} + (10-x)Ca^{2+} + 6H_2PO_4^- + 2F^- \rightarrow (Cd_xCa_{10-x})(PO_4)_6F_2\ (s) + 12H^+ \quad (8.11)$$

However, the immobilization mechanisms of heavy metals in phosphate-amended soils are not very clear. Hamon et al. (2002) argued that until now it has been difficult to assess whether the amendment-induced decrease in metal availability is due simply to increased sorption of the metal or whether it is the result of surface precipitation or other fixation mechanisms, which has made it difficult to predict the long-term stability of such remedial treatments. It is apparent that we are still confused about whether metal-phosphate precipitation dominates the immobilization of heavy metals in contaminated agricultural soils where the contents of heavy metals are much less than that in smelting soils, or whether the immobilization of heavy metals by phosphate materials is essentially different in soils with different surface properties such as variable charge soils and permanent charge soils.

Effects of Different Phosphate Materials on the Mobility and Availability of Heavy Metals in Soils

Immobilization Effectiveness of Different Phosphate Materials

Although *in situ* chemical immobilization of heavy metals, especially Pb, in soils using phosphate materials as a widely accepted technology has been successfully applied to many contaminated soils (Table 8.2), it is obvious that different phosphate sources impact the immobilization effectiveness due to their different solubility, complexing capacity towards heavy metals, mechanism of

TABLE 8.2
Immobilization of Heavy Metals in Soils by Different Phosphate Sources

Amendment	Heavy Metal	Immobilization Effectiveness	Reference
Water-Soluble Phosphates			
DAP	Pb, Zn, Cd, As	Increased retardation (R) 2-fold for Cd, 6-fold for Zn, and 3.5-fold for Pb.	McGowen et al. (2001)
DAP	Pb, Zn, Cu	The concentrations in lettuce was reduced by 91% (Cu), 92% (Pb), and about 70% (Zn).	Khan and Jones (2009)
TSP	Pb	Did not reduce soil exchangeable and plant tissue Pb concentrations.	Aide et al. (2008)
TSP	Zn, Cd	Had no ameliorating effect; on the contrary, increased $BaCl_2$-extractable amounts of Zn.	Filho et al. (2011)
MCP	Pb, Zn, Cd, As	Immobilize Pb and Cd, whereas As and Zn were slightly mobilized. No effect on phytoaccumulation was observed.	Theodoratos et al. (2002)
MCP	Pb, Zn, Cd	Decreased Cd and Zn concentrations 2- and 3-fold, respectively, in leachates, whereas lead content was not significantly changed.	Kucharski et al. (2005)
PDP	Zn, Cd	Decreased solution concentrations of both Cd and Zn from 2.2 to 0.003, 465 to 0.64 mg/kg, respectively.	Hamon et al. (2002)
PDP	Cd	Increased Cd adsorption by soils and reduced Cd phytotoxicity.	Bolan et al. (2003b)
PA	Pb	Reduced bioaccessible Pb by 60%.	Yang et al. (2001)
PA	Pb	Reduced Pb bioaccessibility by 59%, from 1775 to 724 mg/kg.	Yang and Mosby (2006)
PA	Pb, Zn, Cu, Cd	Reduced $Ca(NO_3)_2$ extractable Zn (97%), Cd (99%), Cu (88%), and Pb (95%); decreased plant concentrations of Zn (92%), Cd (87%), Cu (43%), and Pb (93%).	Geebelen et al. (2006)

Water-Insoluble Phosphates			
Natural apatite	Pb, Zn, Cd	100% of Pb, 49% of Cd, and 29% of Zn added were attenuated.	Chen et al. (1997)
Synthetic apatite	Pb	Leachable Pb was reduced from 94 to 3.8 mg/L.	Liu and Zhao (2013)
Natural and synthetic apatite	Pb	Decreased shoot Pb concentration from 170 to 3 mg/kg.	Laperche et al. (1997)
Natural HA	Pb, Cd, Cr, Ni	Decreased bioavailability of Pb, Cd, Cr, and Ni around 22, 36, 26, and 39%, respectively.	Misra and Chaturvedi (2007)
Natural HA	Zn, Ni	Reduced leachable Zn in sandy soil (from 11% to 0.2%), but did not reduce the leachability of Ni.	Zupančič et al. (2006)
Natural HA	Pb, Cu, Zn, Cd, Cr, Ni, Mn, Fe	Immobilization rates were Pb (92.8%) > Cu (91.2%) > Cr (90.4%) > Cd (88.2%) > Fe (78.6%) > Ni (75.8%) > Zn (75.2%) > Mn (74.8%).	Misra and Pandey (2005)
Synthetic HA	Pb	Pyromorphite rose from 19 to 99% when HA/Pb molar ratio increased from 1 to 1000.	Arnich et al. (2003)
PR	Cu, Zn, Cd	Cd, Zn, and Cu concentration in leachates and shoots was decreased by 76%, 71%, and 68%, 63%, 40%, and 65%, respectively.	Elouear et al. (2014)
PR	Pb, Cu, Zn, Cd, Ni	Water-soluble metal reduction: Cu (35%), Cd (38%), Zn (42%), Ni (43%), Pb (66%).	Cajuste et al. (2006)
PR	Pb, Zn, Cd	Reduced uptake in shoots for Cd (19.6%–50%), Pb (21.9%–51.4%), and Zn (22.4%–34.6%).	Chen et al. (2006b)
PR	Pb, Cd, As	Reduced Pb (13.5 to 4.1 mg/kg) and Cd (13 to 3.45 mg/kg) but increased As (608 to 1046 mg/kg) in the fronds.	Fayiga and Ma (2006)
PR	Pb	Extractable Pb was reduced by 93% (85.2%–97.2%).	Lin et al. (2005)
PR	Pb, Cu, Zn	Sorption capacities of Pb, Cu, and Zn were 138, 114, and 83.2 mmol/kg, and surface adsorption/complexation of Pb, Cu, and Zn accounted for 21.7%, 74.5%, and 95.7%, respectively.	Cao et al. (2004)

(Continued)

TABLE 8.2 (CONTINUED)

Immobilization of Heavy Metals in Soils by Different Phosphate Sources

Amendment	Heavy Metal	Immobilization Effectiveness	Reference
Water-Insoluble Phosphates			
PR	Pb, Zn, Cd	Reduced GI-available Pb in both gastric and intestinal solutions (23% and 92%), but did not reduce Zn and Cd phytoavailability.	Basta et al. (2001)
PR	Pb	Nonresidual Pb fraction reduced from 10% to 96%.	Ma and Rao (1997)
Bone meal	Cu, Zn, Ni	Reduced metal release and leachate toxicity, but increased the concentration of DTPA-extractable Zn and Cu.	Hodson et al. (2001)
Bone meal	Pb, Zn, Cd	Leachate Pb, Zn, and Cd concentrations were reduced 57%–67%, 26%–85%, and 88%–95%.	Sneddon et al. (2006)
Bone meal	Pb	Survival times of *E. fetida* increased from 67 h to more than 168 h; LC50 and EC50 increased from 4379 to 5203 μg/g, 1408 to 3334 μg/g, respectively.	Davies et al. (2002)
Mixed Phosphates			
PR, TSP, PR+TSP	Pb, Cu, Zn	Reduced leachability of Pb, Cu, and Zn by up to 89.2%, 24.4%, and 34.3 %, respectively (TSP > TSP+PR > PR).	Cao et al. (2013)
TSP, DAP, PR	Cd	Leachable concentrations of Cd reduced from 306 mg/kg to 140 (PR), 34 (DAP) and 12 mg/kg (TSP) (TSP > DAP > PR).	Thawornchaisit and Polprasert (2009)
Synthetic HA, PR	Co, Ni	Reduced water-soluble concentrations of Co and Ni by about 99% (HA > PR).	Mignardi et al. (2013)

PR, CMP, SSP	Pb, Zn, Cd	Decreased water-soluble and extractable (WE) Pb (22%–81.4%), Cd (1.5%–30.7%), and Zn (11.7%–75.3%); reduced cabbage uptake of Pb (16%–58%), Cd (16.5%–66.9%), and Zn (1.2%–73.2%).	Wang et al. (2008)
Natural HA, PR, TSP, DAP	Pb, Zn, Cd	HA was superior to others and reduced Pb, Zn, Cd uptake by 34%–53%, 31%–47%, and 39%–42%.	Chen et al. (2007)
PA, PR, PA+PR	Pb, Cu, Zn	Reduced Pb water solubility, phytoavailability, and bioaccessibility by 72%–100%, 15%–86%, and 28%–92%; reduced Cu and Zn water solubility by 31%–80% and 40%–69%, but had little effect on the Cu and Zn phytoavailability.	Cao et al. (2009)
K_2HPO_4, $Ca(H_2PO_4)_2$, $Ca_3(PO_4)_2$, $Ca_5(PO_4)_3OH$	Pb, Zn, Cu, Cd, As	Decreased WE and TCLP Pb, Zn, Cu, and Cd, but increased WE and TCLP As; $Ca_5(PO_4)_3OH$ was a better choice to immobilize Pb, Zn, Cu, and Cd.	Wu et al. (2013)
Synthetic HA, MCP, bone meal	Zn, Ni	MCP increased Zn and Ni mobility, whereas HA decreased their mobility; bone meal increased water-soluble Zn and Ni but decreased acetic acid leachable Zn and Ni.	Zupančič et al. (2004)
PA, PA+MCP, PA+PR	Pb	Residual Pb increased by 19%–48% (PA), 22%–50% (PA+MCP), 11%–55% (PA+PR).	Cao et al. (2002)
PDP, TSP	Pb, Zn, Cd	Decreased earthworm bioavailability of Pb (79%), Zn (60%), and Cd (57%) in AR soil; in USDA soil the reduction was 53% (Pb) and 38% (Zn) without reduction for Cd.	Maenpaa et al. (2002)

interaction between P and metal, and soil environment. For example, it was demonstrated that among various phosphate sources used in selected research trials, the most effective amendment was HA (Zhu et al. 2004; Chen et al. 2007; Mignardi et al. 2013), PR (Hettiarachchi et al. 2001; Fang et al. 2012), DAP (Basta and McGowen 2004; Abou Zied 2007), TSP (Thawornchaisit and Polprasert 2009), calcium magnesium phosphate (CMP) (Wang et al. 2008), and PA (Brown et al. 2004), respectively. This means that there is a complex interaction between P and heavy metals depending on the properties of phosphate materials, heavy metals, and soils.

Overall, soluble phosphate materials could provide an abundance of solution phosphorus and increase the efficiency of metal phosphate precipitate/mineral formation (McGowen et al. 2001; Basta and McGowen 2004). PA not only provided solution phosphorus to immobilize heavy metals but also dissolved existing soil minerals (Fe and Mn oxides or carbonate phases) of heavy metals, and thus accelerated their transformation into metal phosphate precipitates/minerals. Because of this, soluble phosphate materials including PA have been widely used to immobilize heavy metals in soils (Table 8.2). Extreme caution should be taken, however, when applying these highly soluble phosphates because of potential environmental risk such as soil acidification (Hettiarachchi et al. 2001; McGowen et al. 2001; Theodoratos et al. 2002; Basta and McGowen 2004) and eutrophication associated with P leaching (Yang et al. 2002; Basta and McGowen 2004). In contrast, application of insoluble phosphate materials such as HA and PR to metal-contaminated soils may reduce this risk (Mignardi et al. 2012). Besides being the main P sources, HA and PR could also act as liming materials due to containing some $CaCO_3$ (Höweler and Woodruff 1968) and to neutralizing soil acidity in dissolution process (Equations 8.6 and 8.7). Insoluble phosphates are more suitable for the immobilization of heavy metals in acidic soils than that in alkaline or calcareous soils where the immobilization is relatively slow because their dissolution is limited (Zhang and Ryan 1998; Chrysochoou et al. 2007).

In general, in order to minimize environmental risk and maximize immobilization effectiveness, a more realistic approach may be to combine soluble phosphates with insoluble phosphates for field application to heavy-metal-contaminated soils so that soluble phosphates could quickly reduce the concentration of heavy metals to below an acceptable level, while insoluble phosphates would supply constant P sources in soils to react with heavy metals for long-term immobilization, as suggested by Cao et al. (2002, 2003). In addition, coapplication of liming materials with soluble phosphate fertilizers such as DAP and PA in nonalkaline soils may also be necessary to offset potential soil acidification (McGowen et al. 2001; Yang et al. 2002; Basta and McGowen 2004). However, Chrysochoou et al. (2007) argued that the use of lime to restore soil pH in acidified soil treatments inhibited further conversion from Pb associated mainly with oxides or carbonates to pyromorphite. It is worth noting that application of biogenic phosphate sources such as bone meal may be a more environmentally sustainable approach to immobilize heavy metals in soils in view of much lower metal concentrations and appreciably more solubility than mined HA or PR (Knox et al. 2006; Chrysochoou et al. 2007).

The optimum application rate for the immobilization of heavy metals in soils has not been specifically studied regardless of phosphate sources. Based on the stoichiometric P to Pb ratio of pyromorphites, a P/M_{total} molar ratio of 3/5, where M_{total} is the sum of the molar content of heavy metals in soils, has been suggested by many researchers (McGowen et al. 2001; Basta and McGowen 2004; Cao et al. 2008; Hashimoto et al. 2009a,b; Szrek et al. 2011). However, a higher P/M_{total} molar ratio was often used to ensure sufficient solution P and thus high immobilization efficiency of heavy metals in soils, such as 2.0 (Thawornchaisit and Polprasert 2009; Liang et al. 2012; Cao et al. 2013), 3.0 (Basta et al. 2001), 4.0 (Cao et al. 2002, 2003; Melamed et al. 2003), 5.0 (Ryan et al. 2001; Chen et al. 2006b), and even more than 10 (Yang et al. 2001). Such a high application rate of phosphates would undoubtedly bring about a series of environmental problems, so the focus should be on both the immobilization efficiency and the environmental risk to determine the optimum application rate of phosphates. For example, McGowen et al. (2001) and Basta and McGowen (2004) found that application of DAP to smelter-contaminated soils at a P/M_{total} molar ratio of 1/15 was not only the most effective for immobilizing heavy metals but could also cause the minimum release of arsenic (As) and P from the contaminated soils. Hence, it was possible to reduce the application rate of phosphate materials in actual remediation.

In addition, because the application of phosphates does not change the total concentrations of heavy metals in soils, it is important to investigate the long-term stability of the immobilization of heavy metals in phosphate-amended soils. With time, continued P consumption by plants may cause the decomposition or activation of heavy metal phosphate precipitates/minerals previously formed; biogeochemistry changes in phosphate-amended soils may also affect the stability of heavy metal immobilization. Houben and Sonnet (2012) found that plants by organic root exudates and secretions could enhance metal release into the environment by accelerating mineral weathering. They therefore suggested that the ability of plants to alter metal phases in soils should be further taken into account. Hashimoto et al. (2009a) provided evidence that organically complexed Pb dominated and was persistent in phosphate-amended soils, indicating that an enriched soil organic carbon may be an inhibitory factor for pyromorphite transformations. Zhang and Zhang (2010) also showed that relatively high rates of P applications could induce a marked increase in dissolved organic matter (DOM) concentrations in the leachates, and thus increase the leaching of Cu, Cd, and Zn from soils due to the formation of metal–DOM complexes. In a word, chemical amendments could introduce elements capable of inducing or affecting the activities of certain lithotrophic microbes that could influence vital geochemical processes such as mineral dissolution and formation, weathering, and organic matter mineralization (Udeigwe et al. 2011), thereby affecting the stability of heavy metal immobilization. Thus, the long-term stability of the immobilized heavy metals in phosphate-amended soils needs to be monitored. Although the immobilization in Pb-contaminated soils using phosphates showed a long-term stability (Cao et al. 2002) without an adverse effect on the environment (Tang et al. 2009), little is known about the long-term stability of phosphate-induced immobilization of heavy metals especially Cu, Zn, and Cd, because very little research has been done to study the effect of time on mobility and availability of heavy metals immobilized in soils.

Mobility and Phytoavailability of the Most Common Heavy Metals in Phosphate-Amended Soils

As shown in Table 8.2, phosphate materials were more widely used to immobilize Pb-contaminated soils, followed by Cd immobilization. There were remarkable differences in the immobilization effectiveness of different heavy metals due to different interaction mechanisms between phosphate and metal. Generally Pb mobility and phytoavailability were reduced markedly, whereas As mobility and phytoavailability were enhanced by phosphate addition; the complex changes in the mobility and phytoavailability of other metals such as Cu, Zn, and Cd occurred. Next, we describe in detail each main heavy metal mobility/phytoavailability affected by phosphate.

Pb

Pb immobilization using different phosphate amendments has been performed by numerous researchers and reviewed by Melamed and Ma (2008) and Miretzky and Fernandez-Cirelli (2008). It is generally accepted that the immobilization effectiveness of Pb by phosphates is very high, with marked reduction in mobility, bioavailability, and toxicity, attributed to the formation of lowly soluble and highly stable pyromorphites. For example, Wang et al. (2008) reported water soluble and exchangeable (WE), TCLP, and plant uptake Pb in the field could be reduced 81.4%, 71.2%, and 58%, respectively; Fang et al. (2012) also showed that P treatments could reduce $CaCl_2$-extractable Pb by 73.1%, dissolved and total leachable Pb by 81.9% and 64.5%, respectively, and plant uptake of Pb by 65.1%. Cao et al. (2002) ascribed the increase in residual Pb (by 19%–48% for PA, 22%–50% for PA+MCP, and 11%–55% for PA+PR) to P-induced formation of chloropyromorphite, identified by XRD, SEM-EDX, TEM, and chemical speciation model.

However, Aide et al. (2008) showed that TSP treatments did not reduce soil exchangeable Pb and plant tissue Pb concentrations, presumably due to sufficient soil Ca that preferentially reacted with TSP. Porter et al. (2004) also concluded that P-added soil could easily form other minerals such as apatite and manganese hydrogen phosphate ($MnHPO_4$) rather than pyromorphites. Thus, they suggest that the only way around this is to add sufficient phosphate to ensure that all the calcium and manganese will change to apatite and

$MnHPO_4$, respectively. According to these researchers, the use of phosphate materials to alleviate Pb may need to be restricted to soil environments that have low Ca activities in addition to the very high concentrations of needed P.

In addition, Hashimoto et al. (2009a) considered that an enriched soil organic carbon may be an inhibitory factor for pyromorphite transformation because organically complexed Pb predominated in soils. Further, Lang and Kaupenjohann (2003) found that organic matter not only strongly inhibited the precipitation of chloropyromorphite at pH ≤ 4 but also markedly diminished the particle size by blocking the surfaces of crystal seeds and impairing crystal growth. That is, organic matter might limit the potential of phosphate to immobilize Pb in soils because it favored the formation of mobile colloids. In short, after application of phosphates, the mobility and phytoavailability of Pb in high organic matter soils may not reduce significantly or even increase due to a marked increase in DOM induced by phosphates, and thus the formation of metal–OM complexes (Zhang and Zhang 2010), which should be carefully considered and avoided.

Cd

Phosphate materials could reduce leachable concentration of Cd from 306 to 12 mg/kg (96%) (Thawornchaisit and Polprasert 2009) and reduce $Ca(NO_3)_2$-extractable Cd and plant uptake of Cd by 99% and 87%, respectively (Geebelen et al. 2006), suggesting that phosphate may immobilize Cd in soils. However, unlike Pb, as pyromorphites immobilized, Cd was predominantly present on particle surfaces where induced-adsorption or coadsorption/coprecipitation occurred, resulting in the formation of mineral–phosphate–Cd ternary surface complex (Agbenin 1998; Bolan et al. 1999; Elzinga and Kretzschmar 2013) or Cd–Ca HA coprecipitate $[(Cd_xCa_{10-x})(PO_4)_6(OH)_2]$ (Xu et al. 1994). Therefore, the immobilization of Cd by phosphate reflected weaker stability and larger variability than that of Pb, depending on soil pH, organic matter, competition cations, and so forth.

Wang et al. (2008) showed that although three commercial phosphate fertilizers in field could reduce significantly WE Cd (1.5%–30.7%) and plant uptake of Cd (16.5%–66.9%), Cd uptake by cabbage was negatively correlated with soil pH rather than with WE Cd, indicating that the impact of phosphate application on Cd availability was complex. It is possible that it was not direct P–Cd interaction but the indirect effect from the P-induced changes of soil environments that dominated the availability of Cd since no Cd–phosphate precipitate/mineral was observed in most soils. First, some phosphate fertilizers, especially DAP and MCP, notably reduced soil pH (e.g., from 7.84 to 6.07) (McGowen et al. 2001; Theodoratos et al. 2002; Basta and McGowen 2004), probably causing the increase in Cd mobility/bioavailability. Wang et al. (2006) reported that shoot Cd concentration in high Cd soil linearly increased with decreasing soil pH, whereas in low Cd soil, it increased with decreasing pH from pH 7.27 to 6.07, and then rapidly decreased in the lower pH treatment. Gray et al. (1999) found that increasing soil pH from 5.5 to 7.0 significantly decreased Cd concentrations in clover, lettuce, carrot, and ryegrass, and to a lesser extent in wheat, although the magnitude of the reduction varied between plant species and soil types. It appears that pH may be a powerful tool in the management of plant Cd concentrations. Second, the competition of accompanying ions such as K^+, NH_4^+, and Ca^{2+} from high rates of phosphate application would also lead to displacement of Cd from adsorption sites, thereby increasing bioavailability. Hayakawa et al. (2011) observed that Ca could promote root uptake of Cd by *Gamblea innovans* but inhibited the transport of Cd from roots through the stem to leaves. However, Lavoie et al. (2012) found that a 50-fold increase in free Ca^{2+} effectively offered some protection and thus significantly decreased Cd uptake and toxicity to green alga (*Chlamydomonas reinhardtii*). Third, DOM induced by phosphates has been shown to increase Cd mobility through the formation of Cd–DOM complex (Zhang and Zhang 2010). Of course, application of phosphate fertilizers led to an increase in crop biomass production and a corresponding increase in soil organic matter, thus enhancing Cd adsorption and complexation.

Cu and Zn

Unlike Pb immobilization where stable pyromorphites formed, and unlike Cd immobilization where P-induced adsorption or coadsorption/coprecipitation occurred, the interactions between P and Cu/Zn were more complex, so the effect of phosphate addition on their mobility and phytoavailability showed different trends that were dependent on soil properties, especially pH and organic matter, and on the types and characteristics of phosphate materials. For example, Geebelen et al. (2006) showed that the extractability for Cu and Zn was reduced by 88% and 97%, respectively, and their phytoavailability was reduced by 43% and 92%, respectively. Cao et al. (2009) found that water solubility of Cu and Zn was reduced by 31%–80% and 40%–69%, respectively, but phosphate had little effect on their phytoavailability. Zupančič et al. (2004) observed that MCP increased Zn mobility, whereas HA decreased mobility, and bone meal increased water-soluble Zn but decreased acetic acid leachable Zn. Spuller et al. (2007) showed that application of phosphate (DAP and MCP) increased Cu mobility and phytotoxicity in shooting range soil.

Compared to Cu–P interaction, with Zn, it was much easier to form phosphate precipitate/mineral such as hopeite (Cotter-Howells and Caporn 1996; McGowen et al. 2001; Lee et al. 2005; Baker et al. 2012; Betts et al. 2013), so the direct effect of phosphates on Zn was greater than on Cu. However, it was well known that there was high affinity of Cu for organic matter. In the presence of DOM, Cu mobility/availability increased as Cu–DOM complexes were formed (Temminghoff et al. 1997; Zhou and Wong 2001); on the other hand, Cu could complex with soil organic matter where stable chelate ring structures formed (Karlsson et al. 2006; Manceau and Matynia 2010) or stable SOM–Cu–mineral (A type) and Cu–SOM–mineral (B type) ternary complexes formed (Liu and Gonzalez 1999; Alcacio et al. 2001; Sheals et al. 2003; Boudesocque et al. 2007), decreasing significantly its mobility/availability. As mentioned in the "Cd" section, application of phosphates would increase DOM as well as SOM, certainly increasing or decreasing or uninfluencing Cu phytoavailability. In general, the indirect effect of phosphates on Cu mobility/availability should be greater than on Zn.

As

Due to strong competition between As and P ions, a high rate of phosphate application would increase As mobility and phytoavailability (McGowen et al. 2001; Theodoratos et al. 2002; Fayiga and Ma 2006; Wu et al. 2013). Therefore, *in situ* chemical immobilization using phosphate addition must be very carefully used to remediate As-contaminated soils. Unfortunately, most Pb/Zn mining/smelter-contaminated soils were usually accompanied by high As content, so the immobilization of these contaminated soils by phosphates would also need to include steps that would minimize As mobility/availability. To address this potential problem, a small amount of insoluble phosphate materials could be applied to these contaminated soils containing high As. As an alternative, application of phosphates in combination with an As hyperaccumulator plant such as *Pteris vittata* L. is an effective remediation measurement to these soils, where on the one hand the mobility and thus bioavailability of Pb, Zn, and Cd were greatly reduced by phosphate, while on the other hand, the P-induced increase in mobility and phytoavailability of As promoted significantly the uptake of As by *Pteris vittata* L. (Fayiga and Ma 2006).

Environmental Risks of the Immobilization of Heavy Metals in Phosphate-Amended Soils

When phosphate materials are introduced into soils for the immobilization of heavy metals, some environmental risks associated with them might occur, such as the accumulation of contaminants from phosphate materials in soils, the eutrophication of water bodies deriving from P leaching, the nutritional imbalance and/or toxicity of As to plants caused by phosphate, and soil acidification.

Accumulation of Heavy Metals in Soils

Phosphate fertilizers around the world generally contain a concentration of most heavy metal(loid)s such as As, Cd, chromium (Cr), Cu, mercury (Hg), Ni, Pb, and Zn. Thus, long-term and frequent application of phosphate fertilizers could result in the accumulation of significant concentrations of some heavy metal(loid)s, especially Cd and Zn in soils. However, it has been observed even in Europe that the input of heavy metals from phosphate fertilizers was much less than that from atmospheric deposition, resulting in improved air quality (Table 8.3), implying that under normal cropping practice, the impact of phosphate fertilizer application on trace element accumulation in receiving soils is not as serious as it might appear. Jiao et al. (2012) also thought the effect of long-term application of phosphate fertilizers on the accumulation of heavy metals in soils has been limited and localized.

Regarding the immobilization of heavy metals in soils by phosphate addition, we must focus on the accumulation of contaminants from phosphate materials in soils by virtue of their high levels of addition. Generally the optimum application rate was a P/M_{total} molar ratio of 3:5 (McGowen et al. 2001; Basta and McGowen 2004; Cao et al. 2008; Hashimoto et al. 2009a,b; Szrek et al. 2011). According to this, we calculate the application rates of phosphates such as DAP (23% P) to be 11,179 kg/ha in a contaminated paddy soil in Zhejiang, China (Pb: 388 mg/kg, Zn: 2286 mg/kg, Cd: 1.6 mg/kg, Cu: 1481 mg/kg) (Zhang et al. 2005), which was 10 times more than the normal amount of DAP applied in China. Further, the accumulation rates of As, Cd, Cr, Cu, Hg, Ni, Pb, and Zn in soil were evaluated to be 176.6, 13.4, 958, 496.3, 9.8, 127.4, 428.2, and 1483.4 g/ha, respectively, based on their average content in phosphates (Luo et al. 2009), which was also 70 times as high as the values in Table 8.3. For heavier smelter-contaminated soil (Cd: 2810 mg/kg; Cu: 455 mg/kg; Pb: 2770 mg/kg; Zn: 45400 mg/kg) (Wei et al. 2014), the application rate up to 137.5 t/ha would be needed, which was 10 and 100 times higher than that in contaminated paddy soil and normal field soil. Obviously, the accumulation of contaminants from phosphate fertilizers in soils would be very serious (almost 1000 times as high as that under normal application of phosphate fertilizers) when they were used to immobilize heavy metals in mining/smelter soils.

P Leaching, Nutrition Imbalance, and Soil Acidification in Phosphate-Amended Soils

Basta and McGowen (2004) showed that total P eluted was only 2.31 mg (< 1% of the total added) for the 10-mg DAP/kg treatment and 335 mg (10.5% of the total added) for the 90-mg DAP/kg treatment. That is, as P application increased, the mass of P eluted from the treated soils increased remarkably. Wei (2010) illustrated that leachable P increased with increasing pore volume for soluble phosphate treatment (TSP), but in insoluble phosphate treatment (PR), it was negligible and was not affected by pore volume. Further, Yang et al. (2002) found that although surface H_3PO_4 treatment resulted in an enhanced leaching of added P, the P leaching under field conditions was very limited (7.2% added P was leached below 30-cm depth). As well, a high rate of phosphate application, especially soluble phosphate materials, may increase the potential risk of surface and groundwater pollution derived

TABLE 8.3

Heavy Metal Addition Rates (g/ha/yr) to Agricultural Soils from Phosphate Fertilizers and Atmospheric Deposition

	Phosphate Fertilizers								Atmospheric Deposition							
Country	**As**	**Cd**	**Cr**	**Cu**	**Hg**	**Ni**	**Pb**	**Zn**	**As**	**Cd**	**Cr**	**Cu**	**Hg**	**Ni**	**Pb**	**Zn**
China[a]	2.4	0.2	13.3	6.8	0.1	1.8	5.8	20.5	27.7	3.7	60.5	106.5	1.5	58.6	196.7	644.2
England and Wales[b]	1.1	1.6	17	4.9	<0.1	3.3	0.5	34	3.1	1.9	7.5	57	1.0	16	54	221
Europe (average)[c]	2.3	1.6	20.7	—	—	3.6	1.0	43.1	2.0	1.9	9.3	—	—	10	38	227
Canada[d]	0.65	0.43	—	0.43	—	—	0.3	5.6	0.05	0.06	—	2.2	—	—	0.36	2.6

[a] Luo, L. et al., *Journal of Environmental Management*, 90, 2524–2530, 2009.
[b] Nicholson, F. A. et al., *Science of the Total Environment*, 311, 205–219, 2003.
[c] Nziguheba G. and Smolders, E., *Science of the Total Environment*, 390, 53–57, 2008.
[d] Sheppard, S. C. et al., *Journal of Environmental Quality*, 38, 919–932, 2009.

from P leaching, which should be noted when phosphate sources are used to immobilize heavy metals in soils.

Boisson et al. (1999) showed that although concentrations of toxic metals in plant leaves decreased after HA application, the uptake of essential trace elements also decreased and probably led to Mn deficiency in corn. In addition, As uptake was found to increase after HA treatment. Theodoratos et al. (2002) also found that phosphate treatment (MCP) had no effect on the uptake of contaminants in plant tissues and in fact had a pronounced negative effect on plant growth, which was combined with a strong deficiency of Ca in the tissue of leaves. These results illustrate that phosphate application for the remediation of heavy-metal-contaminated soils may be effective but not sufficient enough to both reduce pollution and provide food security. During the immobilization of heavy metals using phosphate, nutrition imbalance problems and As toxicity problems may become prominent due to increased PO_4^{2-} concentration in soils, which results in essential nutrient deficiency and increased transfer of As to plants and animals and increases its percolation to ground water (Boisson et al. 1999).

Soil acidic conditions are necessary to provide highly soluble P and dissolve Pb^{2+} from existing minerals and thus successfully transform it in pyromorphites (Equations 8.6 through 8.9), and therefore application of H_3PO_4 or mixed H_3PO_4 with PR is regarded as one of the most effective amendments for immobilization of heavy metals, especially Pb, in soils (Yang et al. 2001, 2002; Cao et al. 2002, 2009; Geebelen et al. 2006). As well, some phosphate fertilizers such as DAP and MCP could also significantly reduce soil pH (McGowen et al. 2001; Theodoratos et al. 2002; Basta and McGowen 2004). Lower pH is beneficial to the formation of heavy metal phosphate precipitates/minerals, but also can increase the mobility and bioavailability of many potentially phytotoxic metals including Mn and Al (Watmough et al. 2007). In addition, after soil acidification, soil microbial activity decreases significantly, leading to lower soil respiration, slower decomposition of OM, and weaker mineralization and fixation of N (Liang et al. 2013). Thus, chemical immobilization of heavy metals using phosphate addition should avoid excessively reducing soil pH and thus deteriorating soil properties. Of course, some phosphate materials could increase soil pH (Boisson et al. 1999; Hodson et al. 2001; Bolan et al. 2003b), which probably increases the mobility of As. In short, appropriate measures must be taken to avoid or alleviate side effects of great pH changes associated with a high rate of amended phosphate on physical, chemical, and biological properties of soils during phosphate-immobilized heavy metals. As a result, it is necessary to regularly monitor soil pH, phosphorus, and contaminant levels in the whole remediation period.

In fact, numerous abiotic and biotic reactions associated with soil chemistry, soil ecology, and soil biology occur after phosphate sources are amended to contaminated soils, where unwanted side effects could arise, as mentioned in this section (e.g., P leaching, As toxicity, nutrition imbalance and soil acidification). The most common situation is that reactions may fix an element and result in unexpected changes in the exposure to another element

(Mench et al. 2006). Most research only focuses on major contaminants and thus ignores those unwanted side effects. Therefore, the use of phosphate amendments should be comprehensively evaluated with respect to potential side effects on, for example, the mobility of P and As, the immobilization of essential trace elements such as Ca, microbial mineralization, and soil acidification, as well as the immobilization effectiveness of contaminants and the cost, which should be of concern. More importantly, high application rates or repeated applications are usually chosen to maintain sufficient solution phosphorus to immobilize heavy metals, potentially increasing the risk of secondary contamination such as eutrophication of the water body, As leaching, and heavy metal accumulation in plants and soils. In this case, the type and rate of phosphate materials and also application method should be carefully studied to maximize the immobilization effectiveness and minimize the side effects.

Conclusions

Phosphate could effectively reduce the mobility, phytoavailability, and toxicity of heavy metals, and therefore phosphate materials as good amendments are usually used to immobilize heavy metals in contaminated soils. There are three main interaction mechanisms between P and metal: (1) phosphate-induced adsorption of metal ions, (2) the formation of metal phosphate precipitates or minerals, and (3) phosphate surface adsorption or complexation of metals or coprecipitation by isomorphous substitution for Ca^{2+} in apatite structure. In general, Pb immobilization proceeds through a pyromorphite precipitation mechanism, whereas phosphate-induced adsorption and coprecipitation are responsible for Cd immobilization. The immobilization of Cu and Zn by phosphate is complex, where the stability of Cu (or Zn)-phosphate precipitates is much weaker than that of pyromorphites, and where Cu and Zn become much more difficult to coprecipitate with Ca than Cd.

Because of very complex interactions between P and metal, phosphate does not always reduce the mobility, phytoavailability, and toxicity of heavy metals in different soils with different compositions and extent of contamination. Sometimes phosphate only reduces metal solubility/mobility but does not affect its phytoavailability, and sometimes phosphate even increases metal mobility and phytotoxicity. In particular, a high rate of phosphate addition, without exception, notably increases As mobility and toxicity in soils due to the strong competition between P and As. Therefore, phosphate materials should be carefully used to remediate heavy-metal-contaminated soils, especially those containing high As concentration. Further, it is emphasized that some environmental risks might occur during the immobilization of heavy metals in soils using phosphate sources in view of a very high application rate of phosphates (several tens of times higher than normal agricultural practice), including the accumulation of contaminants from phosphate materials in soils, P and As leaching into water bodies, nutrition imbalance and essential nutrient deficiency, As toxicity to plants and animals, Al and Mn toxicity derived from soil acidification induced by a high rate of phosphate fertilizers, and so forth. Thus, to maximize the immobilization effectiveness of heavy metals in soils and minimize the side effects that most likely appear during the remediation of soil contaminants by amended phosphate, the type and rate of phosphate sources and also application method should be carefully studied. In particular, the *in situ* chemical immobilization of heavy-metal-contaminated soils using phosphate addition should be comprehensively evaluated.

Two main points should be considered in future research. First, future studies should focus on the transformation and transport of heavy metals in soil–plant systems after phosphate addition by using and integrating various evaluation methods such as SEP, chemical speciation models, and spectroscopic and microscopic techniques, especially XAFS and isotope techniques, in order to distinguish the interaction mechanisms between P and metal, and thus quantitatively evaluate the different immobilization mechanisms in heavy-metal-contaminated soils with different surface properties. Second, risk assessment to predict the effects of a high rate of phosphate amended on the environment should be strengthened. The focus should be on the long-term solubility and mobility of heavy metals immobilized by phosphate and on the long-term *in situ* immobilization of heavy metals by phosphate addition in field-contaminated soils.

References

Abbaspour, A., Golchin, A. 2011. Immobilization of heavy metals in a contaminated soil in Iran using di-ammonium phosphate, vermicompost and zeolite. *Environmental Earth Sciences*, 63, 935–943.

Abou Zied, M. M. A. 2007. Reducing bioavailability of some heavy metals in a contaminated soil using phosphate amendments. *Egyptian Journal of Soil Science*, 47, 9–22.

Adhikari, T., Singh, M. V. 2008. Remediation of cadmium pollution in soils by different amendments: A column study. *Communications in Soil Science and Plant Analysis*, 39, 386–396.

Agbenin, J. O. 1998. Phosphate-induced zinc retention in a tropical semi-arid. *European Journal of Soil Science*, 49, 693–700.

Aide, M., Whitener, K., Westhoff, E., Kelley, J. 2008. Effectiveness of triple superphosphate amendments in alleviating soil lead accumulation in Missouri Alfisols. *Soil and Sediment Contamination*, 17, 630–642.

Alcacio, T. E., Hesterberg, D., Chou, J. W., Martin, J. D., Beauchemin, S., Sayers, D. E. 2001. Molecular scale characteristics of Cu (II) bonding in goethite–humate complexes. *Geochimica et Cosmochimica Acta*, 65, 1355–1366.

Al-Hwaiti, M., Tardio, J., Reynolds, H., Bhargava, S. 2014. Selectivity assessments of a sequential extraction procedure for potential trace metals' mobility and bioavailability in phosphate rocks from Jordan phosphate mines. *Soil and Sediment Contamination*, 23, 417–436.

Alloway, B. J. 1990. *Heavy Metals in Soils.* Blackie Academic & Professional, Glasgow, Scotland.

Anderson, M. A., Rubin, A. J. 1981. *Adsorption of Inorganics at Solid-Liquid Interfaces.* Ann Arbor Science Publishers, Ann Arbor, MI.

Arnich, N., Lanhers, M. C., Laurensot, F., Podor, R., Montiel, A., Burnel, D. 2003. In vitro and in vivo studies of lead immobilization by synthetic hydroxyapatite. *Environmental Pollution*, 124, 139–149.

Baker, L. R., Pierzynski, G. M., Hettiarachchi, G. M., Scheckel, K. G., Newville, M. 2012. Zinc speciation in proximity to phosphate application points in a lead/zinc smelter-contaminated soil. *Journal of Environmental Quality*, 41, 1685–1873.

Basta, N. T., Gradwohl, R., Snethen, K. L., Schroder, J. L. 2001. Chemical immobilization of lead, zinc, and cadmium in smelter-contaminated soils using biosolids and rock phosphate. *Journal of Environmental Quality*, 30, 1222–1230.

Basta, N. T., McGowen, S. L. 2004. Evaluation of chemical immobilization treatments for reducing heavy metal transport in a smelter-contaminated soil. *Environmental Pollution*, 127, 73–82.

Betts, A. R., Chen, N., Hamilton, J. G., Peak, D. 2013. Rates and mechanisms of Zn^{2+} adsorption on a meat and bonemeal biochar. *Environmental Science & Technology*, 47, 14350–14357.

Bi, X. L., Pan, X. B., Zhou, S. W. 2013. Soil security is alarming in China's main grain producing areas. *Environmental Science & Technology*, 47, 7593–7594.

Boisson, J., Ruttens, A., Mench, M., Vangronsveld, J. 1999. Evaluation of hydroxyapatite as a metal immobilizing soil additive for the remediation of polluted soils. Part 1. Influence of hydroxyapatite on metal exchangeability in soil, plant growth and plant metal accumulation. *Environmental Pollution*, 104, 225–233.

Bolan, N. S., Adriano, D. C., Naidu, R. 2003a. Role of phosphorus in (im)mobilization and bioavailability of heavy metals in the soil-plant system. *Reviews of Environmental Contamination and Toxicology*, 177, 1–44.

Bolan, N. S., Adriano, D. C., Duraisamy, P., Mani, A., Arulmozhiselvan, K. 2003b. Immobilization and phytoavailability of cadmium in variable charge soils. I. Effect of phosphate addition. *Plant and Soil*, 250, 83–94.

Bolan, N. S., Khan, M. A. R., Tillman, R. W., Naidu, R., Syers, J. K. 1999. The effects of anion sorption on sorption and leaching of cadmium. *Australian Journal of Soil Research*, 37, 445–460.

Bolland, M. D. A., Posner, A. M., Quirk, J. P. 1977. Zinc adsorption by goethite in the absence and presence of phosphate. *Australian Journal of Soil Research*, 15, 279–286.

Boudesocque, S., Guillon, E., Aplincourt, M., Marceau, E., Stievano, L. 2007. Sorption of Cu (II) onto vineyard soils: Macroscopic and spectroscopic investigations. *Journal of Colloid and Interface Science*, 307, 40–49.

Brown, S., Chaney, R., Hallfrisch, J., Ryan, J. A., Berti, W. R. 2004. *In situ* soil treatments to reduce the phyto- and bioavailability of lead, zinc, and cadmium. *Journal of Environmental Quality*, 33, 522–531.

Cajuste, L. J., Cajuste, L., Garcia, O. C., Cruz, D. J. 2006. Distribution and availability of heavy metals in raw and acidulated phosphate rock-amended soils. *Communications in Soil Science and Plant Analysis*, 37, 2541–2552.

Cao, X. D., Liang, Y., Zhao, L., Le, H. Y. 2013. Mobility of Pb, Cu, and Zn in the phosphorus-amended contaminated soils under simulated landfill and rainfall conditions. *Environmental Science and Pollution Research*, 20, 5913–5921.
Cao, X. D., Ma, L. Q., Chen, M., Singh, S. P., Harris, W. G. 2002. Impacts of phosphate amendments on lead biogeochemistry at a contaminated site. *Environmental Science & Technology*, 36, 5296–5304.
Cao, R. X., Ma, L. Q., Chen, M., Singh, S. P., Harris, W. G. 2003. Phosphate-induced metal immobilization in contaminated site. *Environmental Pollution*, 122, 19–28.
Cao, X. D., Ma, L. Q., Rhue, D. R., Appel, C. S. 2004. Mechanisms of lead, copper, and zinc retention by phosphate rock. *Environmental Pollution*, 131, 435–444.
Cao, X. D., Ma, L. Q., Singh, S. P., Zhou, Q. X. 2008. Phosphate-induced lead immobilization from different lead minerals in soils under varying pH conditions. *Environmental Pollution*, 152, 184–192.
Cao, X. D., Wahbi, A., Ma, L., Li, B., Yang, Y. L. 2009. Immobilization of Zn, Cu, and Pb in contaminated soils using phosphate rock and phosphoric acid. *Journal of Hazardous Materials*, 164, 555–564.
Chen, H. M. 1996. *Heavy Metal Pollution in Soil-Plant System.* Science Press, Beijing (in Chinese).
Chen, G. C., He, Z. L., Stoffella, P. J., Yang, X. E., Yu, S., Yang, J. Y., Calvert, D. V. 2006a. Leaching potential of heavy metals (Cd, Ni, Pb, Cu, and Zn) from acidic sandy soil amended with dolomite phosphate rock (DPR) fertilizers. *Journal of Trace Elements in Medicine and Biology*, 20, 127–133.
Chen, S. B., Zhu, Y. G., Ma, Y. B. 2006b. The effect of grain size of rock phosphate amendment on metal immobilization in contaminated soils. *Journal of Hazardous Materials*, 134, 74–79.
Chen, S. B., Xu, M. G., Ma, Y. B., Yang, J. C. 2007. Evaluation of different phosphate amendments on availability of metals in contaminated soil. *Ecotoxicology and Environmental Safety*, 67, 278–285.
Chen, X. B., Wright, J. V., Conca, J. L., Peurrung, L. M. 1997. Evaluation of heavy metal remediation using mineral apatite. *Water, Air, and Soil Pollution*, 98, 57–78.
Chrysochoou, M., Dermatas, D., Grubb, D. G. 2007. Phosphate application to firing range soils for Pb immobilization: The unclear role of phosphate. *Journal of Hazardous Materials*, 144, 1–14.
Collins, C. R., Ragnarsdottir, K. V., Sherman, D. M. 1999. Effect of inorganic and organic ligands on the mechanism of cadmium sorption to goethite. *Geochimica et Cosmochimica Acta*, 63, 2989–3002.
Conder, J. M., Lanno, R. P., Basta, N. T. 2001. Assessment of metal availability in smelter soil using earthworms and chemical extractions. *Journal of Environmental Quality*, 30, 1231–1237.
Cotter-Howells, J., Caporn, S. 1996. Remediation of contaminated land by formation of heavy metal phosphates. *Applied Geochemistry*, 11, 335–342.
Cotter-Howells, J. D., Champness, P. E., Charnock, J. M. 1999. Mineralogy of Pb-P grains in the roots of Agrostis capillaris L-by ATEM and EXAFS. *Mineralogical Magazine*, 63, 777–789.
Davies, N. A., Hodson, M. E., Black, S. 2002. Changes in toxicity and bioavailability of lead in contaminated soils to the earthworm Eisenia fetida (Savigny 1826) after bone meal amendments to the soil. *Environmental Toxicology and Chemistry*, 21, 2685–2691.

Dudal, Y., Gérard, F. 2004. Accounting for natural organic matter in aqueous chemical equilibrium models: A review of the theories and applications. *Earth-Science Reviews*, 66, 199–216.

Elouear, Z., Bouhamed, F., Bouzid, J. 2014. Evaluation of different amendments to stabilize cadmium, zinc, and copper in a contaminated soil: Influence on metal leaching and phytoavailability. *Soil and Sediment Contamination*, 23, 628–640.

Elzinga, E. J., Kretzschmar, R. 2013. *In situ* ATR-FTIR spectroscopic analysis of the co-adsorption of orthophosphate and Cd (II) onto hematite. *Geochimica et Cosmochimica Acta*, 117, 53–64.

Fang, Y. Y., Cao, X. D., Zhao, L. 2012. Effects of phosphorus amendments and plant growth on the mobility of Pb, Cu, and Zn in a multi-metal-contaminated soil. *Environmental Science and Pollution Research*, 19, 1659–1667.

Fayiga, A. O., Ma, L. Q. 2006. Using phosphate rock to immobilize metals in soil and increase arsenic uptake by hyperaccumulator *Pteris vittata*. *Science of the Total Environment*, 359, 17–25.

Filho, M. R. R., Siqueira, J. O., Vangronsveld, J., Soares, C. R. F. S., Curi, N. 2011. Inorganic materials as ameliorants for soil remediation of metal toxicity to wild mustard (*Sinapis arvensis* L.). *International Journal of Phytoremediation*, 13, 498–512.

Gauglitz, R., Holterdorf, M., Franke, W., Marx, G. 1992. Immobilization of heavy-metals by hydroxylapatite. *Radiochimica Acta*, 58–59, 253–257.

Geebelen, W., Adriano, D. C., van der Lelie, D., Mench, M., Carleer, R., Clijsters, H., Vangronsveld, J. 2003. Selected bioavailability assays to test the efficacy of amendment-induced immobilization of lead in soils. *Plant and Soil*, 249, 217–228.

Geebelen, W., Ruttens, A., Carleer, R., Yperman, J., Bongué-Boma, K., Mench, M., van der Lelie, N., Vangronsveld, J. 2006. Evaluation of cyclonic ash, commercial Na-silicates, lime and phosphoric acid for metal immobilisation purposes in contaminated soils in Flanders (Belgium). *Environmental Pollution*, 144, 32–39.

Gray, C. W., Mclaren, R. G., Roberts, A. H. C., Condron, L. M. 1999. Effect of soil pH on cadmium phytoavailability in some New Zealand soils. *New Zealand Journal of Crop and Horticultural Science*, 27, 169–179.

Guo, G. L., Zhou, Q. X., Ma, L. Q. 2006. Availability and assessment of fixing additives for the *in situ* remediation of heavy metal contaminated soils: A review. *Environmental Monitoring and Assessment*, 116, 513–528.

Hamon, R. E., McLaughlin, M. J., Cozens, G. 2002. Mechanisms of attenuation of metal availability in *in situ* remediation treatments. *Environmental Science & Technology*, 36, 3991–3996.

Hashimoto, Y., Takaoka, M., Oshita, K., Tanida, H. 2009a. Incomplete transformations of Pb to pyromorphite by phosphate-induced immobilization investigated by X-ray absorption fine structure (XAFS) spectroscopy. *Chemosphere*, 76, 616–622.

Hashimoto, Y., Taki, T., Sato, T. 2009b. Sorption of dissolved lead from shooting range soils using hydroxyapatite amendments synthesized from industrial byproducts as affected by varying pH conditions. *Journal of Environmental Management*, 90, 1782–1789.

Hashimoto, Y., Takaoka, M., Shiota, K. 2011. Enhanced transformation of lead speciation in rhizosphere soils using phosphorus amendments and phytostabilization: An X-ray absorption fine structure spectroscopy investigation. *Journal of Environmental Quality*, 40, 696–703.

Hayakawa, N., Tomioka, R., Takenaka, C. 2011. Effects of calcium on cadmium uptake and transport in the tree species *Gamblea innovans*. *Soil Science and Plant Nutrition*, 57, 691–695.
Hettiarachchi, G. M., Pierzynski, G. M., Ransom, M. D. 2000. *In situ* stabilization of soil lead using phosphorus and manganese oxide. *Environmental Science & Technology*, 34, 4614–4619.
Hettiarachchi, G. M., Pierzynski, G. M., Ransom, M. D. 2001. *In situ* stabilization of soil lead using phosphorus. *Journal of Environmental Quality*, 30, 1214–1221.
Hodson, M. E., Valsami-Jones, E., Cotter-Howells, J. D., Dubbin, W. E., Kemp, A. J., Thornton, I., Warren, A. 2001. Effect of bone meal (calcium phosphate) amendments on metal release from contaminated soils—A leaching column study. *Environmental Pollution*, 112, 233–243.
Houben, D., Sonnet, P. 2012. Zinc mineral weathering as affected by plant roots. *Applied Geochemistry*, 27, 1587–1592.
Höweler, R. H., Woodruff, C. M. 1968. Dissolution and availability to plants of rock phosphates of igneous and sedimentary origins. *Soil Science Society of America Proceedings*, 32, 79–82.
Huang, P. M. 1998. *Future Prospects for Soil Chemistry*. Soil Science Society of America, Inc., Madison, WI.
Jalali, M., Moharami, S. 2010. Effects of the addition of phosphorus on the redistribution of cadmium, copper, lead, nickel, and zinc among soil fractions in contaminated calcareous soil. *Soil & Sediment Contamination*, 19, 88–102.
Jiao, W. T., Chen, W. P., Chang, A. C., Page, A. L. 2012. Environmental risks of trace elements associated with long-term phosphate fertilizers applications: A review. *Environmental Pollution*, 168, 44–53.
Kamiishi, E., Utsunomiya, S. 2013. Nano-scale reaction processes at the interface between apatite and aqueous lead. *Chemical Geology*, 340, 121–130.
Karlsson, T., Persson, P., Skyllberg U. 2006. Complexation of copper (II) in organic soils and in dissolved organic matter-EXAFS evidence for chelate ring structures. *Environmental Science & Technology*, 40, 2623–2628.
Kassir, L. N., Darwish, T., Shaban, A., Olivier, G., Ouaini, N. 2012. Mobility and bioavailability of selected trace elements in Mediterranean red soil amended with phosphate fertilizers: Experimental study. *Geoderma*, 189–190, 357–368.
Khan, M. J., Jones, D. L. 2008. Chemical and organic immobilization treatments for reducing phytoavailability of heavy metals in copper-mine tailings. *Journal of Plant Nutrition and Soil Science*, 171, 908–916.
Khan, M. J., Jones, D. L. 2009. Effects of composts, lime and diammonium phosphate on the phytoavailability of heavy metals in copper mine tailing soil. *Pedosphere*, 19, 631–641.
Knox, A. S., Kaplan, D. I., Paller, M. H. 2006. Phosphate sources and their suitability for remediation of contaminated soils. *Science of the Total Environment*, 357, 271–279.
Kucharski, R., Sas-Nowosielska, A., Malkowski, E., Japenga, J., Kuperberg, J. M., Pogrzeba, M., Krzyzak, J. 2005. The use of indigenous plant species and calcium phosphate for the stabilization of highly metal-polluted sites in southern Poland. *Plant and Soil*, 273, 291–305.
Kumpiene, J., Lagerkvist, A, Maurice, C. 2008. Stabilization of As, Cr, Cu, Pb and Zn in soil using amendments—A review. *Waste Management*, 28, 215–225.

Lang, F., Kaupenjohann, M. 2003. Effect of dissolved organic matter on the precipitation and mobility of the lead compound chloropyromorphite in solution. *European Journal of Soil Science*, 54, 139–148.

Laperche, V., Logan, T. J., Gaddam, P., Traina, S. 1997. Effect of apatite amendments on plant uptake of lead from contaminated soil. *Environmental Science & Technology*, 31, 2745–2753.

Lavoie, M., Fortin, C., Campbell, P. G. C. 2012. Influence of essential elements on cadmium uptake and toxicity in a unicellular green alga: The protective effect of trace zinc and cobalt concentrations. *Environmental Toxicology and Chemistry*, 31, 1445–1452.

Lee, Y. J., Elzinga, E. J., Reeder, R. J. 2005. Sorption mechanisms of zinc on hydroxyapatite: Systematic uptake studies and EXAFS spectroscopy analysis. *Environmental Science & Technology*, 39, 4042–4048.

Liang, G. H., Liu, X. Z., Chen, X. M., Qiu Q. Y., Zhang, D. Q., Chu, G. W., Liu, J. X., Liu, S. Z., Zhou, G. Y. 2013. Response of soil respiration to acid rain in forests of different maturity in southern China. *PLoS ONE*, 8, e62207.

Liang, Y., Wang, X. C., Cao, X. D., Zhao, L. 2012. Immobilization of Pb, Cu, and Zn in a multi-metal contaminated soil amended with triple superphosphate fertilizer and phosphate rock tailing. *Advanced Materials Research*, 356–360, 1716–1718.

Lin, C. W., Lian, J., Fang, H. H. 2005. Soil lead immobilization using phosphate rock. *Water, Air, and Soil Pollution*, 161, 113–123.

Lindsay, W. L. 1979. *Chemical Equilibria in Soils*. John Wiley & Sons, New York.

Liu, A., Gonzalez, R. D. 1999. Adsorption/desorption in a system consisting of humic acid, heavy metals and clay minerals. *Journal of Colloid and Interface Science*, 218, 225–232.

Liu, R. Q., Zhao, D. Y. 2007. *In situ* immobilization of Cu(II) in soils using a new class of iron phosphate nanoparticles. *Chemosphere*, 68, 1867–1876.

Liu, R. Q., Zhao, D. Y. 2013. Synthesis and characterization of a new class of stabilized apatite nanoparticles and applying the particles to *in situ* Pb immobilization in a fire-range soil. *Chemosphere*, 91, 594–601.

Liu, Y. L., Wen, C., Liu, X. J. 2013. China's food security soiled by contamination. *Science*, 339, 1382–1383.

Luo, L., Ma, Y. B., Zhang, S. Z., Wei, D. P., Zhu, Y. G. 2009. An inventory of trace element inputs to agricultural soils in China. *Journal of Environmental Management*, 90, 2524–2530.

Ma, L. Q., Choate, A. L., Rao, G. N. 1997. Effects of incubation and phosphate rock on lead extractability and speciation in contaminated soils. *Journal of Environmental Quality*, 26, 801–807.

Ma, L. Q., Rao, G. N. 1997. Effects of phosphate rock on sequential chemical extraction of lead in contaminated soils. *Journal of Environmental Quality*, 26, 788–794.

Maenpaa, K. A., Kukkonen, J. V. K., Lydy, M. J. 2002. Remediation of heavy metal-contaminated soils using phosphorus: Evaluation of bioavailability using an earthworm bioassay. *Archives of Environmental Contamination and Toxicology*, 43, 389–398.

Mahdy, A. M. 2012. Soil-solution speciation by MINEQL+4.6 model and plant uptake of Cd and Zn by barley (*Hordeum vulgare* L.) after application of different phosphate fertilizers. *Journal of Plant Nutrition and Soil Science*, 175, 560–571.

Majs, F. 2011. Potential adverse effects of applying phosphate amendments to immobilize soil contaminants. *Journal of Environmental Quality*, 40, 1135–1142.

Manceau, A., Matynia, A. 2010. The nature of Cu bonding to natural organic matter. *Geochimica et Cosmochimica Acta*, 74, 2556–2580.
Mavropoulos, E., Rossi, A. M., Costa, A. M., Perez, C. A. C., Moreira, J. C., Saldanha, M. 2002. Studies on the mechanisms of lead immobilization by hydroxyapatite. *Environmental Science & Technology*, 36, 1625–1629.
Mavropoulos, E., Rocha, N. C. C., Moreira, J. C., Rossi, A. M., Soares, G. A. 2004. Characterization of phase evolution during lead immobilization by synthetic hydroxyapatite. *Materials Characterization*, 53, 71–78.
McGowen, S. L., Basta, N. T., Brown, G. O. 2001. Use of diammonium phosphate to reduce heavy metal solubility and transport in smelter-contaminated soil. *Journal of Environmental Quality*, 30, 493–500.
Melamed, R., Cao, X. D., Chen, M., Ma, L. Q. 2003. Field assessment of lead immobilization in a contaminated soil after phosphate application. *Science of the Total Environment*, 305, 117–127.
Melamed, R., Ma, L. Q. 2008. Phosphate-induced lead immobilization in contaminated soils: Mechanisms, assessment, and field applications. *In* Violante, A., Huang, P. M., Gadd, G. M. (eds.). *Biophysico-Chemical Processes of Heavy Metals and Metalloids in Soil Environments*. John Wiley & Sons, Inc., Hoboken, NJ, pp. 607–629.
Mench, M., Vangronsveld, J., Lepp, N., Ruttens, A., Bleeker, P, Geebelen, W. 2006. Use of soil amendments to attenuate trace element exposure: Sustainability, side effects, and failures. *In* Hamon, R., McLaughlin, M., Lombi, E. (eds.). *Natural Attenuation of Trace Element Availability in Soils*. CRC Press, Boca Raton, FL, pp. 197–228.
Menzies, N. W., Donn, M. J., Kopittke, P. M. 2007. Evaluation of extractants for estimation of the phytoavailable trace metals in soils. *Environmental Pollution*, 145, 121–130.
Mignardi, S., Corami, A., Ferrini, V. 2012. Evaluation of the effectiveness of phosphate treatment for the remediation of mine waste soils contaminated with Cd, Cu, Pb, and Zn. *Chemosphere*, 86, 354–360.
Mignardi, S., Corami, A., Ferrini, V. 2013. Immobilization of Co and Ni in mining-impacted soils using phosphate amendments. *Water, Air, and Soil Pollution*, 224, doi: 10.1007/s11270-013-1447-y.
Miretzky, P., Fernandez-Cirelli, A. 2008. Phosphates for Pb immobilization in soils: A review. *Environmental Chemistry Letters*, 6, 121–133.
Misra, V., Chaturvedi, P. K. 2007. Plant uptake/bioavailability of heavy metals from the contaminated soil after treatment with humus soil and hydroxyapatite. *Environmental Monitoring and Assessment*, 133, 169–176.
Misra, V., Pandey, S. D. 2005. Immobilization of heavy metals in contaminated soil using nonhumus–humus soil and hydroxyapatite. *Bulletin of Environmental Contamination and Toxicology*, 74, 725–731.
Mulligan, C. N., Yong, R. N., Gibbs, B. F. 2001. Remediation technologies for metal-contaminated soils and groundwater: An evaluation. *Engineering Geology*, 60, 193–207.
Munksgaard, N. C., Lottermoser, B. G. 2011. Fertilizer amendment of mining-impacted soils from Broken Hill, Australia: Fixation or release of contaminants? *Water, Air, and Soil Pollution*, 215, 373–397.
Nicholson, F. A., Smith, S. R., Alloway, B. J., Carlton-Smith, C., Chambers, B. J. 2003. An inventory of heavy metals inputs to agricultural soils in England and Wales. *Science of the Total Environment*, 311, 205–219.

Nriagu, J. O. 1984. Formation and stability of base metal phosphates in soils and sediments. *In* Nriagu, J. O., Moore, P. B. (eds.). *Phosphate Minerals*. Springer-Verlag, Berlin, pp. 318–329.

Nziguheba, G., Smolders, E. 2008. Inputs of trace elements in agricultural soils via phosphate fertilizers in European countries. *Science of the Total Environment*, 390, 53–57.

Padmavathiamma, P. K., Li, L. Y. 2010. Effect of amendments on phytoavailability and fractionation of copper and zinc in a contaminated soil. *International Journal of Phytoremediation*, 12, 697–715.

Pardo, M. T. 2004. Cadmium sorption-desorption by soils in the absence and presence of phosphate. *Communications in Soil Science and Plant Analysis*, 35, 1553–1568.

Parfitt, R. L., Atkinson, R. J., Smart, R. S. C. 1975. The mechanism of phosphate fixation by iron oxides. *Soil Science Society of America Proceedings*, 39, 837–841.

Park, J. H., Bolan, N., Megharaj, M., Naidu, R. 2012. Relative value of phosphate compounds in reducing the bioavailability and toxicity of lead in contaminated soils. *Water, Air, and Soil Pollution*, 223, 599–608.

Pérez-Novo, C., Bermúdez-Couso, A., López-Periago, E., Fernández-Calviño, D. 2011. Zinc adsorption in acid soils: Influence of phosphate. *Geoderma*, 162, 358–364.

Porter, S. K., Scheckel, K. G., Impellitteri, C. A., Ryan, J. A. 2004. Toxic metals in the environment: Thermodynamic considerations for possible immobilisation strategies for Pb, Cd, As, and Hg. *Critical Reviews in Environmental Science and Technology*, 34, 495–604.

Raicevic, S., Kaludjerovic-Radoicic, T., Zouboulis, A. I. 2005. *In situ* stabilization of toxic metals in polluted soils using phosphates: Theoretical prediction and experimental verification. *Journal of Hazardous Materials*, 117, 41–53.

Ramos, L., Hernandez, L. M., Gonzalez, M. J. 1994. Sequential fractionation of copper, lead, cadmium and zinc in soils from or near Donana National Park. *Journal of Environmental Quality*, 23, 50–57.

Ryan, J. A., Zhang, P. C., Hesterberg, D., Chou, J., Sayers, D. E. 2001. Formation of chloropyromorphite in a lead-contaminated soil amended with hydroxyapatite. *Environmental Science & Technology*, 35, 3798–3803.

Sanderson, P., Naidu, R., Bolan, N. 2014. Ecotoxicity of chemically stabilised metal(loid)s in shooting range soils. *Ecotoxicology and Environmental Safety*, 100, 201–208.

Scheckel, K. G., Impellitteri, C. A., Ryan, J. A., McEvoy, T. 2003. Assessment of a sequential extraction procedure for perturbed lead-contaminated samples with and without phosphorus amendments. *Environmental Science & Technology*, 37, 1892–1898.

Scheckel, K. G., Ryan, J. A. 2004. Spectroscopic speciation and quantification of lead in phosphate-amended soils. *Journal of Environmental Quality*, 33, 1288–1295.

Seaman, J. C., Arey, J. S., Bertsch, P. M. 2001. Immobilization of nickel and other metals in contaminated sediments by hydroxyapatite addition. *Journal of Environmental Quality*, 30, 460–469.

Sheals, J., Granström, M., Sjöberg, S., Persson, P. 2003. Coadsorption of Cu (II) and glyphosate at the water-goethite (α-FeOOH) interface: Molecular structures from FTIR and EXAFS measurements. *Journal of Colloid and Interface Science*, 262, 38–47.

Sheppard, S. C., Grant, C. A., Sheppard, M. I., de Jong, R., Long, J. 2009. Risk indicator for agricultural inputs of trace elements to Canadian soils. *Journal of Environmental Quality*, 38, 919–932.

Siebers, N., Kruse, J., Leinweber, P. 2013. Speciation of phosphorus and cadmium in a contaminated soil amended with bone char: Sequential fractionations and XANES spectroscopy. *Water, Air, and Soil Pollution*, 224, doi: 10.1007/s11270-013-1564-7.

Singh, S. P., Ma, L. Q., Harris, W. G. 2001. Heavy metal interactions with phosphatic clay: Sorption and desorption behavior. *Journal of Environmental Quality*, 30, 1961–1968.

Sneddon, I. R., Orueetxebarria, M., Hodson, M. E., Schofield, P. F., Valsami-Jones, E. 2006. Use of bone meal amendments to immobilise Pb, Zn and Cd in soil: A leaching column study. *Environmental Pollution*, 144, 816–825.

Spuller, C., Weigand, H., Marb, C. 2007. Trace metal stabilisation in a shooting range soil: Mobility and phytotoxicity. *Journal of Hazardous Materials*, 141, 378–387.

Strawn, D. G., Hickey, P., Knudsen, A., Baker, L. 2007. Geochemistry of lead contaminated wetland soils amended with phosphorus. *Environmental Geology*, 52, 109–122.

Szrek, D., Bajda, T., Manecki, M. 2011. A comparative study of the most effective amendment for Pb, Zn and Cd immobilization in contaminated soils. *Journal of Environmental Science and Health, Part A: Toxic/Hazardous Substances and Environmental Engineering*, 46, 1491–1502.

Tang, X., Yang, J., Goyne, K. W., Deng, B. L. 2009. Long-term risk reduction of lead-contaminated urban soil by phosphate treatment. *Environmental Engineering Science*, 26, 1747–1754.

Temminghoff, E. J. M., van Der Zee, S. E. A. T. M., de Haan, F. A. M. 1997. Copper mobility in a copper-contaminated sandy soil as affected by pH and solid and dissolved organic matter. *Environmental Science & Technology*, 31, 1109–1115.

Thawornchaisit, U., Polprasert, C. 2009. Evaluation of phosphate fertilizers for the stabilization of cadmium in highly contaminated soils. *Journal of Hazardous Materials*, 165, 1109–1113.

Theodoratos, P., Papassiopi, N., Xenidis, A. 2002. Evaluation of monobasic calcium phosphate for the immobilization of heavy metals in contaminated soils from Lavrion. *Journal of Hazardous Materials*, B94, 135–146.

Tiberg, C., Sjöstedt, C., Persson, I., Gustafsson, J. P. 2013. Phosphate effects on copper (II) and lead (II) sorption to ferrihydrite. *Geochimica et Cosmochimica Acta*, 120, 140–157.

Trevizam, A. R., Villanueva, F. C. A, Muraoka, T., do Nascimento, V. F., Abreu, C. H. 2010. Phosphorus application for the chemical immobilization of cadmium in contaminated soil. *Quimica Nova*, 33, 1235–1241.

Udeigwe, T. K., Eze, P. N., Teboh, J. M., Stietiya, M. H. 2011. Application, chemistry, and environmental implications of contaminant-immobilization amendments on agricultural soil and water quality. *Environmental International*, 37, 258–267.

Wang, A. S., Angle, J. S., Chaney, R. L., Delorme, T. A., Reeves, R. D. 2006. Soil pH effects on uptake of Cd and Zn by *Thlaspi caerulescens*. *Plant and Soil*, 281, 325–337.

Wang, B. L., Xie, Z. M., Chen, J. J., Jiang, J. T., Su, Q. F. 2008. Effects of field application of phosphate fertilizers on the availability and uptake of lead, zinc and cadmium by cabbage (*Brassica chinensis* L.) in a mining tailing contaminated soil. *Journal of Environmental Science-China*, 20, 1109–1117.

Watmough, S. A., Eimers, M. C., Dillon, P. J. 2007. Manganese cycling in central Ontario forests: Response to soil acidification. *Applied Geochemistry*, 22, 1241–1247.

Wei, J., Chen, M. F., Song, J., Luo, F., Han, L., Li, C. P., Dong, M. G. 2014. Assessment of human health risk for an area impacted by a large-scale metallurgical refinery complex in Hunan, China. *Human and Ecological Risk Assessment: An International Journal*, doi: 10.1080/10807039.2014.890479.

Wei, X. X. 2010. Phosphate-induced immobilization of heavy metals in multi-metal contaminated soils. master's thesis (in Chinese with English abstract), Xi'an University of Science and Technology, Xi'an, China.

Wu, W. H., Xie, Z. M., Xu, J. M., Wang, F., Shi, J. C., Zhou, R. B., Jin, Z. F. 2013. Immobilization of trace metals by phosphates in contaminated soil near lead/zinc mine tailings evaluated by sequential extraction and TCLP. *Journal of Soils and Sediments*, 13, 1386–1395.

Xu, J. M., Sparks, D. L. 2013. *Molecular Environmental Soil Science.* Springer, Dordrecht, Netherlands.

Xu, Y., Schwartz, F. W., Tralna, S. J. 1994. Sorption of Zn^{2+} and Cd^{2+} on hydroxyapatite surfaces. *Environmental Science & Technology*, 28, 1472–1480.

Yang, J., Mosby, D. E. 2006. Field assessment of treatment efficacy by three methods of phosphoric acid application in lead-contaminated urban soil. *Science of the Total Environment*, 366, 136–142.

Yang, J., Mosby, D. E., Casteel, S. W., Blanchar, R. W. 2001. Lead immobilization using phosphoric acid in a smelter-contaminated urban soil. *Environmental Science & Technology*, 35, 3553–3559.

Yang, J., Mosby, D. E., Casteel, S. W., Blanchar, R. W. 2002. *In vitro* lead bioaccessibility and phosphate leaching as affected by surface application of phosphoric acid in lead-contaminated soil. *Archives of Environmental Contamination and Toxicology*, 43, 399–405.

Yu, T. R. 1997. *Chemistry of Variable Charge Soils.* Oxford University Press, New York.

Zhang, P., Ryan, J. A. 1998. Formation of pyromorphite in anglesite-hydroxyapatite suspensions under varying pH conditions. *Environmental Science & Technology*, 32, 3318–3324.

Zhang, M. K., Zhang, H. M. 2010. Co-transport of dissolved organic matter and heavy metals in soils induced by excessive phosphorus applications. *Journal of Environmental Sciences-China*, 22, 598–606.

Zhang, X. H., Zhu, Y. G., Chen, B. D., Lin, A. J., Smith, S. E., Smith, F. A. 2005. Arbuscular mycorrhizal fungi contribute to resistance of upland rice to combined metal contamination of soil. *Journal of Plant Nutrition*, 28, 2065–2077.

Zhou, L. X., Wong, J. W. C. 2001. Effect of dissolved organic matter from sludge and sludge compost on soil copper sorption. *Journal of Environmental Quality*, 30, 878–883.

Zhu, Y. G., Chen, S. B., Yang, J. C. 2004. Effects of soil amendments on lead uptake by two vegetable crops from a lead-contaminated soil from Anhui, China. *Environment International*, 30, 351–356.

Zupančič, M., Bukovec, P., Milačič, R., Ščančar, J. 2004. Comparison of various phosphate stabilisation agents for the immobilisation of Ni and Zn in sewage sludge. *Water, Air, and Soil Pollution*, 156, 57–69.

Zupančič, M., Bukovec, P., Milačič, R., Ščančar, J. 2006. Critical evaluation of the use of the hydroxyapatite as a stabilizing agent to reduce the mobility of Zn and Ni in sewage sludge amended soils. *Waste Management*, 26, 1392–1399.

9

Bioavailability of Trace Elements in Soils Amended with High-Phosphate Materials

Grzegorz Siebielec, Aleksandra Ukalska-Jaruga, and Petra Kidd

CONTENTS

Introduction

In many industrial countries, soils are affected by mine waste disposal, acid deposition, and sewage sludge use that could provide large inputs of pollutants and especially trace elements (TEs) to the soils. TEs are defined as elements that are present at low concentrations (mg kg^{-1} or less) in most soils, plants, and living organisms. TEs that have been most extensively studied in soils are those that are essential for the nutrition of higher plants and include copper (Cu), zinc (Zn), iron (Fe), manganese (Mn), molybdenum (Mo), boron (B), cobalt (Co), or selenium (Se). During the last two decades or so, the role of TEs as environmental contaminants has dominated the literature, and TEs have been recognized as one of the most important groups of contaminants. Indeed, 6 out of the 11 most common contaminants at the U.S. National Priority List sites are trace elements: lead (Pb), arsenic (As), chromium (Cr), cadmium (Cd), nickel (Ni), and Zn, in decreasing order of frequency of occurrence.

The bioavailability of contaminants poses a health risk to animals and humans who may be exposed to contaminated sites. Possible exposure pathways include ingestion of contaminated soil or water from the site, direct contact with contaminated soil, inhalation of contaminants adhered to dust in the air, and ingestion of food items that have accumulated contaminants when grown on contaminated soil. Within this context, TE contamination of arable soils through industrial activities is a serious problem worldwide.

The mobilization of metals in soils for plant uptake and leaching to groundwater can, however, be minimized by reducing the bioavailability of metals through chemical and biological immobilization. A variety of *in situ* and *ex situ* remediation methods have been used or proposed for the restoration of soils contaminated with TEs. Recently, there has been increasing interest in the *in situ* gentle immobilization of metals using a range of inorganic compounds or organic materials, including high-phosphate materials. The most important processes involved in this immobilization are the transformation of metals in soils, through precipitation–dissolution, adsorption–desorption, and complexation processes. In addition to reducing TE bioavailability, the incorporation of effective amendments restores soil quality by balancing pH, adding organic matter, increasing water holding capacity, reestablishing microbial communities, and alleviating compaction. As such, the use of soil amendments potentially enables site remediation, revegetation and revitalization, and finally, sustainable reuse.

Materials Containing Phosphates Available at a Practical Scale

Phosphorus (P) is an essential element (macronutrient) required by plants in relatively large amounts. Phosphorus is one of the three major nutrients generally added to soils in fertilizers. As a particle of solid fertilizer comes into contact with the soil, moisture from the soil will begin dissolving the particle. Dissolving of the fertilizer increases the soluble phosphate in the soil solution around the particle and allows the dissolved phosphate to move a short distance away from the fertilizer particle. Movement is slow but may be increased by rainfall or irrigation water flowing through the soil. As phosphate ions in solution slowly migrate away from the fertilizer particle, most of the phosphate will react with the minerals within the soil (He et al. 2005; Liu and Zhao 2007; Withers and Sharpley 1995). Phosphate ions generally react by adsorbing to soil particles or by combining with elements in the soil such as calcium (Ca), magnesium (Mg), aluminum (Al), and

Fe, and forming solid compounds. The adsorbed phosphate and the newly formed solids are relatively available to soil biota (Corami et al. 2008a). Gradually reactions occur in which the adsorbed phosphate and the easily dissolved compounds of phosphate form more insoluble compounds that cause the phosphate to become fixed and unavailable. Over time this results in a decrease in available P in soil. The mechanisms for the changes in phosphate are complex and involve a variety of compounds. In alkaline soils (soil pH greater than 7), Ca is the dominant cation that will react with phosphate. A general sequence of reactions in alkaline soils is the formation of dibasic calcium phosphate dihydrate, octocalcium phosphate, and hydroxyapatite. The formation of each product results in a decrease in solubility and availability of phosphate.

Solubility of P compounds in soil and P availability in soil directly impact processes of inactivation of TEs by phosphate soil amendments. Reactions that reduce TEs and P availability occur in all ranges of soil pH but can be very pronounced in alkaline soils (pH > 7.3) and in acidic soils (pH < 5.5). Maintaining soil pH between 6 and 7 will generally result in the most efficient use of phosphate materials for TE immobilization (Withers and Sharpley 1995).

Phosphates and related materials are required to overcome soil nutrient deficiency, but additionally, in view of their chemical properties, they can be applied to stabilize TE (Pb, Cd, Zn, and Cu) in contaminated soils, sediments, and solid wastes by reducing their phytoavailability, mobility, and subsequently their leaching from the root zone to the groundwater (Liu and Zhao 2007). Stabilization of TEs in soils is based on the modification of contaminant characteristics (e.g., speciation) and soil properties (sorption capacity, buffering potential, etc.) through the addition of amendments. These amendments induce or enhance physicochemical and/or microbial processes, which render TE contaminants less mobile and less bioavailable in soil. They can reduce the negative effects of TEs such as Cr, Cu, Pb, Cd, and Zn, and at the same time improve the quality of contaminated soils (Kumpiene 2008).

Phosphate-containing materials vary widely, but in general, they fall within two major categories: commercial fertilizer sources and organic sources. While it is difficult to make direct comparisons between these two sources, a few loose comparisons can be made.

Distinctions between organic (manure) fertilizers and inorganic (commercial) fertilizers (EPA 2007; Traunfeld and Nibali 2013; Withers and Sharpley 1995):

- *Nutrient analysis.* While commercial fertilizers may have a relatively high content of the major macronutrients (nitrogen, phosphorus, and potassium [NPK]), the nutrient content of manures is much lower. As a result, a larger quantity of manure must be applied to the soil as compared to the addition of commercial fertilizer at an equivalent rate.

- *The nutrient content of manure fertilizers is highly variable.* Factors affecting this nutrient content include animal type and diet, handling, storage, and water content. Since manure is an organic source, the availability of nutrients is also largely influenced by the biological processes of mineralization and immobilization.

Inorganic High-Phosphate Sources

The agronomic effectiveness of inorganic high-phosphate sources depends on the speed of dissolution in the soil, which is usually chemically defined by their solubility in water, citric acid, ammonium citrate, or formic acid, as well as by crop requirements (Withers and Sharpley 1995). The major source of inorganic P fertilizers is rock phosphate. Rock phosphate is a naturally occurring mineral that is mined from the earth. Deposits of rock phosphate occur around the world, such as in the United States, Russia, Morocco, and China.

Rock products (RPs): natural, mined material that contains 27%–41% diphosphorus pentoxide (P_2O_5) and 25% Ca. The minerals that make up RPs are various forms of apatite. The reactivity of RP depends on the type of apatite and its inherent purity. RP is not water-soluble and only becomes available to plants under acidic conditions. RP is most reactive when it is finely ground and incorporated into warm, moist, acidic soils with long growing seasons. Although the availability of RP is slow, it has a great long-term residual effect. RP is loaded with P, but only about 1% to 3% of that P is available; the rest is tightly bound in complex compounds that are slow to break down. On the other hand, colloidal rock phosphate is a fine material left over after processing rock phosphate, which has a lower total P content but slightly higher available P. A selection of those commercially available forms are as follows (EEA 2007; Traunfeld and Nibali 2003):

- *Azomite or rock dust* is an aluminum silicate clay mixed with over 50 minerals from marine deposits (2.5% K).
- *Black rock phosphate,* which is about 30% phosphate rock with calcium oxide (CaO), silicas, and trace minerals. Only 3% of phosphate is immediately available. Slow release builds longer reserves than colloidal phosphate. Best performance in slightly acid soils.
- *Soft rock or colloidal phosphate,* which is phosphate clay with 18%–22% phosphate, 27% calcium oxide, silicates, and 14 trace minerals. Two percent of phosphate is immediately available, and the remaining is slow-released over 3–5 years. Half the liming value of ground lime.
- *Superphosphate,* which is commonly called calcium hydrogen phosphate, is the most important phosphatic fertilizer (Pratiyogita Darpan 2008).

Single superphosphate (SSP) contains 16%–22% P_2O_5, 11%–12% sulfur (S), and 20% Ca. Superphosphate is made by treating phosphate rock, fluorapatite, or bone ash with a calculated quantity of concentration H_2SO_4. This is actually a mixture of calcium hydrogen phosphate and gypsum ($CaSO_4 \cdot H_2O$).

$$Ca_3(PO_4)_2 + 2H_2SO_4 + 5H_2O \rightarrow Ca(H_2PO_4)_2 \cdot H_2O + 2CaSO_4 \cdot 2H_2O$$

$$[3Ca_3(PO_4)_2] \cdot CaF_2 + 7H_2SO_4 \rightarrow 3Ca(H_2PO_4)_2 + 7CaSO_4 + 2HF$$

Triple superphosphate (TSP) contains 44%–52% P_2O_5, 1%–1.5% S, and 13% Ca. TSP is produced by treating phosphate rock or bone ash with phosphoric acid.

$$Ca_3(PO_4)_2 + 4H_3PO_4 + 3H_2O \rightarrow 3Ca(H_2PO_4)_2 \cdot H_2O$$

$$[Ca_5F(PO_4)_3] + 7H_3PO_4 + 5H_2O \rightarrow 5Ca(H_2PO_4)_2 \cdot H_2O + HF$$

Ammonium phosphate is the world's most widely used P fertilizer. It is made from two common constituents—rock phosphate and ammonium sulfate—that are treated with sulfuric acid and ammonium dihydrogen phosphate is obtained. It is popular because of its relatively high nutrient content and its excellent physical properties (Pratiyogita Darpan 2008).

- *Monoammonium phosphate* (MAP) contains 11%–13% N, 48%–62% P_2O_5, is water-soluble, and temporarily lowers the soil pH to 3.5 in areas where MAP initially reacts with the soil.
- *Diammonium phosphate* (DAP) contains 18%–21% N, 46%–53% P_2O_5, is water-soluble, and the soil pH temporarily lowers to 8.5 in areas where DAP initially reacts.

Thomas basic slag is actually obtained as a by-product in the iron industries. This results in the presence of small amounts of P in iron ores. When iron ore is burned with lime stone in the presence of air, calcium phosphate and calcium silicate appear as slag. Thus, the mixture of calcium phosphate and calcium silicate is known as Thomas slag or basic slag. The following reactions are involved in its formation (Pratiyogita Darpan 2008):

$$4P + 5O_2 \rightarrow 2P_2O_5$$

$$3CaO + P_2O_5 \rightarrow Ca_3(PO_4)_2$$

$$CaO + SiO_2 \rightarrow CaSiO_3$$

Calcium oxide is a basic flux that is obtained by adding limestone during the smelting process of Fe extraction. Thomas slag contains 16%–18% P_2O_5 and 40% lime and is useful for lowering the acidity of soil. It is well ground and is used as fertilizer.

Wood ashes generally contain 1%–2% P and 4%–10% K. Hardwood ashes are 45% carbonate-equivalent and are half as effective as lime for raising soil pH. Softwood ashes are less effective than hardwood (Traunfeld and Nibali 2003).

Worm castings are the rich digested *soil* produced from redworm farming. No guaranteed listed analysis is available due to the great variability in feedstock, storage, and handling; however, they represent a concentrated source of Ca, Mg, N, P, and K in a readily available form (Traunfeld and Nibali 2003).

Kelp products are seaweed-based products containing dissolved ocean minerals. Dried kelp will usually contain 1.6% to 3.3% N, 1% to 2% P_2O_5, and 15% to 20% K_2O. Kelp products are also valued as growth stimulants because of their high concentrations of trace minerals (over 60), amino acids, vitamins, and growth hormones, including cytokinins, auxins, and gibberellins (EEA 2007; Traunfeld and Nibali 2003).

Organic Sources of High-Phosphate Materials

Animal manures and bedding materials contain significant amounts of phosphorus in organic forms. After microbial mineralization from the organic forms, the phosphorus applied to soils is subject to the same fates as inorganic fertilizer phosphorus. Rates of land application of animal wastes such as poultry litter should be based on soil tests, nutrient content of the material, and crop needs.

Manures (commercial) are products that carry an NPK fertilizer analysis on their label and will also improve soil structure. Manures vary widely in moisture, nutrient content, and relative stability. Some manures are dewatered or otherwise stabilized for beneficial use, but most are applied directly on nearby agricultural lands as a nutrient and organic matter amendments (EPA 2007).

- *Cow or steer* (dehydrated) manure is exposed to 180°F, dried to 17% moisture, and ground into a fine, soil-like texture. Nutrients are more concentrated and the soluble salt level is probably higher in dehydrated manure than in locally available farm manure.
- *Cockadoodle Doo* is a layer hen manure that has been dehydrated.
- *Cricket* manure is the manure of crickets raised for bait.

Manure (local) is sheep, cattle, horse, and chicken manure widely available from nearby farms. Farm manures usually contain 1% or less each of NPK.

Rabbit, sheep, and chicken manures are higher in these nutrients. Manure mixed with urine-soaked bedding will be higher in N.

Biosolids (sewage sludge) and *biosolids composts* contain approximately 2% to 4% total P. Excess solids from biological processes such as activated sludge may still be referred to as *sludge*, but *biosolids* or *compost* are public relations terms sometimes used to refer to treated human waste (EPA 2007). According to excessive screening of municipal biosolids in Poland (60 biosolids), the median content of P in biosolids was 1.75% with a range of 0.15%–4.1% (Siebielec and Stuczynski 2008). Phosphorus content in composts is highly dependent on the substrates used for composting.

The phosphate in fertilizers and manure is initially quite soluble and available. Most phosphate fertilizers have been manufactured by treating rock phosphate with acid to make it more soluble. Manure contains soluble phosphate, organic phosphate, and inorganic phosphate compounds that are easily available. When the fertilizer or manure phosphate comes into contact with the soil, various reactions begin occurring that make the phosphate less soluble and less available. The rates and products of these reactions and further applicability to TE inactivation are dependent on such soil conditions as pH, moisture content, temperature, and the minerals already present in the soil.

Current engineering remediation technologies are based on thermal and physical–chemical treatments. Chemical precipitation, filtration, complexation, solvent extraction, electrochemical techniques, ion exchange, and adsorption are some of the commonly used processes (Montanher et al. 2005; Pérez-Marín et al. 2007). However, all these techniques are generally costly and highly disruptive as they require the excavation of the contaminated soil, its treatment, and replacement on site or disposal in specific landfills. As an alternative, much research has turned to remediation technologies focusing on the *in situ* immobilization of TEs using metal binding agents (and in particular chemical immobilization). These methods represent environmentally sustainable, nondisruptive, and cost-effective alternatives, especially for contaminated industrial sites. The objective is to immobilize metals primarily through the formation of new minerals with reduced solubility and enhanced geochemical stability in a wide range of environmental conditions (Corami et al. 2008a,b,c). As it will be demonstrated, *in situ* TE immobilization using phosphate is a cost-effective and environmentally positive remediation technology since it transforms reactive and bioavailable TE fractions into more stable forms with reduced mobility and bioavailability (Cao et al. 2009; Mignardi 2012; Yoon et al. 2007), and the United States Environmental Protection Agency (USEPA) recommends this technique for risk-based remediation of Pb-contaminated sites (USEPA 2001).

In this chapter, the effectiveness of high-phosphate amendments of contaminated soils is reviewed in terms of their impact on TE solubility, phytoavailability, leaching, bioaccessibility, and accumulation in soil organisms and mammals.

Concept of TE Bioavailability and Bioaccessibility

TE bioavailability can be defined as the fraction of the total content of the soil that can interact with a biological target (Geebelen et al. 2003). In the soil solution elements are present as free uncomplexed ions, ion pairs, ions complexed with organic anions, and ions complexed with organic macromolecules and inorganic colloids. The most important metal pools in the solid phase include the exchange complex and metals complexed by organic matter, sorbed onto or occluded within oxides and clay minerals, coprecipitated with secondary pedogenic minerals (e.g., Al, Fe, Mn oxides, carbonates and phosphates, sulfides), or as part of the crystal lattices of primary minerals (Adriano 2001). Bioavailability is therefore governed by the pseudoequilibrium between aqueous and solid soil phases rather than by the total metal content. The most important processes controlling the mobility of metals in the environment include dissolution/precipitation, complexation with ligands, sorption/desorption by solids, biotransformation, uptake by soil and aquatic biota, and reduction/oxidation (Reeder et al. 2006).

The replenishment of metals in the soil solution is ultimately controlled by the soil solid-phase components, and these metal fractions can therefore also provide valuable information regarding metal bioavailability or potential availability. Chemical extraction techniques (single extractions or sequential extractions) have been the most widely used methods as indicators of metal availability (Bolan et al. 2014). For example, the content of metal in soil solution (soluble) plus the *weakly adsorbed* content (exchangeable) are generally considered to provide a good measurement of the plant-available amount. Single extractants commonly used include dilute neutral salts (sodium nitrate [$NaNO_3$], ammonium nitrate [NH_4NO_3], calcium chloride [$CaCl_2$], calcium nitrate [$Ca(NO_3)_2$]), dilute acids (hydrochloric acid [HCl], ammonium acetate [NH_4OAc]), and metal chelating agents (ethylenediaminetetraacetic acid [EDTA], diethylene triamine pentaacetic acid [DTPA]). After a comparison of 12 commonly used single-extraction procedures, Meers et al. (2007) found 0.01 M $CaCl_2$ to provide a good indication of phytoavailability of Cd, Cu, Ni, Pb, and Zn in 21 soils differing in soil properties and levels of contamination. However, there is no universal soil extractant that can be used to estimate TE bioavailability to plants, and attempts to correlate extractable soil metal concentrations with plant uptake and bioaccumulation are frequently unsuccessful. Incubating Pb-contaminated soils (shooting range with 346 mg Pb/kg) with rock phosphate or soluble KH_2PO_4 led to a reduction in NH_4NO_3-extractable Pb concentrations (Park et al. 2011). The concentrations of $CaCl_2$-extractable Pb and Zn were significantly reduced in a multimetal-contaminated soil amended with phosphate rock, TSP, or a combination of both (Fang et al. 2012).

Another approach, the sequential extraction procedure, is more related to the chemical forms of metals in soils: specific chemical extractants (of increasing strength) applied in a sequential fashion to the soil sample can theoretically assess different TE fractions of different solubility related to mineralogical phases. The most commonly implemented procedures are the revised Community Bureau of Reference (BCR) three-stage sequential extraction protocol (Rauret et al. 1999) and the five-stage extraction protocol of Tessier et al. (1979). Specific chemical pools measured within the sequential fractionation scheme have been correlated with plant uptake of TEs and in some cases have successfully predicted the plant availability of TEs in soils (Naidu et al. 1997). However, limitations of sequential extractions include the possible redistribution of analytes among phases during extraction, nonselectivity of reagents for target phases, incomplete extraction, and precipitation of *new* mineral phases during extraction (Bacon and Davidson 2008). Application of different P amendments caused TEs in the soil matrix to shift from more available forms (e.g., exchangeable, carbonate, Fe–Mn oxide, organic matter bound) to strongly bound fractions (sulfide or residual). Most authors attribute this increase in the residual fraction to the formation of pyromorphite. For example, Cao et al. (2002) and Melamed et al. (2003) observed a decrease in the carbonate-bound soil Pb fraction, as well as the Fe–Mn oxide bound fraction (albeit to a lesser extent), while the residual fraction increased by up to 60%. This redistribution of Pb led to a greater plant growth and lower metal accumulation in the plant tissues. However, Ryan et al. (2001) recommended that the change of soil Pb to pyromorphite should not be concluded on the basis of sequential extractions alone, but should be confirmed using more sophisticated spectroscopic analyses.

More recently, determining the effectiveness of *in situ* immobilization of TEs (particularly Pb) after addition of phosphate-rich amendments has moved to the use of spectroscopic methods. Instrumental techniques familiar to many geochemists and mineralogists, such as synchrotron-based micro-x-ray fluorescence, micro-x-ray absorption near-edge structure spectroscopy, and micro-x-ray diffraction, are now being used more commonly for determining *in situ* metal speciation in soils and monitoring minute changes at environmentally relevant spatial and temporal scales (Baker et al. 2014). Baker et al. (2014) used synchrotron-based techniques to speciate Pb in a smelter-contaminated soil treated with several fluid and granular P amendments. Moreover, rates of Pb sequestration as Pb phosphates were evaluated by analyzing soils after distinct incubation times. The Pb phosphate mineral plumbogummite was found to be an intermediate phase of pyromorphite formation, which transforms into pyromorphite as P concentrations increase. All fluid (phosphoric acid, ammonium polyphosphate) and granular P sources (phosphate rock, monammonium phosphate, TSP) induced Pb phosphate formation, but fluid phosphoric acid was most effective with time and distance from the treatment. Of the granular P sources, TSP was most effective at greater distances from the point of application. Cao et al. (2011) used x-ray diffraction analysis to demonstrate that biochar derived from dairy manure containing a high amount of available P immobilized Pb in shooting range soil by forming insoluble hydroxypyromorphite ($Pb_5(PO_4)_3(OH)$).

The term bioaccessibility has been introduced in order to characterize oral bioavailability of contaminants. It refers to incidental ingestion of contaminated soil by humans, especially children. The bioacccessible fraction is understood as the pool of a contaminant that might be dissolved in the gastrointestinal tract of the human organism and become available for absorption (Ruby et al. 1999; Schroder et al. 2004). Bioaccessibility is most often expressed as percentage of total content of the contaminant in soil (Scheckel et al. 2009).

It has been reported that soil ingestion was a more significant source of exposure to such metals as Pb in urban areas than crops from urban gardens (Chaney and Ryan 1994).

The original *in vitro* physiologically based extraction test (PBET, Ruby et al. 1996) was designed to mimic the physiology and chemistry of the human gastrointestinal process and to establish the fraction of a metal that would be available for absorption into the body. Since that time, various methods of TE bioaccessibility measurements have been developed: solubility bioaccessibility research consortium (SBRC), *in vitro* gastrointestinal (IVG), the Dutch National Institute for Public Health and the Environment method (RIVM), Deutsches Institut für Normunge. V. (DIN), the Simulator of the Human Intestinal Microbial Ecosystem (SHIME), and TNO intestinal model (TIM) (Grøn 2005; Juhasz et al. 2007; Oomen et al. 2003; Schroder et al. 2003; Wragg et al. 2009). However, all these methods are based on similar principles—they are performed at conditions simulating gastrointestinal situations (pH, temperature, shaking speed). All the methods include stomach and intestinal phases, whereas some of the methods additionally contain a saliva phase. They differ in pH value of the extraction and presence of gastrointestinal enzymes.

Impact of High-P Soil Amendments on TE Bioavailability

TE Solubility and Phytoavailability

Metal immobilization has been studied using a range of P-containing amendments (Kumpiene et al. 2008) such as synthetic and natural apatites and hydroxyapatites (HAs) (Cao et al. 2004; Chen et al. 2007; Corami et al. 2007, 2008b, 2008c), phosphate rock (PR) (Brown et al. 2005; Cao et al. 2004), phosphate-based salts (PBSs) (Cao et al. 2003; Ownby et al. 2005), DAP (McGowen et al. 2001), phosphoric acid (PA) (Brown et al. 2005; Chen et al. 2003; Scheckel et al. 2005), and their combinations. Depending on the source, soil application of phosphate can cause direct adsorption of metals onto these compounds through increased surface charge and enhanced anion-induced metal adsorption. A decrease in TE mobility can be achieved by inducing various sorption processes: adsorption to mineral surfaces, formation of stable complexes with organic ligands, surface precipitation, and ion exchange. Phosphorus fertilizers (such as SSPs, TSPs, and DAP) are acidic phosphate compounds

(Bolan et al. 2003) that lead to a decrease in soil pH and consequent dissolution of both P and Pb and subsequent precipitation of lead phosphate compounds. Precipitation as metal phosphates has been proven to be one of the main mechanisms for the immobilization of metals such as Pb and Zn in soils (McGowen et al. 2001). However, because of the limited knowledge about the relative contribution of each process, it seems that the above four mechanisms could all be involved (Kumpiene et al. 2008). Different P sources and forms (fluid vs. granular) have also been shown to result in different efficacies of Pb stabilization in the same contaminated material (Baker et al. 2014). Park et al. (2011) demonstrated the potential value of both soluble and insoluble P compounds in the immobilization of Pb, thereby reducing its bioavailability and toxicity. These fairly stable metal-phosphate compounds have extremely low solubility over a wide pH range, which makes phosphate application an attractive technology for managing metal-contaminated soils. The formation of the new solid phase (precipitates) occurs when the ionic product in the solution exceeds the solubility product of that phase (Adriano et al. 2004).

Although P amendments have mainly been applied to remediate Pb-contaminated soil, they may also be applicable to other metals that frequently occur simultaneously with Pb in soils (Hamon et al. 2002). The effectiveness of such phosphate materials to immobilize metals other than Pb and the mechanisms involved are still unclear (Chen et al. 2007). Their efficacy varies according to the particular combination of elements at a site as well as the desired end points of concern (Brown et al. 1995): for example, Cu, Zn, and Pb have been shown to have similar responses to the same amendments. On the other hand, in cases of soils cocontaminated with Pb and As, P addition can effectively reduce Pb availability but inadvertently solubilize As. The presence of competing ions usually reduces the treatment efficiency. Competition of Zn for phosphorus amendments was suggested to be responsible for unexpected higher Pb bioavailability in treated soils (Ownby et al. 2005). By contrast, Pb removal by phosphate rock was least affected by the presence of competing metals compared to Cu and Zn (Cao et al. 2004; Kumpiene 2008). According to Porter et al. (2004), stable pyromorphites in soil can be formed only after all soluble Ca and Mn are excluded. Therefore, in addition to the very high concentrations of necessary P, high amounts of precipitated apatite-like minerals can adversely change the soil structure (Kumpiene et al. 2008). In general, high-phosphate materials are considered to be more effective for Pb immobilization than for Zn, Cu, and Cd. The usual stability sequence of metal phosphates is Pb > Cu > Zn (Bolan et al. 2003). It is well documented that lead phosphates are the most insoluble and stable forms of Pb in soils and are thought to form rapidly in the presence of adequate Pb and P. Among all the lead phosphate minerals, chloropyromorphite has the lowest solubility, and thus, it is most stable under favorable environmental conditions. The formation of insoluble pyromorphite-like minerals was responsible for Pb immobilization, whereas Zn, Cu, and Cd immobilization was attributed to coprecipitation and surface complexation mechanisms (Miretzky and Cirelli 2008).

Adsorption of metals onto HA surfaces has been observed for a number of metals including Cd, Cu, Ni, strontium (Sr), and Zn (Aklil et al. 2004; Corami 2007; Gomez del Rio et al. 2004). According to these authors, the sorption process can be divided into two steps: first, rapid complexation on specific sites of the high-phosphate materials surface (e.g., HA); and second, metal diffusion into the structure of the high-phosphate material itself or to a trace-metal-containing high-phosphate material formation (Corami et al. 2007). Rapid metal immobilization indicates that intraparticle diffusion is involved in the overall sorption process. The contribution of the ion exchange process to the metal immobilization is difficult to determine due to HA dissolution produced by the decrease in pH. The substitution of cations in the HA crystal structure is a result of the ion exchange process (Monteil-Rivera and Fedoroff 2002). Several studies suggest that the mechanisms operating in this second step of the sorption process involves either the ion exchange and the partial dissolution of HA followed by the precipitation of a TE-containing HA (Corami et al. 2007, 2008a,b,c; Monteil-Rivera and Fedoroff 2002; Sugiyama 2001).

Sequestration of Cd in phosphates may attenuate the phytoavailability of Cd inputs to soils. However, since the free activities of Cd in soil solutions are well below the potential chemical equilibrium of known solid phases, the solubility and mobility of Cd in soils are therefore likely to be controlled by sorption or coprecipitation mechanisms (Carrillo-Gonzalez et al. 2006; Jiao et al. 2012; Lambert et al. 2007). Munksgaard and Lottermoser (2011) found, among various types of phosphate fertilizers, that potassium orthophosphate fertilizer was the most effective amendment for Cd stabilization, while superphosphate and TSP fertilizers were the most effective amendments for Pb stabilization in mining-impacted soils in Australia. The immobilization of Cd by P depended greatly on the level of Cd in the soil and the level of P applied (Hong et al. 2008; Lambert et al. 2007).

Since the effectiveness of insoluble P sources (such as rock phosphates) on Pb immobilization depends on their rate of dissolution, recent studies have proposed that this dissolution rate can be enhanced by inoculating soils with phosphate-solubilizing bacteria (Park et al. 2011). These bacteria produce organic acids and phosphatases that can increase the solubilization of insoluble P compounds (Park et al. 2010). Bacterial-induced phosphorus dissolution was shown to increase the efficiency of Pb immobilization by rock phosphate in Pb-contaminated soils from a shooting range (Park et al. 2011). A further advantage to the use of these inoculants is that a lower amount of rock phosphate was required, making the process even more cost effective.

The addition of organic amendments to soils has also often been shown to increase the immobilization of TEs through adsorption reactions. However, the ability of such organic amendments to immobilize metals on a long-term basis is unclear, since over time, organic material will decompose. Li et al. (2001) gave evidence for greater affinity for Cd adsorption by the inorganic components of biosolid compost-amended soils, indicating that the increased adsorption of Cd is independent of the added organic matter and of a persistent nature. Increasing evidence seems to indicate a long-term stability of biosolids and limited release of metal(loid)s (Park et al. 2011b). Nonetheless, these organic products can be combined with other materials such as P or Fe-rich products.

It is clear that phosphate can form stable compounds with toxic elements, and particularly with Pb, but the long-term stability of immobilized metals (such as Pb) has been addressed less often. Monitoring the effectiveness of these phosphate amendments in enhancing immobilization and reducing metal bioavailability in contaminated soils on a long-term will therefore be vital. Park et al. (2011) considered the long-term stability of both soluble and insoluble phosphate compounds for Pb immobilization in a contaminated soil from a shooting range. Strong reductions in NH_4NO_3-extractable Pb concentrations were observed after the addition of tricalcium phosphate, HA, rock phosphate, and potassium dihydrogen phosphate. Using the simple bioaccessibility extraction test (SBET), the authors demonstrated that the phosphate-induced reductions in Pb bioavailability remained stable after 2 years of incubation.

Several authors have observed strong reductions in plant phytotoxicity after incorporating phosphate materials to TE-contaminated soils. Brown et al. (2004) evaluated the effects of biosolids in combination with different materials—Fe oxides, TSP, and phosphoric acid—to immobilize Pb in a smelter-contaminated soil. The greatest reductions in the uptake of Pb by lettuce were observed in the TSP and phosphoric acid treatments.

Application of HA significantly reduced Pb in shoots of *Brassica* species (Zhu et al. 2004) and *Sorghum bicolor* (*S. bicolor*) (Chen et al. 2007). Laperche et al. (1997) evaluated the effect of amending Pb-contaminated soil with apatite on the uptake of Pb by *S. bicolor.* The Pb shoot content was reduced drastically after addition of apatite (both natural and synthetic), but Pb and P contents in the plant roots increased as the quantity of added apatite increased. This was attributed to the accumulation of Pb in the roots and formation of pyromorphite on root surfaces. TSP and a combination of rock phosphate and Mn oxide achieved a significant reduction in Pb concentrations in *S. vulgare* and *Beta vulgaris* (Hettiarachchi and Pierzynski 2002). Brown et al. (2004) screened a range of P and Fe amendments for their ability to reduce both phytotoxicity and bioavailability of Pb, Zn, and Cu in a smelter-contaminated soil in the United States. Amending with H_3PO_4 (1%) was the most effective at reducing plant Pb, Zn, and Cd (*Festuca arundinacea* Schreb. cv. Kentucky-31).

Negim et al. (2012) evaluated the use of P-spiked Linz–Donawitz (LD) slag to reduce the phytototoxicity of a Cu-contaminated soil. Bean plants grown on untreated soil showed a high phytotoxicity, while the incorporation of P-spiked LD slag increased the root and shoot dry weight yields. In addition, this amendment increased foliar Ca concentration while reducing Cu, Zn, and Cr concentrations.

Betts et al. (2013) evaluated the effects of biochars produced from meat and bonemeal on Zn adsorption in contaminated soils. These waste materials contain large amounts of calcium phosphate and were proposed as useful sorbents for the remediation of Zn-contaminated soils. Using a commercial bonemeal biochar as an amendment, these authors showed that above a soil pH of 6.1, Zn was quickly adsorbed to the biochar (within 5 h) with a maximum adsorption capacity of 0.65 mmol Zn g^{-1}. Synchrotron-based Zn K-edge extended x-ray absorption fine structure spectroscopy was consistent with a tetrahedral Zn bound to phosphate groups in a monodentate inner-sphere surface complex. With an acidification pretreatment and at a more acidic soil pH, the biochar caused precipitation of a $ZnPO_4$ phase.

Improvements in plant growth after the incorporation of organic amendments due to a reduction in metal phytotoxicity can be found in numerous reports. Pierzynski et al. (2002) applied cattle manure to mine tailings (at a rate of 90 Mg ha^{-1}) and increased growth of *Festuca arundinacea* in the first year after treatment. Farfel et al. (2005) amended Pb-containing yard soils with biosolids compost and showed a significant reduction in bioaccessible Pb fraction and the development of a healthy turfgrass cover at a 1-year follow-up. Composted cattle manure (with and without lime and bentonite) applied to mine wastes led to an increase in waste nutrient content, pH, and available water, and to a decrease in metal availability (Baker et al. 2011). Additional benefits included an increase in microbial biomass and activity.

Soil amendments also proved to be successful in phytostabilization of military-range-contaminated soil from Aberdeen, MD (Siebielec and Chaney 2012). Both Ca-phosphate and Orgro compost substantially reduced Sr-nitrate-extractable Zn, Cd, and Pb; however, only compost treatments fully established tall fescue growth.

TE Leaching

Despite the large number of studies that can be found in the literature demonstrating the potential usefulness of P compounds in TE immobilization, fewer studies have evaluated the effects of these amendments on the leaching of associated metals. The leachability of metals after application of phosphate materials has been evaluated using different methodologies, including the toxicity characteristic leaching procedure (TCLP), or on the basis of leaching column experiments or through the collection of soil pore waters.

In a small-scale field experiment, Melamed et al. (2003) found that different P treatments (combinations of H_3PO_4, $Ca(H_2PO_4)_2$ and rock phosphate) effectively reduced the TCLP-extractable Pb concentrations in the surface soils (0–10 cm) to well below the 5 mg/L critical level for hazardous waste (USEPA 1995). Amending with both H_3PO_4 and $Ca(H_2PO_4)_2$ also reduced TCLP-Pb to below the critical level within subsurface soils (10–20 cm). In contrast, despite an observed reduction in the NH_4NO_3-extractable concentration of Pb, leaching of Pb was increased after amending a Pb-contaminated soil with potassium dihydrogen phosphate. This was attributed to the fact that the soluble P treatment increased the dissolved organic carbon (DOC) concentration of soil, thereby increasing Pb mobility (Park et al. 2011). However, the same authors demonstrated a reduction in this Pb leaching when soils were amended with an insoluble P source (rock phosphate) in combination with phosphate-solubilizing bacterial inoculants.

Most studies evaluating the risks of TE leaching have highlighted the need for caution when amending with H_3PO_4. Phosphoric acid is needed to catalyze the dissolution of metastable Pb, making it available for further immobilization reaction. However, its addition can lead to a decrease in soil pH and consequent leaching of other trace metals. Melamed et al. (2003) stressed that this could be of particular concern in sandy soils with low buffering capacity. As an alternative, they suggested combining H_3PO_4 with either calcium phosphate or rock phosphate, treatments that had less impact on soil pH but effectively induced Pb immobilization. The same authors proposed applying the compounds with less phosphate (such as calcium phosphate or phosphate rock) prior to application of phosphoric acid.

Bonemeal waste materials contain large amounts of calcium phosphate and are potentially useful sorbents for the immobilization of soil metals. In a leaching column experiment, bonemeal additions reduced metal release from a contaminated soil, increased soil and leachate pH, and decreased soil leachate toxicity (Hodson et al. 2001). Minimal P leaching occurred from the soil. The data are consistent with a proton-consuming bonemeal (calcium phosphate) dissolution reaction followed by the formation of metal phosphates. Relatively low (1:50 bonemeal/soil) concentrations of fine (90–500 mm) bonemeal would appear to be an effective treatment for metal-contaminated soils.

In our own study, we examined the effect of various treatments on leachability of TEs from the heavily contaminated soil collected near a former Zn/Pb smelter in Piekary, Poland (Figures 9.1 and 9.2). The following soil amendments were tested: usual rates of NPK, compost GWDA (produced by GWDA company in Pila, Poland) (rate 10% w/w), compost Zabrze (10%), biosolids (10%), Ca-phosphate ($CaHPO_4$ reagent grade) (P = 2%), Fe oxide ($Fe(OH)_3$ reagent grade) (Fe = 2%), clay rock waste (10%), bentonite (5%), and limestone (Ca-carbonate reagent grade) (10%). Leaching of Cd was rather a simple function of pH change after soil treatment (Figure 9.1). In the case of Pb leaching, there were mechanisms other than pH change, such as precipitation of insoluble Pb-phosphates and adsorption and sorption onto organic compounds. Ca-phosphate was the most successful in reducing Pb leaching among all test amendments; Fe-oxide and P-rich Zabrze compost were also effective. In contrast, soil amendment with municipal sewage sludge increased Cd and Pb leaching due to tremendous pH decrease. This proves that sludges should be applied to contaminated soils along with limestone to alleviate risk of TE remobilization resulting from acidification processes present in soil when sludges are mineralized.

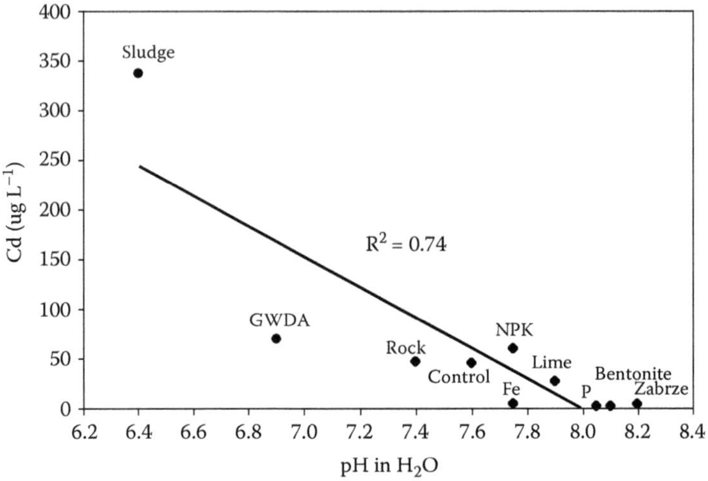

FIGURE 9.1
The impact of various soil amendments on Cd leachability from heavily contaminated soil, near a Zn/Pb smelter. Bentonite = milled bentonite, Control = untreated soil, Fe = iron oxide, GWDA = compost GWDA company, Lime = calcium carbonate, NPK = agricultural NPK rates, P = calcium phosphate, Rock = milled clay rock waste, sludge = municipal sewage sludge, Zabrze = compost municipality of Zabrze.

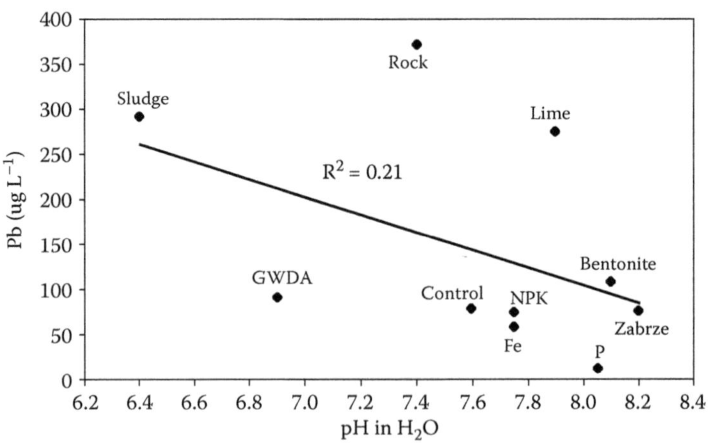

FIGURE 9.2
The impact of various soil amendments on Pb leachability from heavily contaminated soil, near a Zn/Pb smelter. Bentonite = milled bentonite, Control = untreated soil, Fe = iron oxide, GWDA = compost GWDA company, Lime = calcium carbonate, NPK = agricultural NPK rates, P = calcium phosphate, Rock = milled clay rock waste, sludge = municipal sewage sludge, Zabrze = compost municipality of Zabrze.

TE Accumulation in Soil Organisms

Literature data on the impact of soil amendments on TE accumulation in soil organisms are limited to earthworms (*Lumbricus terrestris, Lumbricus rubellus, Eisenia fetida, Eisenia andrei*) and are rather scarce. Some papers report reduction in metal accumulation in earthworm tissues after application of phosphate materials, the most effective of which were usually relatively easily soluble commercial fertilizers or chemical reagents. Ownby et al. (2005) observed significantly reduced accumulation of Pb in *Eisenia fetida* after the treatment of mining waste soil collected in Joplin, MO, with TSP at the rate of 0.5% of soil weight. The combination of TSP and rock phosphate enabled reduced accumulation of both Pb and Zn. On the other hand, monocalcium phosphate was not efficient. Further increasing rates of phosphates and artificial soil acidification caused mortality of the earthworms. Treating the soil from the same area with KH_2PO_4 and TSP at a similar rate of P (0.5% of soil weight) rapidly reduced bioavailability of Pb, Cd, and Zn to earthworms (Maenpaa et al. 2002). The significant effect of the treatments on the metal accumulation had been recorded already after 100 h of the worm presence in the amended soil. No further increase in Pb, Cd, and Zn content in earthworms was observed when the organisms stayed longer than 100 h in the soil. The lower rates of P (0.06% of soil weight) did not have a significant impact on metal accumulation in the *Eisenia fetida*.

Park et al. (2012) documented that the efficiency of rock phosphate materials in reducing metal bioavailability to soil organisms might be enhanced by application of phosphate solubilizing bacteria (*Enterobacter* sp.). Rock phosphate in the presence of the bacteria increased immobilization of Pb in a spiked soil and the metal accumulation in the *Eisenia fetida* tissues at a soil pH of approximately 5.7. A soluble form of phosphates (KH_2PO_4) was slightly more effective than the rock phosphate treated with the bacteria in reducing bioavailability of Pb. In contrast, Morgan et al. (2002) did not observe any effect of phosphate (hydroxyapatite) on Cd, Pb, and Zn accumulation in *Lumbricus rubellus* exposed to heavily contaminated Shipham soil.

The data on effectiveness of organic amendments in reducing metal accumulation in earthworms are scarce and ambiguous. Gonzales et al. (2013) did not observe any compost effect on *Eisenia andrei*, whereas Sizmur et al. (2011) reported reduced Pb accumulation in *Lumbricus terrestris* after green waste compost application to the soil from a former lead mine area at a rate of 10% of soil weight. Unfortunately in both cases, the exact contents of P in the composts are not provided.

TE *In Vitro* Bioaccessibility

Studies on TE *in vitro* bioaccessibility found in the literature usually refer to optimization of methodologies (Juhasz et al. 2007; Oomen et al. 2003; Schroder et al. 2003) or serve as a measure of the TE bioavailability in postindustrial regions (Patinha et al. 2012; Pelfrene et al. 2011). A limited number of papers have reported the effectiveness of various soil amendments in reducing TE bioaccessibility. All of them generally proved effectiveness of P fertilizers in minimizing Pb bioaccessibility; however, data concerning other elements such as Cd or As are not available.

Tang et al. (2004) demonstrated that SSP overperformed rock phosphate in limiting Pb bioaccessibility in the gastric phase, whereas HA was the most effective in the small intestinal phase. Sonmez and Pierzynski (2005) compared the effectiveness of P fertilizers and Mn oxides and their combinations for treating soils from mining area and repository sites in Joplin, MO, or wastes collected in Dearing, KS. Both rock phosphate and TSP significantly minimized Pb bioaccessibility (PBET procedure, Ruby et al. 1996) as single amendments, reducing it to 46%–84% and 58%–88% of untreated soils, respectively. The effectiveness of the P amendments was enhanced by some Mn oxides, especially birnessite (e.g., a TSP and birnessite combined treatment reduced Pb bioaccessibility to 10%–72% of control). The authors attributed this observation to the combined effect of precipitation of insoluble Pb phosphates and adsorption of Pb onto Mn oxides. It is worthwhile to note that soil pH in the experiment was optimized for Mn oxide effectiveness and set at approximately 7.0.

Bosso et al. (2008) confirmed that the speed of Pb immobilization was highly dependent on the solubility of Pb compounds. Sodium phosphate reduced Pb bioaccessibility (PBET) at least 10 times in gastric phase after 1 month of soil incubation. As proved by the scanning electron microscopy, this reduction was attributed to formation of a pyromorphite type of Pb compound. Commercial fertilizers (powder superphosphate and powdered rock phosphate) were less effective and their effectiveness was time-dependent, but they also enabled substantial Pb immobilization, especially after 6 months of soil incubation. This confirms that the P fertilizers can effectively reduce the risk of Pb absorption by humans via ingestion route at a reasonable cost.

Another study by these authors (Bosso and Enzweiler 2008) indirectly proved the role of phosphates in reducing oral bioavailability of Pb. The lowest Pb bioaccessibility among soils and wastes collected in mining areas in Brazil, expressed as percent of total Pb, was measured in wastes where Pb was partly present as phosphates (plumbogummite, $PbAl_3(PO_4)_2(OH)_5{\cdot}(H_2O)$) as detected by x-ray diffraction (XRD).

Brown et al. (2004) compared effectiveness of municipal biosolids compost and inorganic amendments in reducing Pb bioaccessibility (simplified procedure at pH 2.2) in heavily contaminated soil in Joplin, MO. TSP treatment was most effective: at rates of 1% and 3.2%, it reduced bioaccessible Pb by 53% and 87%, respectively. However, these treatments caused unacceptable pH decrease, up to 4.4 and 3.4, respectively, compared to 7.0 in control soil. Bringing soil pH back to the levels before the treatment would be too costly. As an alternative, biosolids compost reduced Pb bioaccessibility by 42% as single amendments and by 65% in combination with lime and Fe and preserved soil pH at an optimal level. Successful application of composts was also demonstrated by Siebielec and Chaney (2012). The compost manufactured using unlimed biosolids as feedstock, containing 1.85% of P, was effective in reducing Pb bioaccessibility (gastric phase at pH 2.2) in military-range-contaminated soil in Aberdeen, MD. The compost treatment at a rate equivalent to 225 Mg ha^{-1} (approximately 11% of soil weight) was especially effective when combined with Ca-carbonate and significantly overperformed the increased agricultural rate of P as $CaHPO_4$ (0.025%, equivalent of 500 kg ha^{-1}).

TE Accumulation in Animals

Feeding trials testing the effectiveness of various soil treatments in reducing TE bioavailability to animals are very limited due to the high cost and controversies related to ethical issues. Brown et al. (2004) demonstrated a positive impact of TSP application at a rate of 1% combined with Fe added at a rate of 2.5% to the Joplin site soil on Pb accumulation in rats fed with the amended soil. Pb contents in bone and blood in animals fed with such soil were substantially lower than in the case of rats fed with control soil or soil treated with compost.

Stuczynski et al. (2007) tested the bioavailability of TEs to young cattle in a feeding trial utilizing the hay collected from the Zn and Pb smelter wasteland revegetated with biosolids and limestone treatment. The municipal biosolids contained 2% of P and were applied at a rate of 300 t DM ha^{-1} combined with by-product lime to extremely toxic slag. The animals fed with the wasteland hay did not accumulate higher levels of Pb and Cd in muscles than cattle fed with control uncontaminated hay. Cadmium that accumulated in the kidney and liver, however, still did not exceed the national maximum permissible levels in these tissues. This study demonstrated that reclamation of such wastelands with biosolids did not pose any substantial risk for wildlife and food safety (Figure 9.3).

FIGURE 9.3
Long-term effect of high-P biosolids plus limestone treatment of smelter wasteland in Piekary, Poland, on establishment of plant growth. Untreated barren control area is shown on the left.

Risk Related to High-P Soil Amendments

A major risk related to the application of high-P materials for the stabilization of contaminated soil is eutrophication of groundwater and surface water. The optimum level of P amendment is a critical issue when soluble P is used as a Pb immobilizing agent due to the potential eutrophication resulting from excessive P leaching to surface and groundwater (Park et al. 2011). Immobilization of TEs such as Pb as insoluble forms (e.g., chloropyromorphite) in soil requires large amounts of P since it has been shown that an important fraction of the added P is itself likely to interact with Ca, Fe, and Al to form various phosphate compounds such as apatite, strengite, and variscite (Porter et al. 2004). There is therefore concern regarding the fate of the excessively added P (Basta and McGowen 2004; Cao et al. 2002; Park et al. 2011).

Application of phosphates for immobilization of Pb or Zn in soils contaminated with a mixture of TEs poses a risk of As leaching. Alkaline high-phosphate materials clearly should be avoided as the mobility of this element usually increases in the alkaline pH range. Acidic soil pH conditions increase metal and phosphate solubility. An increase in pH from 5 to 8, for example, results in a decrease in HA solubility of two orders of magnitude (Shi and Erickson 2001). Zhang and Ryan (1998, 1999) showed that conversion of anglesite, cerussite, and galena to chloropyromorphite ($Pb_5(PO_4)_3Cl$) or hydroxypyromorphite ($Pb_5(PO_4)_3OH$) occurred between pH 4 and 5, which also optimizes the rate of HA dissolution. This element grouping is mainly determined by the dominant TE charge: Cu, Zn, and Pb appear as cations, while As and mobile Cr(VI) appear as oxyanions, and therefore their reaction to pH changes generally follows the reverse pattern. This confirms that pH is one of the most important factors having a very strong impact on the element mobility in soil. Other factors, like redox potential, can greatly modify element mobility, especially of redox-sensitive elements such as As and Cr (Kumpiene 2008; Shi and Erickson 2001; Zhang and Ryan 1998).

Martin and Ruby (2003) evaluated the *in situ* fixation of a Pb- and As-bearing soil. Lead bioavailability was effectively reduced through the use of PO_4-based amendments to induce the formation of insoluble lead-apatite minerals. However, amendment of soils with PO_4 amendments resulted in increased As bioaccessibility and leachability, which was attributed to the competition of PO_4 with As for sorption sites within the soil matrix. These authors suggested that the increase in As leachability and bioaccessibility in response to the phosphate amendment could be offset by including Fe-based chemicals in the amendment formulation (e.g., ferrihydrite and $Ca(H_2PO_4)_2$).

Application of high rates of organic materials causes risk of excessive nutrient transport to groundwater and surface waters. Nutrient-leaching losses from organic soil amendments are an important issue, particularly in the case of N and P from animal manures (Park et al. 2011). Amending soils with highly organic materials can generate large concentrations of soluble organic matter to which free ions (e.g., Cu) can complex and comobilize with organic ligands during leaching events (Bernal et al. 2007). Long-term potential of organic by-products for metal immobilizing materials is questionable.

Examples of Soil Remediation (Metal Immobilization) Using High-P Materials

A number of studies have shown that new practical approaches for *in situ* remediation by addition of amendments that reverse TE risk from soil ingestion have been demonstrated in recent years. Practical, inexpensive methods are available to revegetate such contaminated soil materials and support vegetation that can be safely consumed by wildlife, livestock, and humans. Many laboratory studies have evaluated the effectiveness of phosphate-induced TE immobilization in contaminated soils (Brown et al. 2005; Cao et al. 2002, 2003, 2004; Chen 2007; Chen et al. 2007, 2009; Echeverria et al. 1998; Madrid et al. 2006; Mahmoud et al. 2014; McGowen et al. 2001; Mignardi et al. 2012; Scheckel et al. 2005; Yoon 2007). Nevertheless, the implementation of this method has been very successful in the laboratory, but the immobilization of TE using P amendments on a large scale in field experiments and their evaluation is still limited, perhaps due to the fact that the majority of *in situ* remediation technologies for metals in soils are still primarily in the developmental stage, and site-specific testing is required prior to implementation at a given site.

Examples of Field Studies

Results from the studies of Bert et al. (2012) reported that *in situ* incorporation of Thomas basic slag (TBS) into a landfilled metal-contaminated sediment decreases metal mobility and sediment ecotoxicity. In contrast to what was observed after addition of HA (3% DW), incorporating TBS (5% DW) into a 5-year deposited sediment contaminated with Zn, Cd, Cu, Pb, and As resulted in a decrease in the extractable concentration of Cd and Zn. Shoot Cd and Zn concentration in the dominant plant species (*Calamagrostis epigejos*) also decreased in the presence of TBS. The addition of TBS and HA reduced sediment ecotoxicity and bioavailability.

According to Friesl (2006) the most effective treatment for reducing Cd and Pb uptake and transport in plants is the sludge selected from 15 potential amendments. Results from this experiment identified three of these amendments for use in a field trial. In a pot experiment, a reduction in NH_4NO_3-extractable Cd (<41%) and Pb (<49%) compared to the controls was obtained, with a concurrent reduction in their uptake into barley grain (Cd < 62%, Pb < 68%). In the field, extractable fractions of Cd, Pb, and Zn were reduced by up to 96%, 99%, and 99%, respectively, in amended soils. These authors proposed that the most effective means of reducing Cd transfer to the food chain would be to combine soil amendments with the cultivation of Cd-excluding crop cultivars.

The effectiveness of Pb immobilization using combinations of H_3PO_4, $Ca(H_2PO_4)_2$, and phosphate rock was assessed at a previous battery recycling site. The total soil Pb concentration was 11.6 g/kg, and Pb was primarily associated with the carbonate fraction (50%–70%; as indicated through both a BCR sequential extraction procedure and XRD analyses). In the short term, all P treatments were able to modify the partitioning of Pb from the nonresidual/potentially available fractions (particularly carbonate-bound Pb) to the residual/less-available fraction (60% of total soil Pb was transformed into residual fraction). In surface soils (0–20 cm), the residual Pb fraction increased by up to 60% in comparison with the unamended soil.

Brown et al. (2007) tested a range of soil amendments (DAP fertilizer, municipal biosolids, biosolids compost, and Al- and Fe-based water treatment residues) on Pb-, Zn-, and Cd-contaminated yard soils and tailings at the Tar Creek National Priorities List (NPL) site (Oklahoma) for their effectiveness in restoring a vegetative cover and reducing metal availability. For the yard soils, all amendments reduced bioaccessible (assessed via PBET) Pb and several amendments reduced plant uptake of Zn (*Cynadon dactylon* L.) and NH_4NO_3-extractable concentrations of Cd and Zn. For the tailings, the largest reductions in bioaccessible Pb were achieved after amending with DAP (3%) and a combined treatment of DAP (3%) with biosolids. However, during the first season, the addition of DAP reduced plant growth, while the highest growth was obtained in the treatments that included compost and biosolids. This effect was lost in the second year when growth was vigorous for all treatments. Moreover, plant Zn and Cd content and soil extractable metal concentrations were also reduced.

Cao et al. (2002, 2003) established pilot-scale field experiments comparing the effects of different combinations of P treatments on Pb immobilization. Phosphoric acid was added on its own or in combination with either $Ca(H_2PO_4)_2$ or phosphate rock. Incorporating H_3PO_4 was expected to lead to the solubilization of Pb-bearing minerals and also the phosphate rock, thus increasing readily available Pb in the soil and inducing the formation of insoluble pyromorphite-like minerals. The desired effect was achieved; however, the authors concluded that the combination of H_3PO_4 with phosphate rock produced less impact on soil pH reduction and less leaching of soluble P.

Changes in Pb bioavailability and bioaccessibility at a field test in Joplin, MO, were followed for 3 years after treatments were incorporated. There was a reduction over time in bioavailable Pb but only a small reduction of bioaccessible Pb years after treatment compared to the drop in bioaccessible Pb and bioavailable Pb at the time of treatment (Ryan et al. 2004). This small reduction of bioaccessible Pb over time after the sharp drop at the time of treatment may have resulted from an artifact of bioaccessibility testing in that the chemical reactions to make pyromorphite occur during the extraction as well as in the field (Scheckel and Ryan 2003). Additionally, reduction in soil Pb bioavailability to humans was considerably greater than the reduction measured using rats or pigs (Ryan et al. 2004).

A smelter waste reclamation method based on the application of biosolids and by-product lime combined with a mixture of local grasses was tested in Piekary, Poland (Stuczynski et al. 2007). The pilot implementation involved reclamation of almost 1 ha of the wasteland containing toxic smelter slags. The method enabled successful revegetation of previously barren surface and minimized the risk related to dispersion of Cd and Pb in the local environment. Extensive monitoring of metal solubility, microbial activity, and plant cover confirmed the long-term persistence of the method. The observed metal inactivation processes were the combined effects of pH shift after by-product lime application, sorption potential of biosolids organic matter, adsorption of TEs on biosolids Fe-oxides, and precipitation of TEs with biosolids phosphorus.

Three various P fertilizers and their rates were tested in a field demonstration study in a mining area in Shaoxing, China (Xie et al. 2006). The goal was to set the rates of SSP, RP, and calcium magnesium phosphate effective in the reduction of soil Pb solubility and Pb uptake by cabbage (*Brassica chinensis* L.). All the fertilizers minimized Pb accumulation in the vegetable; however, a substantial effect was found after application of high rates of P, equivalent to 3 and 5 tons ha^{-1}. Such rates would potentially pose a risk for eutrophication of groundwater. The study also revealed that SSP caused significant soil acidification that might result in remobilization of other TEs such as Cd or Zn.

A 2-year field experiment evaluated the use of olive-mill waste compost to promote plant growth and vegetation cover in a TE-contaminated soil (Pardo et al. 2014). Compost addition increased the soil total organic-C concentration, maintained elevated available P concentrations, and stimulated microbial biomass. However, leaf Cu concentrations were increased by amendments, but this was not considered problematic since they remained at nontoxic levels.

Summary

A literature search provides extensive information on testing various high-P materials for remediation of TE-contaminated sites. These materials have been applied especially to reduce Pb solubility in soil and Pb phytoavalability, and therefore the most common parameters related to bioavailability of TEs that have been tested are Pb extractability and Pb uptake by plants. The experiments most commonly involved less soluble forms of phosphorus, such as apatites and other rock phosphates, moderately soluble superphosphates, and for comparison, readily soluble fertilizers such as ammonium phosphates. Less soluble rock fertilizers would be otherwise recommended for practical implementation due to lower cost and lower risk of eutrophication even if the reaction with the contaminants is slower.

Effectiveness of P fertilizers in reduction of Pb solubility is well documented. The impact of P materials on such parameters as TE leaching, TE bioaccessibility, and TE accumulation in soil organisms has been studied relatively rarely. Therefore, it is still difficult to assess the full effects of high-P soil amendments for quality of soil environment and answer the question of whether they can serve as effective single amendments. There is also a lack of experiments comparing P amendments with other types of soil amendments useful in TE inactivation and assessing how to improve the effectiveness of P amendments or avoid risks (e.g., through combining them with other amendments). The risk of using high rates of P materials for soil remediation are well defined: eutrophication of groundwater, As leaching, and nutrient dispersion when high-P organic materials are applied to soil. Furthermore, some P fertilizers, such as superphosphates, can lower soil pH, which creates the risk of remobilization of contaminants and weakens conditions for plant growth and microbially driven processes.

References

Adriano, D., W. Wenzel, J. Vangronsveld, S. Bolan. 2004. Role of assisted natural remediation in environmental cleanup. *Geoderma* 122:121–142.

Adriano, D.C. 2001. *Trace Elements in Terrestrial Environments: Biogeochemistry Bioavailability and Risk Metals*, 2nd Edition, Springer, New York.

Aklil, A., M. Mouflih, S. Sebti. 2004. Removal of heavy metal ions from water by using calcined phosphate as a new adsorbent. *Journal of Hazardous Materials* 112:183–190.

Alkorta, I., J. Hernandez-Alica, J. Becerril, I. Amezaga, I. Albizu, C. Garbisu. 2004. Recent findings on the phytoremediation of soils contaminated with environmentally toxic heavy metals and metalloids such as zinc, cadmium, lead and arsenic. *Reviews in Environmental Science and Bio/Technology* 3:71–90.

Bacon, J.R., C.M. Davidson. 2008. Is there a future for sequential chemical extraction? *Analyst* 133:25–46.

Baker, L.R., G.M. Pierzynski, G.M. Hettiarachchi, K.G. Scheckel, M. Newville. 2014. Micro-x-ray fluorescence, micro-x-ray absorption spectroscopy, and micro-x-ray diffraction investigation of lead speciation after the addition of different phosphorus amendments to a smelter-contaminated soil. *Journal of Environmental Quality* 43:488–497.

Basta, N., J. Ryan, R. Chaney. 2005. Trace element chemistry in residual treated soil: Key concepts and metal bioavailability. *Journal of Environmental Quality* 34:49–63.

Basta, N.T., S.L. McGowen. 2004. Evaluation of chemical immobilization treatments for reducing heavy metal transport in a smelter-contaminated soil. *Environmental Pollution* 127:73–82.

Bernal, M.P., R. Clemente, D.J. Walker. 2007. The role of organic amendments in the bioremediation of heavy metal-polluted soils. In: *Environmental Research at the Leading Edge*, A.B. Gore (ed.), Nova Science Publishers, New York, pp. 1–57.

Bert, V., C. Lors, J. Ponge, L. Caron, A. Biaz, M. Dazy, J. Masfaraud. 2012. Metal immobilization and soil amendment efficiency at a contaminated sediment landfill site: A field study focusing on plants, springtails, and bacteria. *Environmental Pollution* 169:1–11.

Betts, A.R., N. Chen, J.G. Hamilton, D. Peak. 2013. Rates and mechanisms of Zn^{2+} adsorption on a meat and bonemeal biochar. *Environmental Science & Technology* 47:14350–14357.

Bolan, N., A. Kunhikrishnan, R. Thangarajan, J. Kumpiene, J. Park, T. Makino, K. Scheckel. 2014. Remediation of heavy metal (loid)s contaminated soils—To mobilize or to immobilize? *Journal of Hazardous Materials* 266:141–166.

Bolan, N., C. Domy, A. Naidu, R. Naidu. 2003. Role of Phosphorus in (im)mobilization and bioavailability of heavy metals in the soil-plant system. *Reviews of Environmental Contamination and Toxicology* 177:1–44.

Borch, T., R. Kretzschmar, A. Kappler, P. Van Cappelen, M. Ginder-Vogel, A. Voegelin, K. Campbell. 2010. Biogeochemical redox processes and their impact on contaminant dynamics. *Environmental Science & Technology* 44:15–23.

Bosso, S.T., J. Enzweiler. 2008. Bioaccessible lead in soils, slag, and mine wastes from an abandoned mining district in Brazil. *Environmental Geochemistry and Health* 30:219–229.

Bosso, S.T., J. Enzweiler, R.S. Angélica. 2008. Lead bioaccessibility in soil and mine wastes after immobilization with phosphate. *Water, Air, & Soil Pollution*, 195:257–273.

Brown, S., R. Chaney, J. Hallfrisch, J. Ryan, W. Berti. 2004. In situ soil treatments to reduce the phyto- and bioavailability of lead, zinc and cadmium. *Journal of Environmental Quality* 33:522–531.

Brown, S., B. Christensen, E. Lombi, M. McLaughlin, S. McGrath, J. Colpaert, J. Vangronsveld. 2005. An inter-laboratory study to test the ability of amendments to reduce the availability of Cd, Pb, and Zn in situ. *Environmental Pollution* 138:34–45.

Cakmak, D., E. Saljnikov, V. Mrvic, M. Jakovljevic, Z. Marjanovic, B. Sikiric, S. Maksimovic. 2010. Soil properties and trace elements contents following 40 years of phosphate fertilization. *Journal of Environmental Quality* 39:541–547.

Cao, R., L. Ma, M. Chen, P. Singh, W. Harris. 2003. Phosphate-induced metal immobilization in a contaminated site. *Environmental Pollution* 122:19–28.

Cao, X., L. Ma, M. Chen, S. Singh, W. Harris. 2002. Impacts of phosphate amendments on lead biogeochemistry at a contaminated site. *Environmental Science Technology* 36:5296–5304.

Cao, X., L. Ma, Y. Liang, B. Gao, W. Harris. 2011. Simultaneous immobilization of lead and atrazine in contaminated soils using dairy-manure biochar. *Environmental Science and Technology* 45:4884–4889.

Cao, X., L. Ma, D. Rhue, C. Appel. 2004. Mechanisms of lead, copper and zinc retention by phosphate rock. *Environmental Pollution* 131:435–444.

Cao, X., A. Wahbi, L. Ma, B. Li, Y. Yang. 2009. Immobilization of Zn, Cu, and Pb in contaminated soils using phosphate rock and phosphoric acid. *Journal of Hazardous Materials* 164:555–564.
Carrillo-Gonzalez, R., J. Simunek, S. Sauve. 2006. Mechanisms and pathways of trace element mobility in soils. *Advances in Agronomy* 91:111–178.
Cave, W.J., M. Taylor, H. Basta, N. Brandon, E. Casteel, S. Gron, C. Oomen, A.T. Wiele. 2009. Inter-laboratory trial of a unified bioaccessibility procedure. British Geological Survey Open Report, OR/07/027, p. 90.
Chaney, R.L., J.A. Ryan. 1994. *Risk Based Standards for Arsenic, Lead and Cadmium in Urban Soils.* DECHEMA, Frankfurt, Germany, p. 130.
Chen, J., Y. Wang, D. Zhou, Y. Cui, S. Wang, Y. Chen. 2009. Adsorption and desorption of Cu(II), Zn(II), Pb(II), and Cu(II) on the soils amended with nanoscale hydroxyapatite. *Environmental Progress and Sustainable Energy* 29:233–241.
Chen, M., L. Ma, S. Singh, R. Cao, R. Melamed. 2003. Field demonstration of in situ immobilization of soil Pb using P amendments. *Advances in Environmental Research* 8:93–102.
Chen, S., M. Xu, Y. Ma, J. Yang. 2007. Evaluation of different phosphate amendments on availability of metals in contaminated soil. *Ecotoxicology and Environmental Safety* 67:278–285.
Chen, S., Y. Zhu, Y. Ma. 2006. The effect of grain size of rock phosphate amendment on metal immobilization in contaminated soils. *Journal of Hazardous Materials* 134:74–79.
Corami, A., S. Mignardi, V. Ferrini. 2007. Copper and zinc decontamination from single- and binary-metal solutions using hydroxyapatite. *Journal of Hazardous Materials* 146:164–170.
Corami, A., F. D'Acapito, S. Mignardi, V. Ferrini. 2008a. Removal of Cu from aqueous solutions by synthetic hydroxyapatite: EXAFS investigation. *Materials Science and Engineering: B* 149:209–213.
Corami, A., S. Mignardi, V. Ferrini. 2008b. Cadmium removal from single- and multimetal (Cd + Pb + Zn + Cu) solutions by sorption on hydroxyapatite. *Journal of Colloid and Interface Science* 317:402–408.
Corami, A., S. Mignardi, V. Ferrini. 2008c. Removal of lead, copper, zinc and cadmium from water using phosphate rock. *Acta Geologica Sinica* 82:1223–1228.
Echeverria, J., M. Morera, C. Mazkiaran, J. Garrido. 1998. Competitive sorption of heavy metal by soils. Isotherms and fractional factorial experiments, *Environmental Pollution* 101:275–284.
EPA. 2007. *The Use of Soil Amendments for Remediation, Revitalization and Reuse.* Available at http://www.clu-in.org/download/remed/epa-542-r-07-013.pdf, accessed July 31, 2014.
Fang, Y.Y., X.D. Cao, L. Zhao. 2012. Effects of phosphorus amendments and plant growth on the mobility of Pb, Cu and Zn in a multi metal contaminated soil. *Environmental Science and Pollution Research* 19:1659–1667.
Fijalkowski, K., M. Kacprzak, A. Grobelak, A. Placek. 2012. The influence of selected soil parameters on the mobility of heavy metals in soils. *Inżynieria I Ochrona Środowiska* 15:81–92.
Friesl, W., J. Friedl, K. Platzer, O. Horak, M. Grzabek. 2006. Remediation of contaminated agricultural soils neat a former Pb/Zn smelter in Austria: Batch, pot and field experiments. *Environmental Pollution* 114:40–50.

Gadd, J. 2008. Transformation and mobilization of metals, metalloids, and radionuclides by microorganisms. In: A. Violante, P.M. Huang, G.M. Gadd (eds.), *Biophysico-Chemical Processes of Metals and Metalloids in Soil Environments*, Wiley-Jupac Series, Vol. 1, John Wiley & Sons, Hoboken, NY, pp. 53–96.
Geebelen, W, D.C. Adriano, D. Van der Lelie, M. Mench, R. Carleer, H. Clijsters, J. Vansgonsveld. 2003. Selected bioavailability assays to test the efficacy of amendment-induced immobilization of lead in soils. *Plant and Soil* 249:217–228.
Gomez del Rio, J., P. Morando, D. Cicerone. 2004. Natural materials for treatment of industrial effluents: Comparative study of the retention of Cd, Zn, and Co by calcite and hydroxyapatite. Part I: Batch experiments. *Journal of Environmental Management* 71:169–177.
González, V., M. Díez-Ortiz, M. Simón, C.A. van Gestel. 2013. Assessing the impact of organic and inorganic amendments on the toxicity and bioavailability of a metal-contaminated soil to the earthworm Eisenia andrei. *Environmental Science and Pollution Research* 20:8162–8171.
Grøn, C. 2005. Test for bioaccessibility of metals and PAH from soil. Test selection, validation and application. Environment Agency and DHI. Available at http://www.bgs.ac.uk/barge/docs/DH_bioaccess_report_v03.pdf, accessed October 6, 2006.
Gworek, B., A. Barański, K. Czarnowski, J. Sienkiewicz, G. Porębska. Procedura oceny ryzyka w zarządzaniu gruntami zanieczyszczonymi metalami ciężkimi: *Dział Wydawnictw Instytutu Ochrony Środowiska*, Warszawa, 2000.
Hamon, R.E., M.J. McLaughlin, G. Cozens. 2002. Mechanisms of attenuation of metal availability in in situ remediation treatments. *Environmental Science and Technology* 36:3991–3996.
He, Z., X. Yang, P. Stoffella. 2005. Trace elements in agroecosystems and impacts on the environment. *Journal of Trace Elements in Medicine and Biology* 19:125–140.
Hodson, M.E., E. Valsami-Jones, J.D. Cotter-Howells, W.E. Dubbin, A.J. Kemp, I. Thornton, A. Warren. 2001. Effect of bone meal (calcium phosphate) amendments on metal release from contaminated soils—A leaching column study. *Environmental Pollution* 112:233–243.
Hong, C., D. Lee, P. Kim. 2008. Feasibility of phosphate fertilizer to immobilize cadmium in a field. *Chemosphere* 70, 2009–2015.
Juhasz, A., E. Smith, J. Weber, M. Rees, A. Rofe, T. Kuchel, L. Samsom, R. Nadu. 2007a. Comparison of in vivo and in vitro methodologies for the assessment of arsenic bioavailability in contaminated soils. *Chemosphere* 69:961–966.
Juhasz, A.L., E. Smith, J. Weber, M. Rees, A. Rofe, T. Kuchel, L. Sansom and R. Naidu. 2007b. Comparison of in vivo and in vitro methodologies for the assessment of arsenic bioavailability in contaminated soils. *Chemosphere* 69:961–966.
Knox, A.S., D.I. Kaplan, M.H. Paller. 2006. Phosphate sources and their suitability for remediation of contaminated soils. *Science of the Total Environment* 357:271–279.
Kumpiene, J., A. Lagerkvist, C. Maurice. 2008. Stabilization of As, Cr, Cu, Pb, and Zn in soil using amendments—A review. *Waste Management* 28:215–225.
Lambert, R., C. Grant, S. Sauve. 2007. Cadmium and zinc in soil solution extracts following the application of phosphate fertilizers. *Science of the Total Environment* 378:293–305.
Liu, R., D. Zhao. 2007. In situ immobilization of Cu (II) in soils using a new class of iron phosphate nanoparticles. *Chemosphere* 10:1867–1876.

Madrid, F., A.S. Romero, L. Madrid, C. Maqueda. 2006. Reduction of availability of trace metals in urban soils using inorganic amendments. *Environmental Geochemistry and Health* 28:365–373.

Maenpaa, K.A., J.V.K. Kukkonen, M.J. Lydy. 2002. Remediation of heavy metal-contaminated soils using phosphorus: Evaluation of bioavailability using an earthworm bioassay. *Archives of Environmental Contamination and Toxicology* 43:389–398.

Mahmoud, E., N. El-Kader. 2014. Heavy metal immobilization in contaminated soil phosphogyupsum and rice straw compost. *Land Degradation and Development*, doi: 10.1002/ldr.2288.

McBride, M. 2004. Molybdenum, sulfur, and other trace elements in farm soils and forages after sewage sludge application. *Communications in Soil Science and Plant Analysis* 35:517–535.

McGowen, S., N. Basta, G. Brown. 2001. Use of diammonium phosphate to reduce heavy metal solubility and transport in smelter contaminated soil. *Journal of Environmental Quality* 30:493–500.

Meers, E., R. Samson, F.M.G. Tack, A. Ruttens, M. Vandegehuchte, J. Vangronsveld, M.G. Verloo. 2007. Phytoavailability assessment of heavy metals in soils by single extractions and accumulation by Phaseolus vulgaris. *Environmental and Experimental Botany* 60:385–396.

Melamed, R., X. Cao, M. Chen, L.Q. Ma. 2003. Field assessment of lead immobilization in a contaminated soil after phosphate application. *Science of the Total Environment* 305:117–127.

Mignardi, S., A. Corami, V. Ferrini. 2012. Evaluation of the effectiveness of phosphate treatment for the remediation of mine waste soils contaminated with Cd, Cu, Pb, and Zn. *Chemosphere* 86:354–360.

Miretzky, P., A. Cirelli. 2008. Phosphates for Pb immobilization in soils: A review. *Environmental Chemistry Letters* 6:121–133.

Montanher, S., E. Oliveira, M. Rollemberg. 2005. Removal of metal ions from aqueous solutions by sorption onto rice bran. *Journal Hazardous Materials* 117:207–211.

Monteil-Rivera, E., M. Fedoroff. 2002. Sorption of inorganic species on apatites from aqueous solutions. In: *Encyclopedia of Surface and Colloid Science*, A.T. Hubbard (ed.), Marcel Dekker Inc., New York, pp. 1–26.

Morgan, A.J., M. Evans, C. Winters, M. Gane, M.S. Davies. 2002. Assaying the effects of chemical ameliorants with earthworms and plants exposed to a heavily polluted metalliferous soil. *European Journal of Soil Biology* 38:323–327.

Naidu, R., S. Rogers, V. Gupta, R.S. Kookana, N. Bolan. 1997. Bioavailability of metals in the soil–plant environment and its potential in risk assessment: An overview. In: Proc. 4th Int. Conf. the Biogeochemistry of Trace Elements, Berkeley, CA, pp. 757–758.

Oomen, A.G., C.J.M. Rompelberg, M.A. Bruil, C.J.G. Dobbe, D.P.K.H. Pereboom, A.J.A.M. Sips. 2003. Development of an in vitro digestion model for estimating the bioaccessibility of soil contaminants. *Archives of Environmental Contamination and Toxicology* 44:281–287.

Oomen, A., C. Rompelberg, E. Van de Kamp, D. Pereboom, L. De Zwart, A. Sips. 2004. Effect of bile type on the bioaccessibility of soil contaminants in an in vitro digestion model. *Archives of Environmental Contamination and Toxicology* 46:183–188.

Ownby, D.R., K.A. Galvan, M.J. Lydy. 2005. Lead and zinc bioavailability to Eisenia fetida after phosphorus amendment to repository soils. *Environmental Pollution* 136:315–321.

Park, J.H., N. Bolan, M. Megharaj, R. Naidu. 2012. Relative value of phosphate compounds in reducing the bioavailability and toxicity of lead in contaminated soils. *Water, Air, & Soil Pollution* 223:599–608.

Park, J.H., G.K. Choppala, N.S. Bolan, J.W. Chung, T. Chuasavathi. 2011. Biochar reduces the bioavailability and phytotoxicity of heavy metals. *Plant and Soil* 348:439–451.

Patinha, C., A.P. Reis, C. Dias, A. Cachada, R. Adao, H. Martins, E. Ferreira da Silva, A.J. Sousa. 2010. Lead availability in soils from Portugal's Centre Region with special reference to bioaccessibility. *Environmental Geochemistry and Health* 34:213–227.

Pelfrene, A., C. Waterlot, M. Mazzuca, C. Nisse, G. Bidar, F. Douay. 2011. Assessing Cd, Pb, Zn human bioaccessibility in smelter contaminated agricultural topsoils (northern France). *Environmental Geochemistry and Health* 33:477–493.

Pérez-Marín, A., V. Meseguer Zapata, J. Ortuño, M. Aguilar, J. Sáez, M. Lloréns. 2007. Removal of cadmium from aqueous solutions by adsorption onto orange waste. *Journal of Hazardous Materials* 139:122–131.

Porter, A.E., P. Taak, L.W. Hobbs et al. 2004a. Bone bonding to hydroxyapatite and titanium surfaces on femoral stems retrieved from human subjects at autopsy. *Biomaterials* 25:5199–5208.

Porter, S., K. Scheckel, C. Impellitteri, J. Ryan. 2004b. Toxic metals in the environment: Thermodynamic considerations for possible immobilisation strategies for Pb, Cd, As, and Hg. *Critical Reviews in Environmental Science and Technology* 34:495–604.

Pratiyogita Darpan Group. 2008. Phosphatic Fertilizers, *Competition Science Vision* 11(125):611–612.

Rauret, G., J.F. Lopez-Sanchez, A. Sahuquillo, R. Rubio, C. Davidson, A. Ure, P. Quevauviller. 1999. Improvement of the BCR three-step sequential extraction procedure prior to the certification of new sediment and soil reference materials. *Journal of Environmental Monitoring* 1:57–61.

Reeder, R.J., M.A.A. Schoonen, A. Lanzirotti. 2006. Metal speciation and its role in bioaccessibility and bioavailability. *Reviews in Mineralogy and Geochemistry* 64:59–113.

Robinson, B., N. Bolan, S. Mahimairaja, B. Clothier. 2005. Solubility, mobility, and bioaccumulation of trace elements: Abiotic processes in the rhizosphere. In: *Trace Elements in the Environment: Biogeochemistry, Biotechnology, and Bioremediation*, M.N.V. Prasad, K.S. Sajwan, R. Naidu (eds.), CRC Press, Boca Raton, FL., pp. 93–104.

Ruby, M.V., A. Davis, R. Schoof, S. Eberle, C.M. Selstone. 1996. Estimation of lead and arsenic bioavailability using a physiologically based extraction test. *Environmental Science and Technology* 30:422–430.

Ruby, M.V., R. Schoof, W. Brattin, M. Goldade, G. Post, M. Harnois, D.E. Mosby, S.W. Casteel, W. Berti, M. Carpenter, D. Edwards, D. Cragin, W. Chappell. 1999. Advances in evaluating the oral bioavailability of inorganics in soil for use in human health risk assessment. *Environmental Science and Technology* 33:3697–3705.

Ryan, J.A., W.R. Berti, S.L. Brown, S.W. Casteel, R.L. Chaney, M. Doolan et al. 2004. Reducing children's risk to soil Pb: Summary of a field experiment. *Environmental Science and Technology* 38:18A–24A.

Scheckel, K., J. Ryan, D. Allen, N. Lescano. 2005. Determining speciation of Pb in phosphate-amended soils: Method limitations. *Science of the Total Environment* 350:261–272.

Scheckel, K.G., R.L. Chaney, N.T. Basta, J.A. Ryan. 2009. Advances in assessing bioavailability of metal(loids) in contaminated soils. *Advances in Agronomy* 104:1–52.

Schroder, J.L., N.T. Basta, J. Si, S.W. Casteel, T. Evans, M. Payton. 2003. In vitro gastrointestinal method to estimate relative bioavailable cadmium in contaminated soil. *Environmental Science and Technology* 37:1365–1370.

Schroder, J.L., N.T. Basta, S.W. Casteel, T.J. Evans, M.E. Payton, J. Si. 2004. Validation of the in vitro method to estimate bioavailable lead in contaminated soils. *Journal of Environmental Quality* 33:513–521.

Shi, Z., L. Erickson. 2001. Mathematical model development and simulation of in situ stabilization in lead-contaminated soils. *Journal of Hazardous Materials* 87: 99–116.

Siebielec, G., R.L. Chaney. 2012. Testing amendments for remediation of military range contaminated soil. *Journal of Environmental Management* 108:8–13.

Siebielec, G., T. Stuczynski. 2008. Trace metal aspects of biosolids use. CD Proceedings of Protection and Restoration of the Environment IX, Kefalonia, Greece, p. 7.

Sizmur, T., B. Palumbo-Roe, M.E. Hodson. 2011. Impact of earthworms on trace element solubility in contaminated mine soils amended with green waste compost. *Environmental Pollution* 159:1852–1860.

Sonmez, O., G.M. Pierzynski. 2005. Phosphorus and manganese oxides effects on soil lead bioaccessibility: PBET and TCLP. *Water, Air, & Soil Pollution* 166:3–16.

Stuczynski, T., G. Siebielec, W.L. Daniels, G. McCarty, R. Chaney. 2007. Biological aspects of metal waste reclamation with biosolids. *Journal of Environmental Quality* 36:1154–1162.

Sugiyama, S., H. Matsumoto, T. Ichii, H. Hayashi, Y. Hiraga, N. Shigemoto. 2001. Enhancement of Pb-Ba exchangeability of barium hydroxyapatite. *Journal of Colloid and Interface Science* 238:183–187.

Tang, X., Y. Zhu, S. Chen, L. Tang, X. Chen. 2004. Assessment of the effectiveness of different phosphorus fertilizers to remediate Pb contaminated soil using in vitro test. *Environment International* 30:531–537.

Tessier, A., P.G.C. Campbell, M. Bisson. 1979. Sequential extraction procedure for the speciation of particulate trace metals. *Analytical Chemistry* 51:844–851.

Traunfeld, J., E. Nibali. 2013. University of Maryland Extension, Home and Garden Information Center. Available at http://extension.umd.edu/hgic, accessed July 31, 2014.

Valsami-Jones, E., K. Ragnarsdottir, A. Putnis, D. Bosbach, A. Kemp, J. Cressey. 1988. The dissolution of apatite in the presence of aqueous metal cations at pH 2–7. *Chemical Geology* 151:215–233.

Vamerali, T., M. Bandiera, G. Mosca. 2010. Field crops for phytoremediation of metal-contaminated land. A review. *Environmental Chemistry Letters* 8:1–17.

Violante, A., V. Cozzolino, L. Perelomov, A. Caporale, M. Pigna. 2010. Mobility and bioavailability of heavy metals and metalloids in soil environments. *Journal of Plant Nutrition and Soil Science* 10:268–292.

Withers, P., A. Sharpley. 1995. Phosphorus fertilizers. In: *Soil Amendments and Environmental Quality*, J.E. Rechcigl (ed.), CRC Press, Boca Raton, FL.

Wragg, J., M. Cave, H. Taylor, N. Basta, E. Brandon, S. Casteel et al. 2009. Interlaboratory trial of a unified bioaccessibility testing procedure. British Geological Survey, Nottingham, UK, p. 90.

Xie, Z., B. Wang, Y. Sun, J. Li. 2006. Field demonstration of reduction of lead availability in soil and cabbage (Brassica Chinensis L.) contaminated by mining tailings using phosphorus fertilizers. *Zhejiang Univ Science B* 7(1):43–50.

Xu, Y., F. Schwartz, S. Traina. 1994. Sorption of Zn^{2+} and Cd^{2+} on hydroxyapatite surfaces. *Environmental Science and Technology* 28:1472–1480.

Yoon, J., X. Cao, L. Ma. 2007. Application methods affect phosphorus-induced lead immobilization from a contaminated soil. *Journal Environmental Quality* 36:373–378.

Zhang, P., J.A. Ryan. 1998. Formation of chloropyromorphite in anglesite-hydroxyapatite suspensions under varying pH conditions. *Environmental Science and Technology* 32:3318–3324.

Zhang, P., J.A. Ryan. 1999. Transformation of Pb(II) from cerussite to chloropyromorphite in the presence of hydroxyapatite under varying conditions of pH. *Environmental Science and Technology* 33:625–630.

10

Influence of Long-Term Soil Application of Sewage Sludge Rich in Phosphorus on Heavy Metals Bioavailability to Plants

Wanderley José de Melo, Gabriel Maurício Peruca de Melo, and Valéria Peruca de Melo

CONTENTS

Sewage sludge presents risk to the environment and to human health, and therefore its correct disposal is a problem to be solved as quickly as possible. Its use in agriculture as a fertilizer or soil amendment has drawn the attention of producers because sewage sludge contains large concentrations of organic matter and plant nutrients such as nitrogen (N), phosphorus (P), zinc (Zn), and others. However, along with these welcome components, sewage sludge also contains in its composition potentially toxic trace elements and organic substances that are hazardous. Considering the benefits and the risks of applying sewage sludge in agriculture, it is important to know the composition of the residue to be used, the soil properties where it will be used, and the climatic conditions. These factors are important because the trace elements available to plants will also move down the soil profile and contaminate underground water. It is important to prevent the entering of trace elements in the human food chain.

In this chapter, we discuss the behavior of trace elements in soil and in an experiment with a sandy and a clayey soil treated with rates of sewage sludge for 13 successive years and cropped with maize in the thirteenth year. When humans began to live in communities that initially increased without organization, these communities needed food and other goods with each successive increase in size and complexity. In the beginning, food production was done in areas obtained by clear-cutting natural forests. Those areas were cropped with no technology, and most were degraded so that new areas needed to be incorporated into the productive system. Together with the increase in population, production of residues with different characteristics originating from domestic, industrial, and agricultural undertakings also increased. The wastewater that resulted from the use of freshwater in practically all these processes was among the residues, and for many centuries it was contaminated with trace elements, toxic organic substances, and pathogenic microorganisms that returned to rivers, seas, and lakes without any treatment, contaminating the natural resources with each passing day.

Today, the world population exceeds 7 billion and probably will reach 9 billion by 2050. It is no longer acceptable to destroy forests to increase the agricultural areas and to return the untreated wastewater to rivers and lakes. How do we resolve the issues of agricultural production and water use without risks to life and the environment while at the same time supporting sustainability? The answer is treating wastewater and recycling the residue, sewage sludge (SE), or biosolid.

Sewage sludge generally contains large concentrations of phosphorus, an important plant nutrient that takes part in the composition of molecules such as lipids, proteins, and nucleic acids. Phosphorus management is very important in tropical and subtropical soils because it is quickly immobilized by soil components and is then unavailable for plant uptake. On the other hand, phosphate mines are being exhausted around the world, and discovering new potential sources of the element to produce fertilizers will be a challenge in the coming decades. The phosphorus contained in SE perhaps will be the way of the future.

In this chapter, the behavior of trace elements in soil and the effect of long-term application of a SE rich in P on the availability of some trace elements to plant absorption are discussed.

SE

SE is the residue generated in the treatment of municipal wastewater with the objective to obtain a water that can be returned to the rivers with minimal risk. In some countries, this residue is also called biosolid. The composition of SE varies greatly, depending on the origin of the wastewater and

TABLE 10.1

Trace Element Concentration in SE Produced by Brazilian Wastewater Treatment Stations

	mg kg^{-1}. Dry Weight Basis								
Wasterwater Treatment Station	**As**	**Cd**	**Pb**	**Cu**	**Hg**	**Mo**	**Ni**	**Se**	**Zn**
Barueri (SP)	12	18	189	850	2	13	349	1	1870
Suzano (São Paulo [SP])	39	6	345	733	24	19	227	0.8	1873
Franca (SP)	0.2	7	31	160	0.14	5.5	34	0.3	1560
Lavapés (SP)	na	6	2	26	na	na	32	na	25
Belém (PR)	na	na	123	439	1	na	73	na	824
RALF (PR)	na	na	64	89	0.5	na	40	na	456
Brasília (DF)	na	10	50	186	4	na	34	na	1060

Note: As = arsenic, Cu = copper, DF = Brazilian capital, Hg = mercury, Mo = molybdenum, na = not available, Ni = nickel, PR = Paraná state, Se = selenium, SP = São Paulo state.

the process used in its treatment. It is generally rich in organic matter and plant nutrients such as N, P, sulfur (S), and Zn, but it may also contain trace elements such as lead (Pb), cadmium (Cd), and chromium (Cr); organic toxic substances such as pesticides; and pathogen agents such as virus, fecal coliforms, and helminthes eggs. When originating from a domestic sewer, SE is normally poor in trace elements but may be considerably contaminated by pathogenic agents depending on the health status of the population. On the other hand, when the wastewater comes from industrialized regions, it is normally rich in trace elements and poor in pathogenic agents (Table 10.1).

Agricultural Use of SE

Along with such plant nutrients as N and P, sewage sludge may also contain elements, substances, and organisms potentially hazardous to the environment and to humans so that its disposition in soil must be well studied. In order to minimize the risks, many countries have complex laws regulating its disposal (Table 10.2). The American legislation was published in 1993 by the Regulations Federal Committee (Title 40, parts 257, 403, and 503). It includes nine trace elements and the maximum concentrations they may reach in the soil and the annual rate that is added to the soil through sewage sludge (Table 10.3). The United States Environmental Protection Agency is now studying modifications in the legislation for the use of sewage sludge in agriculture.

Brazilian legislation for use of sewage sludge in agriculture was published in 2006 (CONAMA 2006). It includes 11 trace elements, the maximum concentration in the residue, and the accumulated addition of the metals in successive applications (Table 10.4). The legislation also includes plants that may not be fertilized with sewage sludge: pasture, vegetables, roots and tubers, and flooded crops.

TABLE 10.2

Maximum Concentration of Trace Elements in Sewage Sludge Permitted for Use in Agriculture in the Legislation of Some Countries

	mg kg^{-1} (dry weight basis)									
Metal	**Sweden**	**Switzerland**	**Scotland**	**Belgium**	**Denmark**	**Germany**	**France**	**Italy**	**Spain**	**Holland**
As	ni	ni	ni	ni	ni	ni	ni	ni	ni	15
Cd	15	30	20	12	0.8	10	20	20	20	10
Cr	1000	1000	2000	500	100	900	1000	ni	1000	500
Cu	3000	1000	1500	750	1000	800	1000	1000	1000	600
Pb	300	1000	1500	600	120	900	800	750	750	100
Hg	8	10	ni	10	0.8	8	10	10	16	ni
Ni	500	200	25	100	30	200	200	300	300	100
Zn	1000	3000	2500	2500	4000	2500	3000	2500	2500	2000

Source: Mathews, P., International perspective. Workshop biosolids: Current and emerging issues, WWEFTEC'99, New Orleans, LA, 1999.

Note: ni = not included.

TABLE 10.3

Maximum Trace Element Concentration Permitted in Sewage Sludge for Use in Agriculture, Maximum Concentration That May Be Reached in Soil, and Maximum Annual Rate That May Be Added

	mg kg^{-1}		kg ha^{-1} ano^{-1}
Element	**Sewage Sludge**[a]	**Soil**	**Annual Rate**
As	75	41	2
Cd	85	39	1.9
Pb	840	300	15
Cu	4300	1500	75
Hg	57	17	0.85
Mo	75	ni	ni
Ni	420	420	21
Se	100	100	5
Zn	7500	2800	140

Source: USEPA (United States Environmental Protection Agency). Title 40 CFR. Part 503. Final rules standards for the use or disposal of sewage sludge. Federal Register 58-9387-9415, 1999.

[a] Dry weight basis.

TABLE 10.4

Maximum Trace Element Concentration Permitted in Sewage Sludge for Use in Agriculture and the Accumulated Rate of Trace Elements Reached after Successive Applications

Element	Maximum Concentration mg kg^{-1} (dwb)	Accumulate Rate kg ha^{-1}
Arseniun	41	30
Barium	1300	265
Cadmium	39	4
Lead	300	41
Copper	1500	137
Chromium	1000	154
Mercury	17	1.2
Molybdenum	50	13
Nickel	420	74
Selenium	100	13
Zinc	2800	445

Source: CONAMA (Conselho Nacional Do Meio Ambiente). Resolução 375. Critérios e procedimentos para o uso agrícola de lodos de esgoto gerados em estações de tratamento de esgoto sanitário e seus produtos derivados. Brasília, 2006.

Note: dwb = dry weight basis.

Phosphorus Content

Phosphorus is an essential element for plant nutrition. It takes part in the composition of molecules involved in the cell bioenergetic systems, such as adenosine triphosphate/adenosine diphosphate (ATP/ADP) and guanosine triphosphate/guanosine diphosphate (GTP/GDP), in the structure of membranes such as phospholipids and also in several metabolic pathways (Mengel and Kirkby 1987).

Soil-soluble P, whose source may be phosphate rocks (apatite), organic and mineral fertilizers, and organic wastes, reacts with calcium, aluminum, and iron and forms substances with low solubility. It may also link to clay minerals, iron, and aluminum oxides and may become unavailable for plant uptake (Stevenson 1986).

Sorghum P uptake increased linearly when soil was treated with rates of SE (Oliveira et al. 1995).

Trace Elements in Soil

Trace elements are very important in human activities and include plant micronutrients such as Cu, iron (Fe), manganese (Mn), Zn, Mo, and boron (B). Some authors also consider Ni as a plant micronutrient (Malavolta 2006). Other trace elements, such as gold (Au), Pb, tin (Sn), Fe, and antimony (Sb), are important in the production of equipment and other goods. Trace elements are also important in the generation of energy as uranium (U) and for human adornment in the form of Au, platinum (Pt), and diamond. However, if trace elements are taken up in high quantities in plants, animals, and microorganisms, they may be toxic and even lethal. Soils have trace elements in their composition in different concentrations.

The natural origin of soil trace elements is the rock on which it was formed, since most of the rocks contain trace elements in their composition. However, trace element concentrations in soils may increase by anthropic activities such as application of fertilizers, pesticides, irrigation, oil and charcoal burning, motor fuel combustion, urban and industrial waste burning, mining, metal melting and purification, and residue disposition.

Sewage sludge may be a source of trace elements in soil. A soil amended with 180 t ha^{-1} sewage sludge presented 75.5 mg kg^{-1} of total lead, 1.6 mg kg^{-1} as exchangeable + adsorbed, 3.9 mg kg^{-1} as organic, 51.00 mg kg^{-1} as carbonate, and 17.00 mg kg^{-1} as sulfide. The residue is also an import source of Zn. The risk of high concentrations of soluble trace elements in soil is their uptake by plants, which serves as a means of their entry into the animal food chain.

Each essential trace element has a specific function in cell metabolism, and the concentration that is needed varies according to the element and the organism. A lack or an excess causes alterations in the metabolism, and symptoms of deficiency or toxicity appear in specific organs. Peanut plants submitted to an excess of lead exhibited symptoms of Fe deficiency (Figure 10.1).

FIGURE 10.1
Peanut plants exhibiting Fe deficiency when cropped in a soil with high concentration of available Pb. (From Melo, W.J. et al., O uso agrícola do biossólido e as propriedades do solo. In: Tsutyia, M.T. et al. (eds.), *Biossólidos na Agricultura*, Chapter 11. São Paulo, ABES/SP, pp. 289–363, 2002.)

Lead interferes with a variety of cell processes and is toxic to the heart, bones, intestines, kidneys, and reproductive and nervous systems, mainly in children. The Centers for Disease Control in the United States has set blood lead levels as 10 ($\mu g\ dl^{-1}$) for adults and 5 (μg/dl) for children (ACCLPP 2012). The primary cause of lead's toxicity is its interference with a group of enzymes that has as their composition the sulfhydryl groups (HS) because Pb binds to this functional group, causing inhibition of enzyme activity. The reaction with the group HS also occurs with other metal trace elements.

Mobility in Soil Profile

Soil trace elements may be in different forms and, depending on the form, may or may not be available for plant uptake or to move in the soil profile.

Some forms are not available to plants, such as those taking part in the crystalline structure of primary minerals and oxides, and also in secondary minerals that originate from weathering, such as clay minerals. During the weathering, trace elements may be solubilized and then be uptaken by plants. Some soil trace elements are in the composition of hydroxides, such as hydrated manganese oxide ($MnO_2 \cdot nH_2O$). The binding of the element to the oxide may be so strong that it is not available to plants. Depending on the soil pH and the presence of certain anions, metal trace elements can precipitate and become unavailable for plant uptake or soil percolation. Cadmium can precipitate in the presence of carbonates and bicarbonates as cadmium carbonate or cadmium bicarbonate. Soil organisms use organic substances as a source of carbon and energy. Trace elements are also absorbed, and while they are a component of the organisms, they are not available to the plant. When those organisms die, other soil organisms promote trace element cycling.

In the interface between a solid phase and the solution in which it is involved, the trace element suffers an attraction strength. They are known as the adsorbed trace elements, and their availability to plants depends on the attraction strength. When the interaction is between a hydrated metal ion and colloidal particles, the attraction force is of an electrostatic nature, a weak strength, and is seen as having no specific adsorption and forms an external sphere (Ji and Li 1997). An ion that has a large hydrated ionic ray is attracted with lower intensity (Camargo et al. 2001) and is called a changeable cation. When the attraction strength is strong (covalent or ionic), the adsorption is called specific and forms an internal sphere complex.

The main soil components participating in the formation of internal sphere complexes are the oxides, in particular those of aluminum (Al), Fe, and Mn (McBride 1994). In Brazilian oxisols, Pb^{2+} is not very mobile in soil since it forms internal sphere complexes with Al, Fe, and Mn oxides, while Cd^{2+} and Zn^{2+}, which form external sphere complexes, are relatively mobile (Pierangeli et al. 2001). Zinc adsorption is stronger in amorphous Al oxides than in amorphous Fe oxides, and for the same type of oxide, the adsorption of metals varies with the grade of crystallinity of the oxide because weathering causes changes to the form of the crystal in its superficial area and chemical properties (Yu 1997). Iron and manganese oxides have more capacity to adsorb metals than aluminum oxides and other clay minerals (Brown and Parks 2001).

The more electronegative cations tend to form covalent bindings with the oxygen on a mineral surface. In the case of divalent cations, the strength of the binding decreases in the order Cu > Ni > Co > Pb > Cd > Zn > Mg > Sr. Trivalent cations such as Cr^{3+} and Fe^{3+} are preferentially adsorbed compared to the bivalents (McBride 1994). In seven Brazilian soils, the metals more strongly adsorbed were Cr, Pb, and Cu and the most weakly adsorbed were Cd, Zn, and Ni (Gomes et al. 2001).

Trace elements adsorbed in a colloidal system are in dynamic equilibrium with the same ion in soil solution. The trace element soluble in soil solution when uptaken by a plant is replaced by adsorbed ions establishing a new equilibrium. The two forms are considered available to plants. When the plant removes a cation from the soil solution, a new equilibrium is established (Figure 10.2).

The concentration of trace elements in soil solution is very low, generally in the range of 1–1000 mg L^{-1}, and sometimes below 1 μg L^{-1} (McBride 1994). Soil organic matter, mainly humic substances (fulvic and humic acids), presents in its composition hydroxyl groups in carboxylic, phenolic, and other chemical functions, which may be ionized according to soil pH. Trace elements may form complexes with those hydroxyls. Humic substance complexes reduce the availability of trace element metals to plants (Sposito 2008).

The movement of trace elements in a soil profile is regulated by their chemical nature and by soil properties, whose interactions act on the sorption/desorption, precipitation/dissolution, chelates and complex formations, and redox potential (Oliveira and Mattiazzo 2001). Among the soil properties that may affect trace element mobility in a soil profile is pH, organic matter, redox potential, type and content of clay minerals, texture, and plant

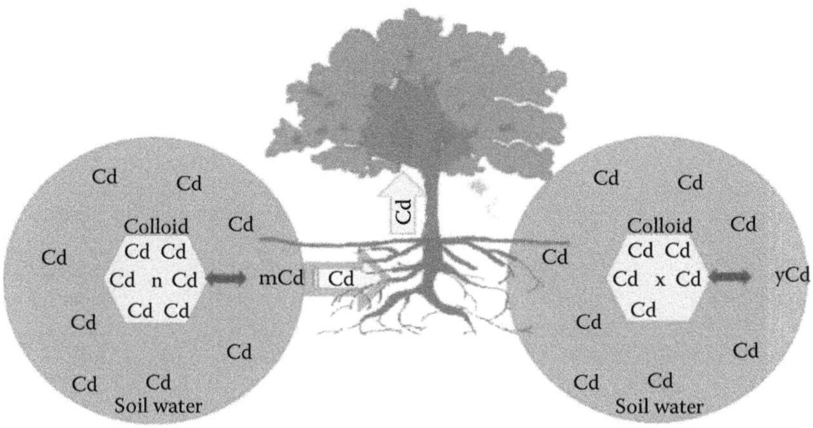

FIGURE 10.2
Available trace element equilibrium in soil.

exudates (Rieuwerts et al. 2006). Iron oxides and fulvic acids are the most active in the mechanisms of sorption and desorption (Covelo et al. 2007).

Authors have suggested that Cr, Cu, Ni, and Pb are not very mobile in the soil and are accumulated in the soil surface, while Cd and mainly Zn are much more mobile (Oliveira et al. 2005; Nogueira et al. 2008; Merlino et al. 2010).

When a trace element is added to a soil surface as an organic waste or fertilizer, once mineralized, it will be initially in the upper layers adsorbed, precipitated, or coprecipitated or in the constitution of an organism or mineral (Figure 10.3). If it is soluble in soil water, it may move down the soil profile and concentrate in deep layers. If it originated from rock, the trace element tends to concentrate in the deep layers, but if soluble in water, it can move to the upper layers with the water movement. Therefore, the mobility of trace elements in the soil profile depends on the dynamics of the water in the soil profile. The interaction between trace elements and the soil solid phase is also very important in their mobility (Ross 1994; Camargo et al. 2001; Cornu et al. 2001) as shown in Figure 10.2.

Generally the movement of trace elements is higher in sandy and acid soils, which have low cation exchange capacity or higher oxidation levels (Ross 1994). Copper added to the soil surface stays there for a long time, since the element is strongly adsorbed by the soil organic matter, Fe, Al, and Mn oxides, and by clay minerals. It is one of the less mobile trace elements in soil. On the other hand, Cd is relatively mobile in the soil profile and may contaminate underground water. Iron produces chelates and complexes with soil organic matter that are soluble and responsible for the migration of the element to the deeper layers.

Inconsistent results have been observed for Cr distribution in the soil profile. In some works, it has been found uniformly distributed in the soil profile, but in another, it has been found accumulated in the soil surface.

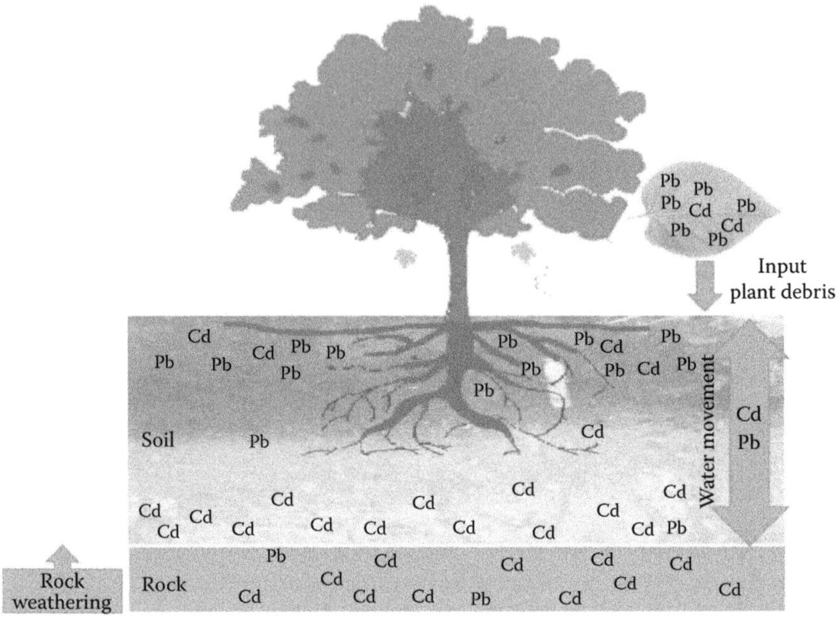

FIGURE 10.3
Movement of trace elements in soil profile.

Lead generally has been found in the upper soil layer because plant debris and atmospheric deposition are the main sources of this metal, but it has also been found in depths of 30–45 cm (Levy et al. 1992). In soils treated with sewage sludge for 6 consecutive years, Pb was found in the first 15 cm (Chang et al. 1984).

Availability to Plants

The availability of trace elements to plants is affected by a number of factors. Texture and structure affect the number of micropores and macropores and the relation to air/water. They also affect the soil cation exchange capacity (CEC). The more clayey the soil, the higher the CEC and the availability of trace elements to plants. The soil CEC is affected by the concentration and the type of clay mineral. Clay minerals with a 1:1 crystalline structure, such as kaolinite, have lower CECs than those with a 2:1 crystalline structure, such as montmorillonite. In this way, soils that are more weathered, such as the soils located in the tropics and subtropics, have lower CECs in relation to the clay minerals.

Perhaps pH is the most important soil property affecting trace element solubility. The soil management affecting pH is an effective way to decrease the availability of trace elements to plants and their mobility in the soil profile. Therefore, soil amendment with alkaline substances such as carbonates (CO_3^{-2}), silicates, and complexing agents (ethylenediaminetetraacetic [EDTA] acid, diethylenetriaminepentaacetic [DTPA] acid, plant residues, organic wastes) is important in order to attenuate the effects of soil contamination with trace elements. The anions SiO_3^{-2} present in calcium silicates ($CaSiO_3$) react with the ion H^+ present in the soil solution in a similar way as the anion CO_3^{-2}, promoting an increase in soil pH and a decrease in the ion activity (mobility in water, adsorption to soil colloids, uptake by plants) by the precipitation of the metals as hydroxides (Pierangeli et al. 2001). Furthermore, $CaSiO_3$, mainly at high rates, promotes the polymerization of silicates, which are potential ligants for the formation of insoluble complexes with trace elements (Shriver 2003).

Trace elements are affected differently by soil pH. Some are more soluble when pH increases, while others have their solubility increased by a decrease in pH. Manganese may occur in soil with three valences: Mn^{2+} (adsorbed and in soil solution), Mn^{3+} (as the highly reactive Mn_2O_3), and Mn^{4+} (as MnO_2). The three species are in a dynamic equilibrium. The species 3^+ and 4^+ are favored by an elevation in pH and oxidant conditions. In order for Mn^{3+} to be transformed to Mn^{2+}, it is necessary to have an acid and reductor environment. When soil pH increases, Pb produces insoluble substances such as hydroxides, phosphates, and carbonates. It also forms insoluble complexes with organic matter.

Cadmium concentration in soil solution is influenced by the content of organic matter and by the soil pH. If Cd concentration in soil is low, organomineral complexes are produced, and if the concentration is very high, cadmium carbonates and phosphates are formed. In acid soils, the factors that most affect Cd solubility are the formation of complexes with organic matter and the reactions with sesquioxides.

Soil organic matter influences trace element availability because it increases soil CEC and acts in the formation of soluble and insoluble complexes. In a very old experiment, in which $PbCO_3$ was added to soil up to 5568 kg ha^{-1}, only 3.2 mg kg^{-1} soluble Pb was detected at barley sowing, a lower value (1.1 mg kg^{-1}) at harvesting, and the crop growing was not affected by the trace element (Keaton 1937). Lead carbonate is very insoluble, and the sorption on soil organic matter and the formation of insoluble complexes contributed to the low Pb solubility. Iron is another trace metal that forms chelates and complexes with soil organic matter, which are responsible for the migration of Fe from the top to the deep layers of the soil and for the availability of the element to plants. In tropical soils, the amount of Fe in soil is very high, but the available Fe is in little concentration as the ion Fe^{2+}, which is the ion uptaken by the plants.

The concentration of O_2 in the soil atmosphere is responsible for the oxidation level of the trace elements, transforming soluble to insoluble species. In environments rich in O_2, reduced ions of Mn (Mn^{2+}) and Fe (Fe^{2+}), which are soluble, are oxidized to Mn^{3+} and Fe^{3+}, which are insoluble and not available to plants. For example, in very reducing conditions, as in flooded crops of rice, Cd can become cadmium sulfide, which is very insoluble.

Interactions of Trace Elements with Phosphorus

Metal trace elements may react with the anion phosphate and produce insoluble or low soluble compounds, making P unavailable to plants. This is also a way to decrease the risks of the presence of metal trace elements in soil. The anion di-hydroxyphosphate $\left(H_2PO_4^-\right)$ is able to form low soluble complexes with the metals Zn, Cu, Pb, and Cd (Carvalho et al. 2001). In this way, soil phosphorus fertilization may contribute to the decrease in the availability of those metals to plants and their mobility in soil profile. In a study with soil contaminated by trace metals, the application of phosphorus as $Ca(H_2PO_4)_2$ was efficient in decreasing their availability (Paim et al. 2006).

Case Study

Long-Term Experiments with Annual Applications of Sewage Sludge to Brazilian Oxisols

Experiments were installed in two oxisols, a Typic Eutrorthox (TE) and Typic Haplorthox (TH) under field conditions in Jaboticabal, SP, Brazil (21° 14′ 05″ S, 48° 17′ 09″ W, 600 m altitude, climate Aw according to Koppen classification), in the growing season 1997–1998. The experiment design was a randomized block with four treatments (control, with mineral fertilization based on soil chemical analysis, 5, 10, and 20 t ha^{-1} sewage sludge, dry weight basis) and five replications. The plots that received sewage sludge were also fertilized with nitrogen, phosphorus, and potassium (NPK) mineral fertilizer, if necessary, so that they received the same amounts of NPK as the control. Sewage sludge was obtained from three sewage treatment plants operated by Companhia de Saneamento Básico do Estado de São Paulo (SABESP), situated in Barueri, Franca, and Monte Alto, all of them in the state of São Paulo. It was spread on soil surface before sowing and incorporated to 0–10 cm with a disc harrow.

Before the experiment installation (September 1997), the area was plowed, harrowed, and limed with 2.5 t ha^{-1} of dolomitic limestone to raise the bases saturation (BS) to 70%. Thirty days after liming, the plots were marked out (6 × 10 m), sewage sludge was applied, the soil was furrowed (six lines with 10 m and 0.90 cm between lines), the mineral fertilizers were distributed into the furrows, and maize was seeded. Nitrogen and potassium fertilizers were applied in two stages: at sowing and 45 days later. Soil and plants were sampled for chemical, biochemical, and biological analysis. Soil samples were taken 60 days after seeding. In each plot, 10 simple samples were taken from a depth of 0–20 cm and mixed to form a composite sample. Leaf samples for leaf diagnosis were collected on the same day as the soil samples. In each plot, the first leaf in front and below the first cob of five plants were collected. At about 120 days after sowing, five plants were cut just on the soil surface to evaluate dry mass production and element absorption. When the maize grains had about 13% moisture, cobs were sampled for estimating grain production.

Annually, when the raining period began, herbicide was applied to kill the weeds, the soil was harrowed, and the experiment was installed again. In the growing season 2009–2010, the experiment was installed for the thirteenth time. The granulometry and chemical analysis of soil samples taken in August, four months before maize sowing, are shown in Tables 10.5 and 10.6.

In this growing season, sewage sludge was supplied by SABESP operating in Barueri, SP, whose chemical analysis is shown in Table 10.7.

Liming was carried out in order to raise the BS to nearly 70%, using a dolomitic limestone (Table 10.8). Sewage sludge was spread on the soil surface in December 2009, incorporated to 10 cm depth and 20 days after the soil was fertilized (Table 10.9), and maize (hybrid Agroceres–AG 5020) was sowed (7–8 seeds per linear meter).

Sixty days after seedling emergence, when the female flower had appeared in more than 50% of the plants, soil and leaf were sampled as described above. Soil samples were air-dried, sieved to 2 mm, and used for the analysis. The middle third of each leaf was removed, rinsed with tap water, distilled water, and deionized water, dried at 60°C in an oven with forced air circulation, weighed, milled in a mortar, and used for the analysis. The shoot vegetal material was washed with tap water, distilled water, and deionized water, dried at 60°C in an oven with forced air circulation, weighed, and crushed in a Willey type mill with a stainless chamber and knives.

At 125 days after seedling emergence, eight ears from each plot were sampled. The ears were husked and the grains were prepared as described above. Plant and soil samples were extracted according to the SW 846–3051A

TABLE 10.5

Granulometry of the Soils Typic Eutrorthox and Typic Haplorthox

	Typic Eutrorthox			**Typic Haplorthox**		
	0–0.1 m	**0.1–0.2 m**	**0.2–0.3 m**	**0–0.1 m**	**0.1–0.2 m**	**0.2–0.3 m**
Fraction			g kg^{-1}			
Clay	485	508	525	245	278	285
Silt	297	281	273	68	62	63
Total sand	219	212	202	687	661	652
Gross sand	90	86	77	388	349	356
Fine sand	129	126	125	299	312	296

Source: Melo, V.P. et al., *Pesq. Agropec. Bras.* 39:67–72, 2004.

TABLE 10.6

Chemical Properties of a Typic Eutrorthox and a Typic Haplorthox Four Months before Maize Sowing

Treatments	pH $CaCl_2$	OM g dm^{-3}	P resin mg dm^{-3}	K	Ca	Mg	H + Al	SB	CEC	V %
				mmol$_c$ dm^{-3}						
Typic Eutrorthox										
T-0	4.62	21.4	31.0	3.66	22.6	7.2	50.4	33.46	83.86	39.8
T-5	4.60	24.0	34.8	3.42	24.8	7.6	56.4	35.88	92.22	39.2
T-10	4.74	26.0	80.8	4.16	32.2	9.0	54.8	45.36	100.16	45.2
T-20	4.40	26.2	86.0	2.98	22.4	5.8	70.4	31.18	101.58	30.6
Typic Haplorthox										
T-0	4.66	19.4	47.8	2.14	18.6	4.8	39.0	25.54	64.54	40.0
T-5	4.98	19.8	55.2	2.46	24.6	6.8	32.2	33.86	66.06	51.2
T-10	4.84	21.2	95.6	2.50	26.2	6.2	40.0	34.90	74.90	47.0
T-20	4.50	20.2	94.0	2.00	19.2	4.6	49.8	25.80	75.60	35.2

Source: Andrade, M.G. et al., *Semin.-Ciênc. Agrár.* 35:135–148, 2014.
Note: T-0, T-5, T-10, T-20 = control, 5, 10, and 20 t ha^{-1} sewage sludge, dry weight basis. OM = organic matter, SB = sum of bases, V = bases saturation.

TABLE 10.7

Chemical Properties of the Sewage Sludge Used in the Thirteenth Year of Experimentation

Property	Value	Trace Element	Value
Organic carbon	246.75 g kg^{-1}	As	6.71 mg kg^{-1}
P total	20.36 g kg^{-1}	Ba	2238.89 mg kg^{-1}
N Kjeldahl	24.8 g kg^{-1}	Cd	5.10 mg kg^{-1}
pH water	5.80	Cr	671.08 mg kg^{-1}
pH $CaCl_2$	5.54	Cu	629.34 mg kg^{-1}
K total	2.38 g kg^{-1}	Hg	4.37 mg kg^{-1}
Na total	1.08 g kg^{-1}	Mo	4.28 mg kg^{-1}
Ca total	15.87 g kg^{-1}	Ni	277.54 mg kg^{-1}
Mg total	4.23 g kg^{-1}	Pb	63.13 mg kg^{-1}
Moisture	81.3%	Se	<0.05 mg kg^{-1}
		Zn	1330.55 mg kg^{-1}

Source: Andrade, M.G. et al., *Semin.-Ciênc. Agrár.* 35:135–148, 2014.

methodology (USEPA 2007) and the content of As, Ba, Cd, Cr, Cu, Hg, Mo, Ni, Pb, Se, and Zn was estimated in the extracts by inductively coupled plasma optical emission spectrometry (ICP-OES).

According to the Brazilian legislation (CONAMA 2009), the soil Typic Eutrorthox presented concentrations of Cd, Cr, and Cu higher than the prevention value (PV) and the soil Typic Haplorthox presented Cr concentration

TABLE 10.8

Rates of Dolomitic Limestone Applied to the Soils in Order to Raise the Bases Saturation to Nearly 70%

Treatments	Typic Eutrorthox	Typic Haplorthox
t ha^{-1}	Limestone (t ha^{-1})	
T-0	2.52	1.97
T-5	2.86	1.23
T-10	2.48	1.74
T-20	4.00	2.71

Source: Andrade, M.G., Elementos-traço As, Ba, Cd, Cr, Cu, Hg, Mo, Ni, Pb, Se e Zn em Latossolos e plantas de milho após treze aplicações anuais de lodo de esgoto. PhD thesis, Jaboticabal, Faculdade de Ciências Agrárias e Veterinárias, 2011.

Note: T-0, T-5, T-10, T-20 = control, 5, 10, and 20 t ha^{-1} sewage sludge, dry weight basis.

TABLE 10.9

Mineral Fertilization in the Treatments of the Experiment with the Thirteenth Annual Applications of Sewage Sludge

	N		P	K	
	Sowing	Side Dressing	Sowing	Sowing	Side Dressing
Treatments			kg ha^{-1}		
Typic Eutrorthox					
T-0	30.0	140.0	70.0	50.0	40.0
T-5	—	96.0	—	32.5	40.0
T-10	—	52.0	—	15.0	40.0
T-20	—	—	—	—	20.0
Typic Haplorthox					
T-0	30.0	140.0	50.0	50.0	40.0
T-5	—	96.0	—	32.5	40.0
T-10	—	52.0	—	15.0	40.0
T-20	—	—	—	—	20.0

Note: T-0, T-5, T-10, T-20 = control, 5, 10, and 20 t ha^{-1} sewage sludge, dry weight basis. Nitrogen and potassium side dressing was split into two applications 25 and 40 days after seedling emergence. N as urea, P as simple superphosphate, and K as potassium chloride.

higher than the PV (Table 10.10). Considering the actual concentrations of trace elements in these soils and in the sewage sludge, it is possible that the first element to get the investigation value (IV) for agronomic use is Cr. On the other hand, it is important to consider that the metal trace elements immobilized in organic matter may become available to plants if the application

TABLE 10.10

Trace Elements in Typic Eutrorthox and Typic Haplorthox Treated with Sewage Sludge for 13 Consecutive Years

	As	Ba	Cd	Cr	Cu	Hg	Mo	Ni	Pb	Se	Zn
Treaments	mg kg^{-1} Soil										
Typic Eutrorthox											
T-0	5.80 [a]	110.77 [a]	1.70 [a]	135.38 [a]	121.57 [a]	0.08 [a]	<0.001	22.29 [a]	5.83 [a]	<0.05	97.91 [b]
T-5	5.86 [a]	113.58 [a]	1.42 [a]	138.78 [a]	122.66 [a]	0.09 [a]	<0.001	23.70 [a]	6.77 [a]	<0.05	107.24 [ab]
T-10	5.66 [a]	133.69 [a]	1.67 [a]	143.78 [a]	135.74 [a]	0.11 [a]	<0.001	28.82 [a]	7.63 [a]	<0.05	129.26 [ab]
T-20	5.69 [a]	109.57 [a]	1.60 [a]	144.14 [a]	126.66 [a]	0.11 [a]	<0.001	24.52 [a]	8.00 [a]	<0.05	135.46 [a]
CV%	23.5	17.1	12.8	4.1	11.5	43.3		20.9	16.7		15.4
Typic Haplorthox											
T-0	3.06 [a]	32.55 [a]	0.48 [a]	76.01 [c]	10.27 [c]	0.15 [a]	<0.001	4.88 [c]	4.32 [b]	<0.05	24.51 [b]
T-5	3.56 [a]	24.30 [a]	0.56 [a]	82.77 [bc]	15.32 [c]	0.20 [a]	<0.001	6.73 [bc]	5.10 [b]	<0.05	38.01 [b]
T-10	3.64 [a]	29.76 [a]	0.62 [a]	95.41 [ab]	22.76 [b]	0.24 [a]	<0.001	9.27 [ab]	6.36 [a]	<0.05	61.10 [a]
T-20	3.76 [a]	44.78 [a]	0.66 [a]	99.00 [a]	29.40 [a]	0.19 [a]	<0.001	11.01 [a]	7.02 [a]	<0.05	66.63 [a]
CV%	22.9	39.4	20.1	7.7	17.6	54.2		19.1	9.0		17.8
PV value	15	150	1.3	75	60	0.5	30	30	72	5	300
IV value	35	300	3	150	200	12	50	70	180	—	450

Source: PV and IV values are obtained from CONAMA (Conselho Nacional Do Meio Ambiente). Resolução 420. Critérios e valores orientadores de qualidade do solo quanto à presença de substâncias químicas e estabelece diretrizes para o gerenciamento ambiental de áreas contaminadas por essas substâncias em decorrência de atividades antrópicas. Brasília, 2009.

Note: T-0, T-5, T-10, T-20 = control, 5, 10, and 20 t ha^{-1} sewage sludge, dry weight basis. CV = coefficient of variation. Means followed by the same letter are not different according to the Tukey test ($P < 0.05$). PV = prevention value and IV = investigation value.

of sewage sludge is stopped and the soil organic matter decomposed by soil organisms (McBride 1994).

Comparing As, Ba, Cu, Hg, Ni, Pb, and Zn concentrations in the thirteenth year with the concentrations in the twelfth year, a small decrease may be observed in some treatments that received sewage sludge. This is also observed for Cd and Cr in all the treatments that received the residue. This decrease may be attributed to the formation of compounds that are not solubilized by the SW 846–3051A methodology as observed by Melo et al. (2007) for Ni.

Maize plants did not show any phytotoxicity symptoms caused by trace elements in all the treatments. Copper concentration in the leaves ranged in the interval of 6–20 mg kg^{-1} (Figure 10.4), which is considered normal (Malavolta et al. 1997; Kabata-Pendias and Pendias 2001). This concentration was affected by the treatments in both the soils, and the lowest value occurred in the highest rate of sewage sludge. The low Cu concentration in the leaves may be due to its low mobility in the plant (Martins et al. 2003). According to Faquin (2001), Cu uptake by plants is inhibited when Zn concentration is high (6.65, 13.30, and 26.61 kg ha^{-1}).

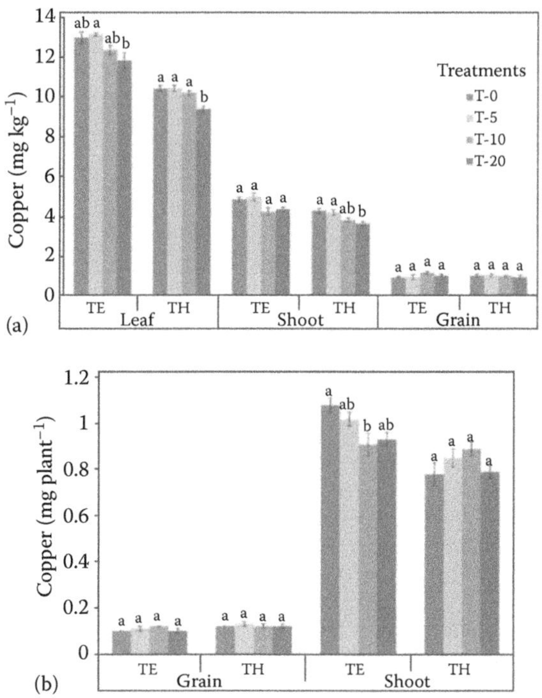

FIGURE 10.4
Copper concentration (a) and accumulated (b) in maize plant parts when cropped in Typic Eutrorthox (TE) and Typic Haplorthox (TH) treated with sewage sludge for 13 consecutive years. T-0, T-5, T-10, T-20 = control, 5, 10, and 20 t ha^{-1} sewage sludge, dry weight basis. Bars followed by the same letter in the same soil are not different according to the Tukey test ($P < 0.05$).

Some authors have observed that there is not always a positive relation between Cu concentration in soil and in leaf (Logan et al. 1997). The total concentration of Cu in soil is not a useful indicator of the potential availability of the element to plants because it may be strongly immobilized by the soil organic matter (Kabata-Pendias and Pendias 2001; McBride 2001).

Zinc concentration in shoot and leaves increased with the rate of sewage sludge (Figure 10.5). This is due to the high concentration of the element in the residue and to its solubility in soil, but no Zn toxicity symptoms were exhibited. This behavior was also observed by other authors (Rappaport et al. 1988; Berti and Jacobs 1996; Schimidt 1997; Nascimento et al. 2004).

Higher Cu and Zn concentrations in leaves than in grains were also observed by other authors (Martins et al. 2003; Rangel et al. 2006). In order to protect the species, seeds accumulate fewer trace elements than the other parts of the plant (Jing and Logan 1992). For use in the diet, the maximum allowed contents of Cu and Zn in grains (dry matter) are 30 and 50 mg kg^{-1}, respectively (ABIA 1985), which are concentrations not reached in this experiment.

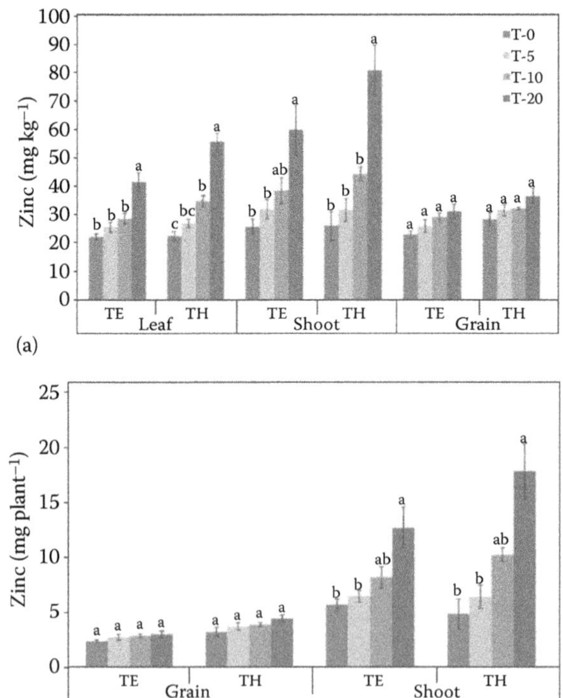

FIGURE 10.5
Zinc concentration (a) and accumulated (b) in maize plant parts when cropped in Typic Eutrorthox (TE) and Typic Haplorthox (TH) treated with sewage sludge for 13 consecutive years. T-0, T-5, T-10, T-20 = control, 5, 10, and 20 t ha^{-1} sewage sludge, dry weight basis. Bars followed by the same letter in the same soil are not different according to the Tukey test ($P < 0.05$).

In the fifth year of experimentation, Oliveira et al. (2005) found a content of Cu in maize grains similar to that obtained in the thirteenth year. However, Zn concentration in the fifth year ranged from 14.6 to 21.65 mg kg^{-1} (soil TE) and from 11.76 to 17.19 mg kg^{-1} (soil TH), which are concentrations lower than those obtained in the thirteenth year. This finding may be explained by an increase in the accumulated rates of sewage sludge or by the climatic conditions during the two growing seasons, increasing the availability of Zn to plants in the preceding growing season. Zinc concentration in parts of maize plants was generally higher when the plant was cropped in the soil TH. This may be attributed to a higher Zn immobilization in the soil TE, which has more clay particles and organic matter.

Barium concentration in the diagnosis leaf is below the toxic level of 500 mg kg^{-1} (Figure 10.6).

The ions Ba^{2+} in soil solution may form an insoluble salt ($BaSO_4$) instead of soluble complexes with fulvic and humic acids (USEPA 1984), which

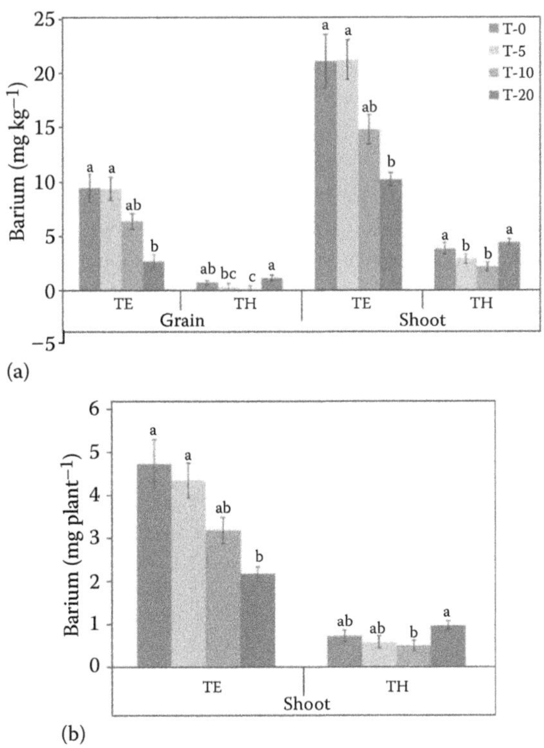

FIGURE 10.6
Barium concentration (a) and accumulated (b) in maize plant parts when cropped in Typic Eutrorthox (TE) and Typic Haplorthox (TH) treated with sewage sludge for 13 consecutive years. T-0, T-5, T-10, T-20 = control, 5, 10, and 20 t ha^{-1} sewage sludge, dry weight basis. Bars followed by the same letter in the same soil are not different according to the Tukey test ($P < 0.05$).

decreases its uptake by the plants. They are also specifically adsorbed by Fe and Mn oxides (Choudhury and Carey 2001; Liu et al. 2002), so that only a small fraction of the soil Ba is available to plants.

The data suggest that maize plants cropped in sandy soils with low organic matter content uptake more trace elements, particularly Zn and Cd, as compared to plants cropped in clayey soils with more organic matter (Alloway 1995).

Grain production in both the soils was not affected by the treatments (Figure 10.7). Similar results were obtained by other authors (Anjos and Mattiazzo 2000; Galdos et al. 2004; Nogueira et al. 2008). However, grain production was higher in the sandy soil (soil TH), probably due to a lower CEC associated with better physical properties. The same behavior was observed by Anjos and Mattiazzo (2000) when comparing the production in sand and clayey soils. Silva et al. (2002) obtained higher maize grain production in the treatments that received sewage sludge as compared to the treatment receiving NPK mineral fertilization.

Shoot dry matter production was not affected by the treatments in the clayey soil, but this was observed in the sandy soil (Figure 10.7), where the highest production occurred in the treatment with 20 t ha^{-1} sewage sludge (dry weight basis). This finding corroborates the data obtained by Martins et al. (2003), Simonete et al. (2003), and Nascimento et al. (2004).

The maize shoot dry matter production was higher in the plants cropped in the clayey soil, which was the inverse of the grain production. According to Prado (2008) and Cakmak (2008), plants growing in soil having high availability of nutrients, mainly nitrogen, may produce more amino acids, proteins, enzymes, nucleic acids, and other nitrogen compounds, including chlorophyll, but a low production of carbohydrates. In these conditions, plants present vigorous growing but low grain production.

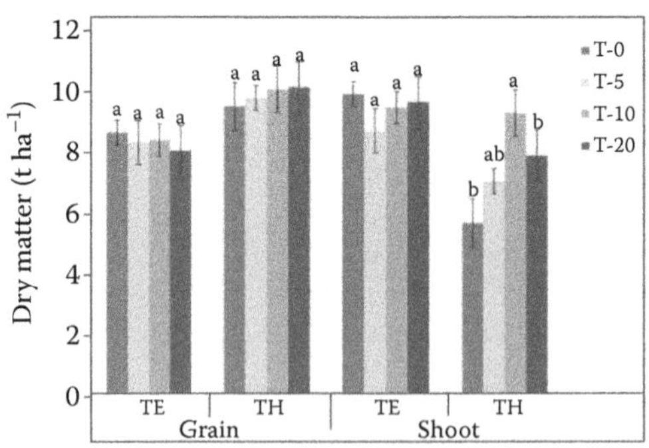

FIGURE 10.7
Grain and shoot dry matter production by maize plants cropped in Typic Eutrorthox (TE) and Typic Haplorthox (TH) treated with sewage sludge for 13 consecutive years. T-0, T-5, T-10, T-20 = control, 5, 10, and 20 t ha^{-1} sewage sludge, dry weight basis. Bars followed by the same letter in the same soil are not different according to the Tukey test ($P < 0.05$).

Summary and Future Directions

The environmental and economic disposition of sewage sludge is a problem that must be solved urgently. The residue is attractive for use in agriculture as an organic fertilizer because of its contents of organic matter and plant nutrients; however, it also contains components that have a negative impact on the environment and human health.

To make a decision about the use of sewage sludge in agriculture that carries minimum risk, it is important to develop long-term field experiments evaluating different soils, plants, climates, and rates of the residue in order to obtain data that support that disposition. The data will supply support for the establishment of more secure legislation regulating the use of sewage sludge in agriculture.

In trials to evaluate rates of sewage sludge related in this chapter, the residue was applied for 13 consecutive years in sandy and clayey soils, using maize as the main crop.

The data showed clearly that even after 13 annual applications of a sewage sludge obtained from the treatment of a polluted sewer to sandy and clayey soils, the concentration of trace elements in the soils did not reach the investigation value adopted by Brazilian legislation, and maize plants did not present toxicity symptoms caused by trace elements.

The rates of sewage sludge did not affect maize grain production, but the grain production was higher in the sandy soil.

Copper and zinc accumulated mainly in the leaves and shoot. Barium also accumulated in the shoot, but a significant concentration was also observed in the seeds.

The highest concentration of trace elements was found in the clayey soil.

References

ABIA (Associação Brasileira das Indústrias da Alimentação). Compêndio da legislação dos alimentos. São Paulo, 1985.

ACCLPP (Advisory Committee on Childhood Lead Poisoning Prevention). CDC, 2012.

Adriano, D.C. *Trace Elements in the Terrestrial Environment*. New York, Springer-Verlag, 1986.

Alloway, B.J. Soil processes and the behaviour of metals. In: Alloway, B.J. (ed.), *Heavy Metals in Soils*. Glasgow, Blackie Academic and Professional, 1995, pp. 7–28.

Andrade, M.G. Elementos-traço As, Ba, Cd, Cr, Cu, Hg, Mo, Ni, Pb, Se e Zn em Latossolos e plantas de milho após treze aplicações anuais de lodo de esgoto. PhD thesis, Jaboticabal, Faculdade de Ciências Agrárias e Veterinárias, 2011.

Andrade, M.G., Lima, A.S.T., Melo, W.J., Santos, E.J., Herrmann, A.B. Elementos-traço em dois Latossolos após aplicações anuais de lodo de esgoto por treze anos. *Semin.-Ciênc. Agrár.* 35:135–148, 2014.

Anjos, A.R.M., Mattiazzo, M.E. Metais pesados em plantas de milho cultivadas em Latossolos repetidamente tratados com biossólido. *Sci. Agric.* 57:769–776, 2000.

Berti, A.M., Jacobs, L.W. Chemistry and phytotoxicity of soil trace elements from repeated sewage sludge applications. *J. Environ. Qual.* 25:1025–1032, 1996.

Brown, G.E. Jr., Parks, G.A. Sorption of trace elements on mineral surfaces: Modern perspectives from spectroscopic studies and comments on sorption in the marine environment. *Int. Geol. Rev.* 43:963–1073, 2001.

Cakmak, I. Enrichment of cereal grains with zinc: Agronomic or genetic biofortification? *Plant and Soil* 302:1–17, 2008.

Camargo, O.A., Alleoni, L.R.F., Casagrande, J.C. Reações dos Micronutrientes e Elementos Tóxicos no Solo. In: Ferreira, M.E., Cruz, M.C.P., Raij, B. van, Abreu, C.A. (eds). *Micronutrientes e elementos tóxicos na agricultura.* Jaboticabal, CNPq/FAPESP/POTAFOS, 2001, pp. 89–124.

Carvalho, R., Neto, A.E.F., Curi, N., Fernandes, L.A., Oliveira, A.C. Interações silício-fósforo em solos cultivados com eucalipto em casa de vegetação *Pesqui. Agropecu. Bras.* 36:557–565, 2001.

Chang, A.C., Warneke, J.E., Page, A.L., Lund, L.J. Accumulation of heavy metals in sewage sludge-treated soils. *J. Environ. Qual.* 13(1):87–90, 1984.

Choudhury, H., Carey, R. Barium and barium compounds. Concise International Chemical Assessment Document 33, International Programme on Chemical Safety, World Health Organization, Geneva, Switzerland, 2001, http://inchem.org/documents/cicads/cicads/cicad33.htm, accessed August 8, 2011.

CONAMA (Conselho Nacional Do Meio Ambiente). Resolução 375. Critérios e procedimentos para o uso agrícola de lodos de esgoto gerados em estações de tratamento de esgoto sanitário e seus produtos derivados. Brasília, 2006.

CONAMA (Conselho Nacional Do Meio Ambiente). Resolução 420. Critérios e valores orientadores de qualidade do solo quanto à presença de substâncias químicas e estabelece diretrizes para o gerenciamento ambiental de áreas contaminadas por essas substâncias em decorrência de atividades antrópicas. Brasília, 2009.

Cornu, S., Neal, C., Ambrosi, J.P., Whitehead, P., Neal, M., Sigolo, J., Vachier, P. The environmental impact of heavy metals from sewage sludge in Ferrasols (São Paulo, Brazil). *Sci. Total Environ.* 271:27–48, 2001.

Covelo, E.F., Vega, F.A., Andrade, M.L. Competitive sorption and desorption of heavy metals by individual soil components. *J. Hazard. Mater.* 140:308–315, 2007.

Faquin, V. Nutrição mineral de plantas. Lavras, Brazil, Universidade Federal de Lavras, 2001.

Galdos, M.V., De Maria, I.C., Camargo, O.A. Atributos químicos e produção de milho em um Latossolo Vermelho eutroférrico tratado com lodo de esgoto. *R. Bras. Ci. Solo* 28:569–577, 2004.

Gomes, P.C., Fontes, M.P.F., Silva, A.G., Mendonça, E.S., Netto, A.R. Selectivity sequence and competitive adsorption of heavy metals by Brazilian soils. *Soil Sci. Soc. Am. J.* 65:843–852, 2001.

Ji, G.L., Li, H.Y. Electrostatic adsorption of cations. In: Yu, T.R. (ed.), *Chemistry of Variable Charge Soils.* New York, Oxford University Press, 1997, pp. 65–79.

Jing, J., Logan, T.J. Effects of sewage sludge cadmium concentration on chemical extractability and plant uptake. *J. Environ. Qual.* 21:73–81, 1992.

Kabata-Pendias, A., Pendias, H. *Trace Elements in Soils and Plants.* Boca Raton, FL, CRC Press LLC, 2001.

Keaton, C.M. Influence of lead compounds on the growth of barley. *Soil Sci.* 43:401–411, 1937.

Levy, D.B., Barbaric, K.A., Siemer, E.G., Sommers, L.E. Distribution and portioning of trace metals in contaminated soils near Leadville, Colorado. *J. Environ. Qual.* 21:185–195, 1992.

Liu, F., Colombo, C., Adamo, P., He, J.Z., Violante, A. Trace elements in manganese–iron nodules from a Chinese Alfisol. *Soil Sci. Soc. Am. J.* 66:661–670, 2002.

Logan, T.J., Lindsay, B.J., Goins, L.E., Ryan, J.A. Field assessment of sludge metal bioavailability to crops: Sludge rate response. *J. Environ. Qual.* 26:534–550, 1997.

Malavolta, E., Vitti, G.C., Oliveira, S.A. *Avaliação do estado nutricional das plantas: Princípios e aplicações,* 2. ed., Piracicaba, Brazil, Associação Brasileira para Pesquisa da Potassa e do Fosfato, 1997.

Malavolta, E. *Manual de nutrição mineral de plantas.* Piracicaba. Ed. Agronômica Ceres. 2006.

Martins, A.L.C., Bataglia, O.C., Camargo, O.A., Cantarella, H. Produção de grão e absorção de Cu, Fe, Mn e Zn pelo milho em solo adubado com lodo de esgoto, com e sem calcário. *R. Bras. Ci. Solo* 27:563–574, 2003.

Mathews, P. International perspective. Workshop biosolids: Current and emerging issues, WWEFTEC'99, New Orleans, LA, 1999.

McBride, M.B. *Environmental Chemistry of Soils.* New York, Oxford University Press, 1994.

McBride, M.B. Cupric ion activity in peat soil as a toxicity indicator for maize. *J. Environ. Qual.* 30:78–84, 2001.

Melo, W.J., Marques, M.O., Melo, V.P. O uso agrícola do biossólido e as propriedades do solo. In: Tsutyia, M.T. et al. (eds.), *Biossólidos na Agricultura,* Chapter 11. São Paulo, ABES/SP, 2002, pp. 289–363.

Melo, V.P., Beutler, A.N., Souza, Z.M., Centurion, J.F., Melo, W.J. Atributos físicos de Latossolos adubados durante cinco anos com biossólido. *Pesq. Agropec. Bras.* 39:67–72, 2004.

Melo, W.J., Aguiar, P. de Stefani, Melo, G.M.P., Melo, V.P. Nickel in a tropical soil treated with sewage sludge and cropped with maize in a long-term field study. *Soil Biol. Biochem.* 39:1341–1347, 2007.

Mengel, K., Kirkby, E.A. *Principles of Plant Nutrition.* Bern, Switzerland, International Potassium Institute, 1987.

Merlino, L.C.S., Melo, W.J., Macedo, F.G., Guedes, A.C.T.P., Ribeiro, M.H., Melo, V.P., Melo, G.M.P. Bário, cádmio, cromo e chumbo em plantas de milho e em Latossolo após onze aplicações anuais de lodo de esgoto. *R. Bras. Ci. Solo* 34:2031–2039, 2010.

Nascimento, C.W.A., Barros, D.A.S., Melo, E.E.C., Oliveira, A.B. Alterações químicas em solos e crescimento de milho e feijoeiro após aplicação de lodo de esgoto. *R. Bras. Ci. Solo* 28:385–392, 2004.

Nogueira, T.A.R., Oliveira, L.R., Melo, W.J., Fonseca, I.M., Melo, G.M.P., Melo, V.P., Marques, M.O. Cádmio, cromo, chumbo e zinco em plantas de milho e em Latossolo após nove aplicações anuais de lodo de esgoto. *R. Bras. Ci. Solo* 32:2195–2207, 2008.

Oliveira, F.C., Marques, M.O., Bellingieri, P.A., Precin, D. Lodo de esgoto como fonte de macronutrientes para a cultura do sorgo granífero. *Sci. Agríc.* 52:360–367, 1995.

Oliveira, F.C., Mattiazzo, M.E. Mobilidade de metais pesados em um Latossolo Amarelo distrófico tratado com lodo de esgoto e cultivado com cana-de-açúcar. *Sci. Agric.* 58:807–812, 2001.

Oliveira, K.W., Melo, W.J., Pereira, G.T., Melo, V.P., Melo, G.M.P. Heavy metals in oxisols amended with biosolids and cropped with maize in a long-term experiment. *Sci. Agric.* 62:381–388, 2005.

Paim, L.A., Carvalho, R., Abreu, C.M.P., Guerreiro, M.C. Estudos dos efeitos do silício e do fósforo na redução da disponibilidade de metais pesados em área de mineração. *Quím. Nova.* 29:28–33, 2006.

Pierangeli, M.A.P., Guilherme, L.R.G., Curi, N., Silva, M.L.N., Oliveira, L.R., Lima, J.M. Teor total e capacidade máxima de adsorção de chumbo em Latossolos brasileiros. *R. Bras. Ci. Solo* 25:279–288, 2001.

Prado, R.M. *Nutrição de plantas*. São Paulo: Editora UNESP, 2008, p. 407.

Rangel, O.J.P., Silva, C.A., Bettiol, W., Dynia, J.F. Efeito de aplicações de lodos de esgoto sobre os teores de metais pesados em folhas e grãos de milho. *R. Bras. Ci. Solo* 30:583–594, 2006.

Rappaport, B.D., Martens, D.C., Reneau, Jr., R.B., Simpson, T.W. Metal availability in sludge-amended soils with elevated metals levels. *J. Environ. Qual.* 17:42–47, 1988.

Rieuwerts, J.S., Ashnore, M.R., Farago, M.E., Thornton, I. The influence of soil characteristics on the extractability of Cd, Pb and Zn in upland and moorland soils. *Sci. Total Environ.* 366:64–875, 2006.

Ross, S.M. *Toxic Metals in Soil-Plant Systems*. Chichester, England, John Wiley & Sons, 1994.

Schimidt, J.P. Understanding phytotoxicity thresholds for trace elements in land-applied sewage sludge. *J. Environ. Qual.* 26:4–10, 1997.

Shriver, D.F. *Química Inorgânica*, 3ª ed., Porto Alegre, Brazil, Bookman, 2003.

Silva, J.E., Resck, D.V.S., Sharma, R.D. Alternativa Agronômica para o biossólido produzido no Distrito Federal. I–efeito na produção de milho e na adição de metais pesados em Latossolo no cerrado. *R. Bras. Ci. Solo* 26:487–495, 2002.

Simonete, M.A., Kiehl, J.C., Andrade, C.A., Teixeira, C.F.A. Efeito do lodo de esgoto em um Argissolo e no crescimento e nutrição de milho. *Pesq. Agropec. Bras.* 38:1187–1195, 2003.

Sposito, G. *The Chemistry of Soils*. New York, Oxford University Press, 2008.

Stevenson, F.J. *Cycles of Soil Carbon, Nitrogen, Phosphorus, Sulfur, Micronutrients*. New York, John Wiley & Sons, 1986, pp. 231–284.

USEPA (United States Environmental Protection Agency). Health effects assessment for barium. EPA 540/1-86-021. USEPA, Cincinnati, OH, 1984.

USEPA (United States Environmental Protection Agency). Title 40 CFR. Part 503. Final rules standards for the use or disposal of sewage sludge. Federal Register 58-9387-9415, 1999.

USEPA (United States Environmental Protection Agency), Method 3051A–Microwave assisted acid digestion of sediments, sludges, soils, and oils. Washington, DC, Revision 1, February 2007, http://www.epa.gov/epawaste/hazard/testmethods/sw846/pdfs/3051a.pdf, accessed October 7, 2009.

Yu, T.R. *Chemistry of Variable Charge Soils*. New York, Oxford University Press, 1997.

11

Effectiveness of Lime and Glauconite-Phosphorite Containing Organo-Mineral Ameliorants in Heavy-Metal-Contaminated Soils

Maya Benkova and Irena Atanassova

CONTENTS

This chapter refers to a case study involving aspects of soil acidity and amelioration of copper (Cu)-, zinc (Zn)-, and lead (Pb)-contaminated soils around nonferrous industrial plants in Bulgaria. This chapter first outlines the various soil physicochemical characteristics that influence Cu, Zn, and Pb behavior, and then discusses the effects of applying organomineral mixtures containing lime, phosphorite-glauconite, peat, coal, and iron (Fe) hydroxides on the solubility, speciation, and bioavailability of Cu, Zn, and Pb to ryegrass plants. Although the soils contained low amounts of organic carbon (C) (0.5%–1.7%), the major species of Cu, Zn, and Pb in the pH range of 5.7–7.4 were organic complexes with dissolved organic carbon (DOC). The organomineral amendment effectiveness to the ryegrass was proved by significant increases in the biomass yields and >70% for Zn and >90% for Cu and Pb decrease in metal solution concentration as compared to the control variants. However, no significant difference in metal uptake among treated soils and control soils could be discerned. Plant tissues took up normal levels of Cu, Zn, and Pb, and no correlation existed between solution/plant concentrations. The mineral nutrition and biogeochemical processes in the rhizosphere of ryegrass plants could also influence the pH, ionic composition, and metal speciation in solution and bioavailability.

Numerous ameliorants such as lime, peat, composts, phosporus (P)-based amendments, coal ash, zeolites, and Fe oxides have long been used to raise soil pH and lower metal toxicity to plants (Gworek 1992; Naidu et al. 1994; Shende et al. 1994; Hooda and Alloway 1996; Chlopecka and Adriano 1997; Edwards et al. 1999; McBride and Martinez 2000; Oste et al. 2002; Bolan and Duraisamy 2003; Bolan et al. 2003). Aside from carbonates and bicarbonates, organic matter has an important role in buffering the soil acidity and increasing the pH of acid soils. It has been suggested that organic amendments (sewage sludge, composts, manures, coals, and peat) may react similarly to calcium carbonate ($CaCO_3$) by precipitating or complexing aluminum (Al) and Fe (Pierzynski and Schwab 1993; Mench et al. 1994; Hooda and Alloway 1996; Iyamuremye et al. 1996; Shestopalov and Gorbov 1998; Smith 2009). Phosphorus-bearing materials such as soluble P-compounds, phosphate fertilizers, hydroxyapatites, and phosphate rock have been widely applied in immobilization and reducing phytoavailability of Cu, Zn, cadmium (Cd), and arsenic (As) (Boisson et al. 1999; Bolan et al. 2003; Miretzky and Fernandez-Cirelli 2008; Hashimoto et al. 2009; Waterlot et al. 2011; Siebers et al. 2013). The phosphorus-induced immobilization is mainly due to formation of metal–phosphate precipitates (Cao et al. 2003; Miretzky and Fernandez-Cirelli 2008). The efficiency and mechanism of nanohydroxyapatite (nHA) in immobilizing Pb and Cd in contaminated soil were assessed by He et al. (2013). The nHA application could significantly reduce water-soluble and bioavailable Pb and Cd by 65.3% and 64.6% in contaminated soil, respectively. Diammonium phosphate (DAP) has not been extensively tested as a chemical immobilization treatment, but it was the most effective treatment for immobilizing heavy metals and was superior to limestone and rock phosphate (McGowen et al. 2001; Basta and McGowen 2004). A potential negative consequence of soluble phosphate amendments is phosphate leaching from soils and runoff that can promote eutrophication in water bodies.

A method of increasing amelioration effectiveness was developed based on addition of organomineral lime-containing materials that lead to strengthening interactions between soil colloids and heavy metals (Raytchev et al. 1999; Raichev et al. 2000; Arsova and Raytchev 2001; Raychev et al. 2002; Benkova et al. 2005). According to a theoretically elaborated design, a combined treatment of polluted acid soils with lime materials and organic products (anthracite coal), containing compounds with a high degree of humification (humic acids), increased soil pH and decreased heavy metal mobility due to association of metals in stable organomineral complexes (Raytchev et al. 1999; Benkova 2005). The priority of using anthracite coals over brown coals is due to the higher content of humic acids caused by an advanced stage of humification. Waste carbonate deposit and anthracite coal powder were tested in laboratory and pot experiments, and their applicability for soil acidity neutralization and heavy metal immobilization was confirmed (Arsova and Raychev 1999). A compost of carbonate deposit and anthracite coal powder was patented as an ameliorant for copper-polluted acid soils (Raytchev et al. 1999). Detailed investigation was carried out to determine the optimal quantitative ratios between the waste products providing maximal effectiveness. The amount of carbonate deposit and coal powder applied in the polluted soils was considered in parallel with the toxic soil acidity and the sorption capacity of the humus components (Arsova and Raytchev 2001). In a study involving zeolite, glauconite, and phosphorite in different ratios (Petkova et al. 2000), it was established that soil acidity was neutralized both by using $CaCO_3$ and phosphorite–glauconite. The Pb content of radishes decreased by approximately three times with these ameliorant mixtures compared to radish grown on Pb-contaminated soil alone.

In this chapter, we present results on a specifically designated scheme of organomineral amelioration of acid heavy metal polluted soils including lime, glauconite–phosphorite, and organic amendments and their effect on the solubility, speciation, and bioavailability of Cu, Zn, and Pb.

Study Sites

Soil samples were taken from areas located near the nonferrous metallurgical plants, which are major sources of pollution with heavy metals in Bulgaria (Figure 11.1). Site 1 is close to the Kremikovtsi Steel Plant in Sofia; 6000 ha of agricultural land near Kremikovtsi is now contaminated with heavy metals even though the plant stopped functioning in 2008. Site 2 is close to the Umicor Copper smelter AD, Pirdop. Soil properties and land use in combination with lithogenic enrichment with heavy metals are prerequisites for elevated contents of metals exceeding the maximum permissible loads of the national standards (State Gazette, No. 71, 12.08.2008).

Ameliorative materials and their composition in organomineral mixtures were selected based on greenhouse experiments (Raychev et al. 2002; Benkova 2005). Organomineral mixtures involving glauconite–phosphorite, lime, and organic amendments were proposed. The following mineral and organic materials were used as ameliorants:

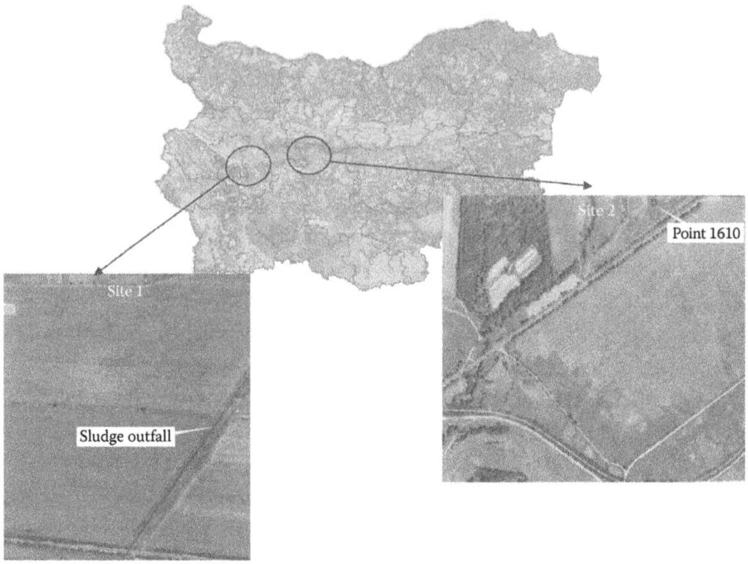

FIGURE 11.1
General scheme of Site 1 (near Kremikovtsi iron smelter) and Site 2 (near Umicor copper smelter, Bulgaria).

- Lime material, calcium oxide (CaO), which has a neutralizing ability of 3500 cmol/1000 g material (nBM = 100/3500 ≈ 0.03 g/1 mequ).
- Glauconite–phosphorite ameliorant, which is a sedimentary homogeneous phosphorite from North Bulgaria (Sanidinovo) containing fluorapaptite–glauconite concretions ($Ca_5(PO_4)_3F$) (Table 11.1). Glauconite has beneficial effects on soil structure contributing to soil water-holding and cation exchange capacity.
- Ferric hydroxide, a waste product containing 55% $Fe(OH)_3$ or 25 g Fe/100 g.
- Coal powder, an anthracite waste powder from a coal mine near Sofia (pH = 7.74, ash content 15%–20%).

TABLE 11.1
Chemical Composition of Glauconite–Phosporite

Mineral Composition (%)									
SiO_2	Al_2O_3	Fe_2O_3	CaO	MgO	MnO	Na_2O	K_2O	TiO	As
45.30	5.70	10.64	16.00	1.82	0.03	0.44	3.77	0.12	<0.001
Available Forms (mg/kg)									
pH_{H_2O}	N	P	K	Zn	Fe	Cu	Ca	Mg	Na
7.92	7.00	16.30	184.26	41.00	0.39	0.20	4000.70	187.20	16.10

- Peat, from the area of Venets, a village in Bulgaria.
- Manure containing 5% nitrogen (N), 2.8% diphosphorus pentoxide (P_2O_5), 3.4% potassium oxide (K_2O), 4.9% CaO, 0.7% magnesium oxide (MgO), 3.2% silicon dioxide (SiO_2), 8.0% water (H_2O), and 28% ash.

The experiments were carried out in pots of 5-kg contaminated soil. The appropriate amount of fertilizer was mixed with the soil in each pot (except in the control variant). After 2 months of composting the tested materials at 70% field capacity, two samplers were installed to extract soil solution in each pot. Soil solution was obtained by vacuum application of Rhizon soil moisture samplers (plastic capillary tubes with a filter at one end), allowing for collection of soil solution for laboratory analysis. The low sorption capacity of the plastic material makes them suitable for suction at high pressure (Wenzel and Wieshammer 1995). A layer of ~1–1.5 cm of perlite was placed at the bottom of each pot, on top of which an antirooting mat was positioned to prevent roots from growing out of the pots. Then the pots were filled with a layer of ~15 cm of soil; two samplers were placed parallel to each other and spaced about 10 cm apart from each other, and then were filled to a few centimeters below the edge of the pot (Figure 11.2).

FIGURE 11.2
Design of the greenhouse experiment for biomass and solution analyses using Rhizon soil solution samplers.

In each pot, 5 g seeds of *Lolium perenne* L. ryegrasses were sown. Soil solution was sampled twice, before planting and before collecting the biomass.

Design of the Greenhouse Experiments

The experimental setup and amounts of ameliorants were carried out in three replicates.

Kremikovtsi (Site 1)

1. K, control soil without ameliorants
2. Ca = lime material
3. CaGp = 1/4 CaO + 3/4 glauconite–phosphorite
4. CaGpPFe = 1/4 CaO + 3/4 glauconite–phosphorite + peat + Fe
5. CaGpPCFe = 1/4 CaO + 3/4 glauconite–phosphorite + 1/2 peat + 1/2 coal + $Fe(OH)_3$
6. CaGpFe = 1/4 CaO + 3/4 glauconite–phosphorite + $Fe(OH)_3$

Pirdop (Site 2)

1. K, control soil without ameliorants
2. Ca, lime material
3. CaGp, 3/4 CaO + 1/4 glauconite–phosphorite
4. CaGpP, 3/4 CaO + 1/4 glauconite–phosporite + peat
5. CaCGpPC, 3/4 CaO + 1/4 glauconite–phosphorite + 1/2 peat + 1/2 coal
6. CaGpC, 3/4 CaO + 1/4 glauconite–phosphorite + coal

The amount of lime material and glauconite–phosphorite in the organo mineral mixtures was calculated on the basis of an equation determining lime requirement for acid soils (Ganev 1987):

$$LM_{H8.2}\ (g/100g) = n_{LM} \times [H_{8.2} + 0.8 \times (\%HS) \times K]$$

where

n_{LM} = neutralizing ability of the lime material, cmol/kg
%HS = organic C in soil
K = exch.Al/[exch.Al + Σ(exch.Ca + exch.Mg)]

The applied amounts of peat (PE) and coal powder (CP) were calculated according to the optimal rate for organomineral mixtures (Raychev et al. 2002; Benkova 2005):

$$[OM]^1\ (g/100g) = (50 \times \Sigma Al^{exch})/\%HA$$

$$[OM]^2\ (g/100g) = (25 \times \Sigma Al^{exch})/\%HA$$

where

%HA = humic substances, g/100g, determined by the ash content: %HA = 100% ash

ΣAl^{exch} = exchange acidity, cmol/kg

The dose of ferric hydroxide for amelioration of As toxicity in the Kremikovtsi soil was calculated on the basis of As content (138.3 mg/kg > maximum permissible load [MPL] <25 mg/kg, based on the Bulgarian National Standards) (Boldirev 1983). Mineral fertilization was applied at a dose of 0.1 g/100 g of soil with the following composition of salts (in g/100 g): NH_4NO_3 0.022, Ca $(H_2PO_4)_2$ 0.034, $KHPO_4$ 0.037, $MgSO_4$ 0.0066, and a ratio of 1N:1P:0.2S:0.5Ca:0.5K:0.2Mg.

Soil solution was taken twice in the experimental year (before sowing and before harvesting the biomass). The liquid phase was analyzed for the following parameters: pH, conductivity, DOC, K^+, Na^+, Ca^{2+}, Mg^{2+} Cu^{2+}, Zn^{2+}, Pb^{2+}, Cd^{2+}, SO_4^{2}, PO_4, HCO_3^-, NO_3^-, and Cl^-.

After harvesting, plant samples were obtained by cutting the grass in each pot, drying at 70°C, and grinding to powder using a mill. Plant material was digested using concentrated HNO_3 and H_2O_2 (Methods of Soil Analysis 1982). The metal concentration was determined by ICP-AES.

Soil Analysis

Cation exchange properties of the soil were determined using the methods of Ganev and Arsova (1980), and exchangeable acidity ($_{exch}Al$) was assessed by titration after displacement of Al^{3+} ions with 1N $CaCl_2$, hydrolytic acidity ($H_{8.2}$), and exchangeable bases (Ca + Mg) by titration; cation exchange capacity (CEC) was calculated as the sum of the total acidity and exchangeable bases. Soil reaction was measured in 1:2.5 soil–water suspensions.

The content and composition of soil organic matter were determined according to Turin's modified method (acid–dichromate digestion, 120°C, 45 min in a thermostat, catalyst Ag_2SO_4) and Kononova–Belchicova's method (Kononova 1966).

The total heavy metal content in the soils was determined by aqua regia digestion (1 g soil + mixture of concentrated acids $HCl:HNO_3 = 3:1$), diluted with deionized water to 50 mL after digestion (ISO 11466:1995). The extracts were filtered and analyzed for Cd, Cu, Pb, and Zn by inductively coupled plasma atomic emission spectroscopy (ICP-AES) to obtain the total heavy metal contents.

Bioavailable forms of heavy metals in the soil were determined by extraction with 0.43N nitric acid (1:10 soil/solution ratio) using the work of Houba et al. (1988) and 0.01 M $CaCl_2$ (1:10 soil/solution ratio) using the work of Houba et al. (1999).

Soil Solution Analysis

Samples of the liquid phase of the soil were analyzed for pH and conductivity, ions of potassium, sodium by flame photometry, calcium, magnesium, and heavy metals (copper, zinc, lead, and cadmium) by ICP-AES, and HCO_3^- was determined by titration with 0.02 H_2SO_4 to pH = 4.4. Nitrates, sulfates, phosphates, chlorides, and DOC were analyzed by ultraviolet–visible spectroscopy (UV–VIS) (Spectroquant Pharo 100). DOC was analyzed by digestion with sulfuric acid and peroxidisulfate. The CO_2 generated reacted with the indicator solution, whose color was photometrically determined at 596 nm using standard phthalate acid solutions. Metal speciation was calculated by using the Visual Minteq 3.0 through the Stockholm Humic Acid (SHM) model and inorganic solution chemistry.

Measurements were made of total yield dry weight. One gram of dried plant material was digested using concentrated HNO_3 and H_2O_2. The metal concentration (Cu, Zn, Pb, and Cd) was determined by ICP-AES (Methods of soil analysis 1982). Results were treated statistically by one-way analysis of variance (ANOVA), multiple-variable analysis, and multiple regression analysis. A STATGRAPHICS Centurion Version XV was used.

Physicochemical Characteristics of Soils

The soil from Kremikovtsi is moderately acidic (pH 5.3) and possesses lower CEC than the Pirdop soil (22.3 $cmol{\cdot}kg^{-1}$). The Pirdop soil is highly acidic pH 4.6 with an organic matter content of 2.2%, is exchangeable (Al 15.7% of CEC_{sa}), and has hydrolytic soil acidity (46.64% of CEC) (Table 11.2). According to the classification of Orlov and Grishina (Orlov 1985), Kremikovtsi soil (Site 1) has a low humus content (1.5%), a humic type of humus (Ch/Cf–0.87), a low content of Ca-bound humic acids, and a low degree of humification (Table 11.3). In this soil, 90% of humic acids are presented as free and sesquioxide R_2O_3 complexed forms. The soil from Pirdop Site 2 has an average humus content of 2.2% and a humic–fulvic type of humus (Ch/Cf–0.51). Humic acids have a low degree of condensation, and 100% are presented as free and R_2O_3-bound.

TABLE 11.2

Physicochemical Parameters and Values of the Strongly Acidic and Weakly Acidic Cation Exchange Sites on the Soil Adsorbents for Kremikovtsi (Site 1) and Pirdop (Site 2) Soil

Soil Sample	pH (H_2O)	CEC	CEC_{sa}	CEC_a	exch.$H_{8.2}$	$_{exch}Al$	Ca	Mg	Base Saturation V%
		cmol·kg^{-1}/% of $CEC_{pH8.2}$							
Soil 1	5.3	19.3	*14.3*	*5*	*5.3*	*0.34*	*12.5*	*1*	72.5
Kremikovtsi			74.09	25.91	27.46	1.76	64.77	5.18	
Soil 2	4.7	22.3	*15.5*	*6.8*	*10.4*	*3.5*	*9.5*	*2.8*	53.4
Pirdop			69.51	30.49	46.64	15.7	42.62	12.56	

TABLE 11.3

Content and Composition of Organic Matter

			Organic C (%) Extracted with 0.1 M $Na_4P_2O_7$ + 0.1 M NaOH				C in Humic Acids Fractions (%)		
Site	Total C%	SOM	Total	Humic Acid	Fulvic Acids	C_h/C_f	Free and R_2O_3-Complexed	Ca-Complexed	
Kremikovtsi	0.87	1.5	*0.39*[a]	*0.19*	0.2	0.95	*0.17*	*0.02*	*0.48*
			44.83[b]	21.84	22.99		89.47[c]	10.53	55.17
Pirdop	1.28	2.2	*0.71*	*0.24*	*0.47*	0.51	100	0	*0.57*
			55.47	18.75	36.72				44.53

[a] Percent of soil.
[b] Percent of total C.
[c] Percent of total humic acids; humic to fulvic acids ratio (C_h/C_f).

Organic Carbon and Cation Exchange Properties of the Soils Following Ameliorant Application

After 60 days of ameliorant application in all the variants, pH increased in comparison with control soils (Table 11.4).

ANOVA of the pH values showed that four homogeneous groups were formed. Each homogeneous group combines data that are not statistically different at $p < 0.05$. The control variants formed a statistically different homogeneous group than the other variants. The highest pH values were observed when lime and glauconite–phosphorite were added as single amendments at both sites. Soil amelioration with organic matter in combination with lime material resulted in a pH increase to ~6.0. Following ameliorant application, pH of the Pirdop soil increased enough to neutralize soil acidity. Variants 4 and 5 at Site 1 and variants 4, 5, and 6 at Site 2 are in a group that is significantly different from the others. In all the variants, exchangeable acidity was

TABLE 11.4

ANOVA of the Average Values of pH at Site 1 (Kremikovtsi) and Site 2 (Pirdop)

Site 1						Site 2					
Variant	Average	Homogeneous Groups				Variant	Average	Homogeneous Groups			
1	5.53	X				1	4.07	X			
5	6.03		X			5	6.03		X		
4	6.03		X			4	6.07		X		
6	6.13			X		6	6.17		X		
3	6.20			X		3	6.37			X	
2	6.77				X	2	6.70				X

neutralized after the amendment additions. Cation exchange capacity did not increase; however, exchangeable Ca occupying the strongly acidic cation exchanger rose (Table 11.5). The treatments with organomineral mixtures of lime and glauconite–phosphorite with coal and peat had a smaller effect on cation exchange properties compared to single treatments with CaO at both sites. Activation of the organic components in the ameliorative mixtures was not achieved during the experiment because it required higher pH. Organic amendment addition led to an increased content of organic C of the mixtures (Table 11.5).

Total, Reactive, and Bioavailable Forms of the Heavy Metal in the Ameliorated Treatments

Table 11.6 shows the average values of total, reactive (extracted by 0.43 M HNO_3), and bioavailable (0.01 M $CaCl_2$) metal fractions. The total aqua regia extractable heavy metals copper, lead, and zinc in the soils studied exceeded the MPLs stated in the National Standard (2008).

The total contents of the elements varied slightly by variants. The amounts of metals extracted with 0.43 M HNO_3, the so-called reactive pool, are considered the fraction that controls the concentration in the soil solution in relation to other soil properties such as pH, organic matter, and clay content (Gooddy et al. 1995; Temminghoff et al. 1997; Cances et al. 2003; Römkens et al. 2004; de Vries et al. 2007). The addition of ameliorants influenced the bioavailable fraction of Cu and Zn and the reactive fractions of Cu in the more acidic Pirdop soil (Table 11.6) but not the Kremikovtsi soil. In the Pirdop soil (Site 2), a significant decrease was observed in the *specifically* adsorbed (dilute acid-extracted) Cu (McLaren et al. 1990; Atanassova and Okazaki 1997) as a result of amelioration. The difference between the total amount

TABLE 11.5

Average Contents of Organic Carbon and Cation-Exchange Properties of the Analyzed Soils Following Ameliorant Application

Kremikovtsi										
	C	pH	CEC	CEC_{CA}	CEC_A	exch.$H_{8.2}$	Al	Ca	Mg	V
Variant	(%)	(H_2O)	cmol·kg^{-1}							(%)
1	0.49	5.5	19.4	*14.4*	*4.9*	*5.3*	*0.5*	*12.8*	*1.07*	72.8
				74.3	25.6	27.4	2.41	65.91	5.5	
2	0.43	6.8	19.6	*16.7*	*2.87*	*2.4*	0.0	*15.9*	*1.07*	87.6
				85.3	14.7	12.4		51.1	5.45	
3	0.57	6.2	19.7	*15.7*	*4.0*	*3.5*	0.0	*15.3*	*1.07*	82.4
				79.7	23.3	17.6		77.5	5.4	
4	0.99	6.0	19.8	*15.8*	*3.9*	*3.8*	0.0	*15.1*	*1.00*	80.9
				80.0	20.0	19.0		76.3	5.05	
5	1.03	6.0	19.6	*15.9*	*3.7*	*3.5*	0.0	*15.2*	*1.07*	82.3
				81.2	18.7	17.6		77.6	5.62	
6	0.96	6.1	19.4	*16.3*	*3.07*	*2.7*	0.0	*15.63*	*1.03*	86.3
				84.2	15.8	13.7		80.6	5.33	

Pirdop										
	C	pH	CEC	CEC_{CA}	CEC_A	$H_{8.2}$	Al	Ca	Mg	V
Variant	(%)	(H_2O)	cmol·kg^{-1}/%CEC							(%)
1	1.05	4.07	20.6	*14.5*	*6.0*	*8.7*	*2.8*	*9.4*	*2.3*	57.4
				70.7	29.3	42.5	13.61	45.7	11.0	
2	1.03	6.7	20.7	*16.4*	*4.4*	*4.0*	0.0	*14.4*	*2.2*	80.9
				78.9	21.1	19.1		69.6	10.6	
3	1.04	6.4	20.3	*14.7*	*5.5*	*5.1*	0.0	*13.2*	*2.2*	75.0
				72.7	27.3	25.0		64.9	10.9	
4	1.18	6.1	20.7	*14.8*	*5.8*	*5.7*	0.0	*12.6*	*2.3*	72.6
				71.8	28.2	27.4		60.8	10.9	
5	1.68	6.0	20.6	*14.8*	*5.7*	*5.6*	0.0	*12.6*	*2.1*	72.8
				72.1	27.9	27.2		61.4	10.4	
6	1.73	6.2	20.63	*15.10*	*5.5*	*5.3*	0.0	*13.2*	*2.2*	73.5
				73.17	26.83	25.69		63.97	10.83	

of heavy metals and the reactive fraction (i.e., the operationally called non-reactive fraction) is not directly available to plants or leached down the soil profile. Transfer functions were derived relating total soil metal concentrations with reactive soil metal concentrations and soil characteristics similar to those in de Vries et al. (2007).

Multiple regression analysis was carried out to find transfer functions relating the reactive metal pool (extracted with 0.43 M NHO_3) and the total

TABLE 11.6

Average Values of Total and Reactive and Bioavailable Heavy Metals Determined by Extraction with 0.43 M HNO_3 for the Two Soils

	Cu mg/kg		Zn mg/kg			Pb mg/kg	
Kremikovtsi	**TM**	**0.43 M HNO_3**	**TM**	**0.43 M HNO_3**	**0.01 M $CaCl_2$**	**TM**	**0.43 M HNO_3**
1	56.4	15.1	421	40.3	1.13	138	30.8
2	55.0	14.8	292	42.3	0.02	139	32.1
3	55.9	14.4	330	44.7	0.02	141	32.1
4	58.9	14.4	295	40.3	0.01	138	32.1
5	58.2	14.9	294	50.3	0.01	137	32.1
6	62.1	16.7	308	42.3	0.02	141	32.9
SD	3.40	1.21	70.65	6.8	0.01	4.11	0.98

	Cu mg/kg			Zn mg/kg			Pb mg/kg	
Pirdop	**TM**	**0.43 M HNO_3**	**0.01 M $CaCl_2$**	**TM**	**0.43 M HNO_3**	**0.01 M $CaCl_2$**	**TM**	**0.43 M HNO_3**
1	382	217	26.3	135	4.93	1.90	55.67	27.60
2	353	183	1.00	118	6.07	0.02	57.33	30.47
3	354	174	0.80	115	6.37	0.02	57.00	29.17
4	352	168	0.83	140	5.97	0.02	52.33	29.50
5	358	172	0.73	118	5.33	0.03	53.67	28.50
6	353	159	0.70	118	5.87	0.17	57.33	28.57
SD	13.3	20.7	9.82	21.0	17.8	0.72	2.62	1.34
MPL pH < 6	<80			<200			<60	

metal content as well as other soil characteristics, such as CEC and soil organic matter (SOM).

$$\log HM_{reactive} = a + b^{x}\log HM_{tot} + c^{x}\log SOM + d^{x}\log CEC \qquad (11.1)$$

where

HM_{re} = reactive heavy metal concentration in soil ($mg{\cdot}kg^{-1}$)
HM_{tot} = total heavy metal concentration in soil ($mg{\cdot}kg^{-1}$)
SOM = soil organic matter (%)
CEC = cation exchangeable capacity (%)
a, *b*, *c*, and *d* = regression coefficients

The relationships between the total and reactive metal contents according to Equation 11.1 are presented in Table 11.7. Results show that for Cu and Pb, the relationships between the reactive and total contents were significantly high: R^2 = 99.8 and 99.2, respectively. For Zn, this relationship is less significant at 82.4. The *b* coefficients were close to 1 for Cu and Zn, which means

TABLE 11.7

Values for the Regression Coefficients in Multiple Regressions between Total HM Concentrations, Reactive HM Concentrations (Extracted by 0.43 M HNO_3), and Other Soil Characteristics

Metal	Intercept a	TM_{tot} b	SOM c	CEC d	R^2 %	SE
Cu	−1.01	1.40	−0.13	0.21	99.8	0.027
Zn	−2.88	1.78	−0.028	0	82.4	0.184
Pb	−4.10	2.61	0.018	0	99.2	0.048

that the reactive heavy metal (HM) content in soil increases with the total HM content. For Pb this relationship indicates that the reactive metal content in soil samples is low compared to the total contents. The values of *c* showed that the reactive HM contents are affected strongly by SOM for some metals. For Cu the sign of the *c* coefficient is negative. This may be due to the strong binding by OM, especially for copper. For the Zn and Pb reactive concentration, the OM variable is not significant.

Composition and Properties of the Soil Solution from the Greenhouse Experiment with Heavy Metal–Contaminated Soils

The chemical composition of soil solution is a sensitive indicator for the impact of the contaminants and the role of the ameliorants in the fixation and release of heavy metals.

Statistical analysis of the data as well as the time for collection of soil solution is very important. Both the impact of the amendments and the time for collection of the soil solution have significant influence on the measured parameters at $p = 0.05$, and therefore interpretation of the data will consider the average values of the components of the soil solution at the two sampling times.

Changes in Soil Solution pH and Electrical Conductivity

Table 11.8 shows that variation of the acidity of the Kremikovtsi and Pirdop soil solution sampling is insignificant and coefficient of variation (CV) < 3%. The data for the electrical conductivity of the solution varied significantly by variants (CV from 6.94% to 46%) for soil solution sampling 2 (SS II) in comparison with 11%–20% for sampling 1 (SS I). At all the variants, however, the electrical conductivity are >1 in all the treatments, indicating high

TABLE 11.8

pH and Electrical Conductivity of the First (SS I) and Second (SS II) Soil Solution Samplings at Kremikovtsi (Site 1) and Pirdop (Site 2)

	pH (H_2O)					Conductivity (mS/cm)				
Variant	Min	Mean	Max	SD	CV%	Min	Mean	Max	SD	CV%
Soil Solution I										
Site 1										
1	5.06	5.13	5.17	0.059	1.14	0.826	1.016	1.270	0.229	22.50
2	6.56	6.65	6.71	0.081	1.22	2.794	3.302	4.064	0.672	20.35
3	5.79	5.85	5.94	0.081	1.39	3.175	3.852	4.572	0.699	18.16
4	5.66	5.75	5.83	0.085	1.48	4.191	4.614	5.207	0.529	11.46
5	5.83	5.91	5.98	0.076	1.29	2.413	3.133	3.556	0.626	20.00
Site 2										
1	3.60	3.62	3.64	0.021	0.57	0.254	0.288	0.330	0.039	13.46
2	6.39	6.42	6.45	0.031	0.48	1.143	1.418	1.651	0.257	18.10
3	6.20	6.29	6.44	0.129	2.04	1.080	1.228	1.334	0.132	10.76
4	5.89	6.00	6.05	0.092	1.54	2.286	2.582	2.921	0.320	12.38
5	5.90	6.04	6.31	0.231	3.82	1.588	2.053	2.286	0.403	19.63
Soil Solution II										
Site 1										
1	6.41	6.45	6.5	0.046	0.71	0.699	0.800	0.876	0.091	11.40
2	7.35	7.40	7.46	0.057	0.77	2.032	2.392	2.731	0.350	14.63
3	6.80	6.94	7.04	0.125	1.80	1.270	1.905	2.921	0.889	46.67
4	6.56	6.66	6.78	0.111	1.66	1.588	2.604	3.747	1.085	41.67
5	6.66	6.85	6.94	0.162	2.36	1.905	2.307	2.604	0.361	15.66
Site 2										
1	3.61	3.73	3.83	0.111	2.97	0.203	0.267	0.356	0.080	29.87
2	5.94	6.04	6.13	0.095	1.57	1.524	2.074	2.921	0.744	35.87
3	5.93	5.99	6.07	0.072	1.20	1.27	1.990	2.54	0.652	32.75
4	6.14	6.26	6.47	0.185	2.96	2.159	2.307	2.477	0.160	6.94
5	6.08	6.25	6.42	0.170	2.72	1.397	2.582	3.429	1.057	40.95

mineralization in the liquid phase due to the inorganic components in the ameliorants compared to the untreated soil at Site 2 (0.254–0.654 mS/cm).

The ANOVA of the pH values for both soil solution samplings SS I and SS II showed that four homogeneous groups for Kremikovtsi soil and three homogeneous groups for the Pirdop soil were obtained. Each homogeneous group contains data of the tested indicator that are not statistically different at $p \leq 0.05$.

The control treatments formed single homogeneous groups that were statistically different from all the other variants. Application of the ameliorants and their combinations resulted in a significant increase in the pH of the soil solution to neutral values for Pirdop soil for the two samplings ranging from 5.99 to 6.42. The highest pHs were observed in the variants with single lime treatments (pH = 6.65) at Site 1. There is an increase in the pH values at the second sampling for Kremikovtsi soil (1 pH unit at each variant), which is probably related to the properties of the soil from this site and the effect of the ameliorants, while for the Pirdop soil, no increase in the pH of the soil solution (SS II) was observed. It has been shown that rhizosphere has a direct acidifying effect through the release of H^+ during the excess uptake of cations over anions. Solubilization of lime is possible through the release of low molecular weight organic acids as root exudates (Marschner 1991) enhancing microbial activity and stimulating the decomposition of native SOM, production of H_2CO_3, and soil respiration, leaving Ca^{2+} and other cations in the soil solution (i.e., the so-called priming effect) (Haynes and Naidu 1998; Cheng 2009; Dumale et al. 2011).

Heavy Metal Concentrations in the Soil Solution

Copper was <1 mg/L in all the variants of the Kremikovtsi soil because of the smaller concentration load in that soil and the higher soil and solution pH (pH 5.13). With the other ameliorants, pH ranged between 5.8–6.5 (SS I) and 6.5–7.4 (SS II) (Table 11.9). No statistical difference was observed in Cu concentrations. The concentrations of zinc ranged from 0.17 to 0.58 mg/L for both dates of sampling and were below the MPL (<1 mg/L). The application

TABLE 11.9

ANOVA of the pH Values for the Two Soil Solution Samplings SS I and SS II at Sites 1 and 2

Kremikovtsi						Pirdop					
Variant	Average	Homogeneous Groups				Variant	Average	Homogeneous Groups			
SS I						*SS I*					
1	5.13	X				1	6.45	X			
4	5.75		X			4	6.66		X		
3	5.85		X	X		5	6.85		X	X	
5	5.91			X		3	6.94			X	
2	6.65				X	2	7.40				X
SS II						*SS II*					
1	3.62	X				1	3.73	X			
4	6.00		X			3	5.99		X		
5	6.04		X			2	6.04		X	X	
3	6.29			X		5	6.25			X	
2	6.42			X		4	6.26			X	

of different ameliorative materials including the phosphorite–glauconite did not influence zinc concentration (Figure 11.3) for SS I, but Zn was significantly mobilized at SS II in the control variant. At pH < 4 for the Pirdop soil, Cu concentrations in the soil solution of the control soil reached high values (2.6–4.7 mg/L) at both dates of sampling (Figure 11.4). Ameliorative mixtures led to a significant decrease in Cu in all the other variants. The concentration of the soluble Cu in these variants was 10–20 times the acceptable standard of water (0.2 mg/L, State Gazette 87). The concentration of zinc in the soil solution was similarly affected and lower concentrations were observed, especially following the second sampling. We ascribe this effect to the role of plants in the soil from this site.

Soluble copper, zinc, and lead were significantly decreased in the ameliorated soils—about 70% of Zn, 80%–90% of Cu, and >90% of Pb—but statistical differences between the data of the different ameliorative mixtures

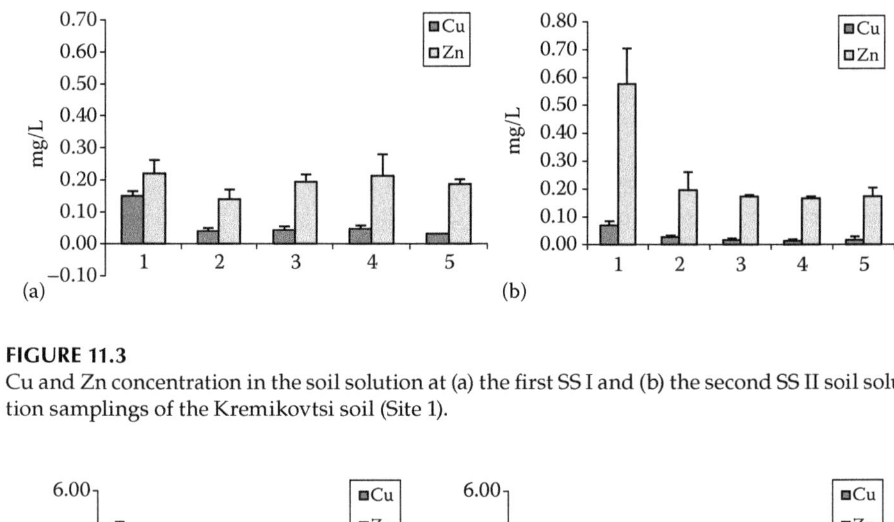

FIGURE 11.3
Cu and Zn concentration in the soil solution at (a) the first SS I and (b) the second SS II soil solution samplings of the Kremikovtsi soil (Site 1).

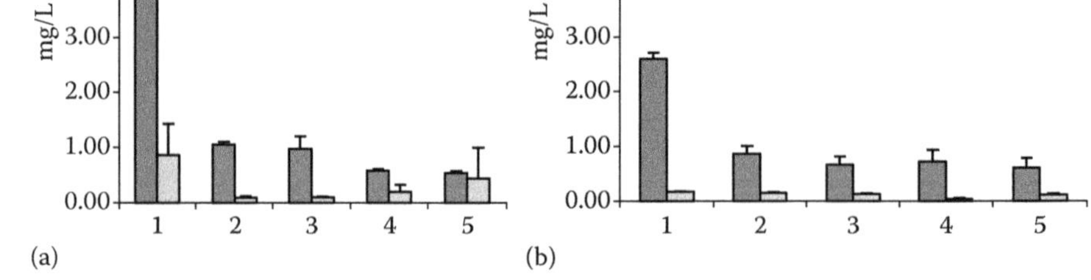

FIGURE 11.4
Cu and Zn concentration on the soil solution at (a) the first SS I and (b) the second SS II soil solution samplings of the Pirdop soil (Site 2).

containing organic amendments and the inorganic ones containing glauconite–phosphorite and lime could not be discerned.

Ionic and DOC Composition of the Soil Solution

The concentration of the basic cations in the soil solution increased significantly in all the amended variants compared to the controls (Table 11.10). The cation concentration at the first solution sampling (SS I) for the two soils was higher than the second, which was probably related to the consumption of nutrients for the growth of ryegrass. In general, the concentration of the major cations and anions in the soil solution for the two soils increased with the application of ameliorants, contributing to the input of inorganic and organic compounds into the soil solution. Glauconite–phosphorite (pH = 7.92) has a high content of calcium (Ca), magnesium (Mg), and potassium (K). With the exception of binding metals, the organic ameliorants applied supplied the soil with additional nutrient elements such as nitrogen, phosphorus, and potassium (NPK). The organomineral amendment provoked an increase in DOC and metal–organic complexation in the ameliorated variants of raised pH 5.75–7.4 for SS I and SS II, especially in the Pirdop soil for V2 and 3 SS II soil solution sampling (Figure 11.5). However, statistically significant differences of DOC concentrations between ameliorated variants

TABLE 11.10

Basic Cations in Soil Solution

	Site 1					Site 2			
	K	Na	Ca	Mg		K	Na	Ca	Mg
Variant	(mg/L)				Variant	(mg/L)			
SS I									
1	17.18	6.33	252	45.33	1	18.45	6.74	0.53	8.53
2	45.45	27.76	818	113	2	54.61	22.79	158	12.40
3	58.65	35.30	474	120	3	50.77	22.56	202	12.27
4	62.24	40.70	651	197	4	61.21	30.46	605	53.00
5	46.73	25.09	791	160	5	52.88	23.74	475	29.00
SD	17.66	13.13	258	68.01	SD	17.48	8.84	236	17.67
CV%	38.35	48.57	43.22	53.55	CV%	36.73	41.59	81.84	76.70
SS II									
1	1.60	2.39	13.25	3.03	1	13.85	9.7	12.6	8.40
2	7.79	5.65	663	77.67	2	12.72	13.24	463	26.67
3	7.93	8.75	524	72.67	3	5.74	6.99	605	28.67
4	12.51	11.93	745	106	4	6.54	6.32	767	49.33
5	8.16	9.35	640	93.33	5	5.27	23.58	763	50.00
SD	4.24	4.25	360	44.49	SD	6.72	42.27	323	19.33
CV%	55.85	55.76	69.59	63.07	CV%	76.16	64.34	61.94	59.28

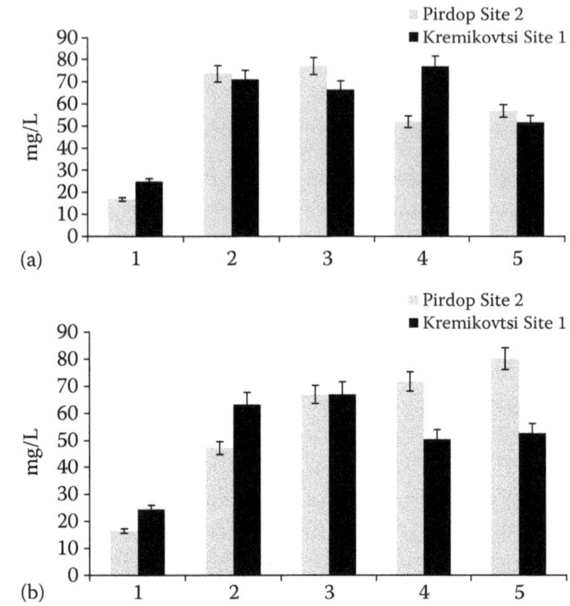

FIGURE 11.5
DOC concentration in the first soil solution sampling SS I (a) and the second soil solution sampling SS II (b).

were not detected. It has been found that lime treatment enhances the leaching of DOC because of stimulated biological activity, SOM mineralization (Biasi et al. 2008), and formation of low molecular organic compounds that are prone to leaching (Andersson et al. 1994). In this study, increased solution concentrations of DOC associated with raised pH were accompanied by a sharp decrease in the concentration of Cu, Zn, and Pb.

We speculate that enhanced C mineralization as a result of microbial activity associated with increased soil pH caused higher DOC concentrations in the soil solution. We also observed higher concentrations of dissolved inorganic carbon as HCO_3^-. Ahmad et al. (2013) have found an increase in the cumulative CO_2-C efflux from 7 to 98 days after planting in a limed soil, suggesting that liming enhanced SOM decomposition. Higher DOC present in the soil solution can coat the ferric hydroxide and phosphoritic surfaces, thus inhibiting the sorption of metals on these materials.

Speciation of Cu, Zn, and Pb in the Soil Solution

In the unamended acid Pirdop soil at the very acidic pH of 3.62, 73.4% of Cu, 74.5% of Zn, and 20.1% of Pb were present as free Me^{2+} ions as calculated by the Visual Minteq speciation program (Tables 11.11a, b, c, data for soil solution SS I, before planting the ryegrass). A decrease in the percentage of Cu^{2+}

TABLE 11.11a

Speciation of Cu in Solution (SS I) from the Pirdop Soil (Site 2) and Kremikovtsi Soil (Site 1)

	Pirdop	Krem	Pirdop	Krem	Pirdop	Krem	Pirdop	Krem	Pirdop	Krem
Species Name	**V1**		**V2**		**V3**		**V4**		**V5**	
Cu^{+2}	73.4	0.22					0.03		0.02	
FA-Cu^{+}+2G (aq)	4.1	30.0	18.7	12.4	19.6		19.8		19.9	
$CuCl^{+}$	0.04									
$CuSO_4$ (aq)	2.9	0.02					0.02			
FA_2Cu (aq)	2.9	57.5	70.9	86.6	70.7	30.8	69.8	8.04	71.3	89.4
$FACu^{+}$ (aq)	16.5	12.2	9.7	0.90	9.2		10.04		8.3	0.1
FA_2CuOH (aq)		0.04	0.68		0.5	69.2	0.33	92.0	0.3	10.5

TABLE 11.11b

Speciation of Zn in Solution (SS I) from the Pirdop Soil (Site 2) and Kremikovtsi Soil (Site 1) in % of Total Concentration

	Pirdop	Krem	Pirdop	Krem	Pirdop	Krem	Pirdop	Krem	Pirdop	Krem
Species Name	**V1**		**V2**		**V3**		**V4**		**V5**	
Zn^{2+}	74.5	3.3		0.07			0.2	0.01	0.3	0.2
$FAZn^{+}$+2G (aq)	19.4	88.4		97.6			94.0		89.9	0.01
$ZnSO_4$ (aq)	3.0	0.32		0.01			0.1		0.12	0.05
$ZnCl^{+}$	0.05									
$Zn(OH)_2$ (aq)			0.43		0.21	0.07		0.05		
$Zn(OH)_3^-$			0.02							
$FAZn^{+}$ (aq)	3.1	6.8	0.2	1.5	0.3	2.9	3.2	3.8	5.9	30.6
FA_2Zn (aq)	0.04	1.1	99.4	0.8	99.5	96.9	2.5	96	3.8	69.1

TABLE 11.11c

Speciation of Pb in Solution (SS I) from the Pirdop Soil (Site 2) and Kremikovtsi (Site 1) in % of Total Concentration

	Pirdop	Krem	Pirdop	Krem	Pirdop	Krem	Pirdop	Krem	Pirdop	Krem
Species Name	**V1**		**V2**		**V3**		**V4**		**V5**	
Pb^{2+}	20.1	0.80						0.01		
$FAPb^{+}$+2G (aq)	57.5	50.7	76.9	70.1	74.1	66.8	76.5	68.3	79.8	73.1
$PbCl^{+}$	0.2	0.02								
$PbSO_4$ (aq)	1.80							0.01		
$FAPb^{+}$ (aq)	16.7	23.8	0.7	0.7	1.1	2.9	1.4	3.0	1.2	2.0
FA_2Pb (aq)	3.8	24.4	22.4	29.0	24.8	30.3	22.1	28.6	19.0	24.9

free ions in the soil solution was observed in response to increasing pH and the presence of inorganic (lime + glauconite–phosphorite) and organic (peat + coal) amendments, while the percentage of copper complexes with dissolved organic matter rose. The degree of soil copper contamination and soil type had no influence on the percentages of Cu complexes in the soil solution. In all the other variant Cu–organic complexes, both mono- and bidentate account for >90%. Our finding corresponds to those of Temminghoff et al. (1997) who observed ~30% of the Cu in the solution bound by DOC at pH 3.9, while at pH 6.6, Cu–DOC complexes comprised >99%. It has been shown that fulvic acid usually dominates (75%–95%) in the DOC of natural waters (Oliver et al. 1983). Similar to our study, they found that DOC mobility was very sensitive to pH and calcium concentration. For Cu, mainly bidentate complexes were found to be formed in the soils studied, while some Cu species were present as monodentate complexes in the Pirdop soil with higher Cu concentration. Bidentate complexes were dominant at lower Cu concentrations in the soil solutions of pH 5.75–6.65. For the Kremikovtsi soil, Zn monodentate fulvic acid (FA) species dominate in the variants without organic amendments, and bidentate FA species predominate when organic ameliorants were added. In the Pirdop soils, monodentate FA–Zn complexes are the dominant species in the variants with organic amendments, in which lower DOC concentrations were encountered. For the more strongly bound Pb ions, the speciation model resulted in a substantial difference in the relative contribution of monodentate and bidentate complexes. Lead formed stronger monodentate complexes in the soils tested (Table 11.11c). These results indicate that Pb is more strongly bound to DOC complexes than Cu and Zn at the acid pH = 3.62 and at higher equilibrium concentrations, whereas Cu binds much stronger than Pb at the higher pH. The share of carbonate, sulfate, chloride, hydroxyl, and bicarbonate complexes was negligible. Similar results on metal speciation were obtained for the second soil solution sampling (SS II) before collecting the plant biomass.

Yields and Heavy Metal Contents in Shoots

As a result of the statistical analyses, weights were arranged into three homogeneous groups (Table 11.12). The first homogeneous group included the control variants, which were with the lowest or zero amounts of biomass. The application of ameliorants led to an increase in yields for all the variants loading the second and third homogeneous groups.

One of the best variants at Site 1 was treatment 5, in which the biomass increased by 52% compared to the control (Figure 11.6). No plants developed on the Pirdop control soil at Site 2 because of the high acidity of that soil. The most effective were variants 2 and 3 with lime and glauconite–phosphorite. In

TABLE 11.12
ANOVA of Shoot Biomass (g) by Variants

Site 1				Site 2			
Variants	Average	Homogeneous Groups		Variants	Average	Homogeneous Groups	
1	5.38	X		1	0.00	X	
6	9.05	X		6	9.27	X	
2	10.13	X	X	5	10.42	X	X
4	10.58	X	X	4	10.53		X
3	10.58	X	X	3	11.60		X
5	11.25		X	2	11.67		X

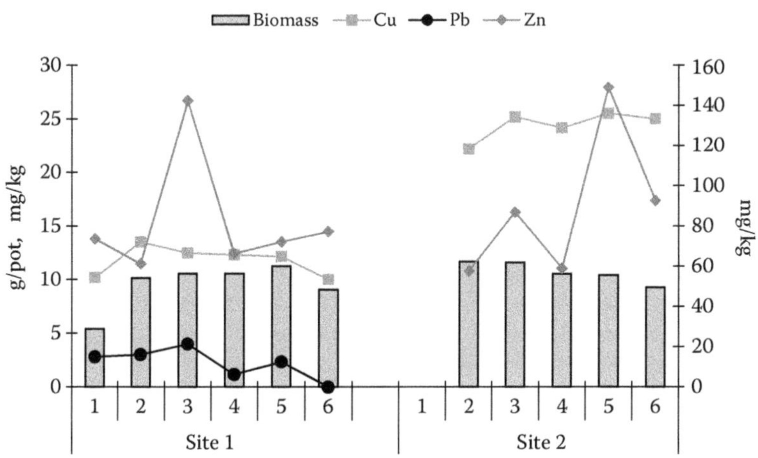

FIGURE 11.6
Cu, Zn, and Pb contents (mg/kg) in plant biomass (g) by variants (V1–V5).

general, addition of ameliorants to the acid soils led to increased crop yields. The addition of mixtures of lime, glauconite–phosphorite, peat, and coal resulted in considerable improvement of soil properties and higher yields. There are no significant differences (at 95% confidence interval) among average Cu concentrations at the Kremikovtsi site. In variant 3 comprising CaO and glauconite–phosphorite, the Zn content in the plant biomass exhibited an increase of 90% as compared to the control (i.e., no immobilization by glauconite–phosphorite minerals of bioavailable Zn took place under these experimental conditions). The content of Pb was very low in all the variants, including the control. Copper and lead are stored in roots, and the transport to above-ground shoots is low, especially for Pb, which had a lower accumulation factor (AF) compared to Zn and Cu (Figure 11.7).

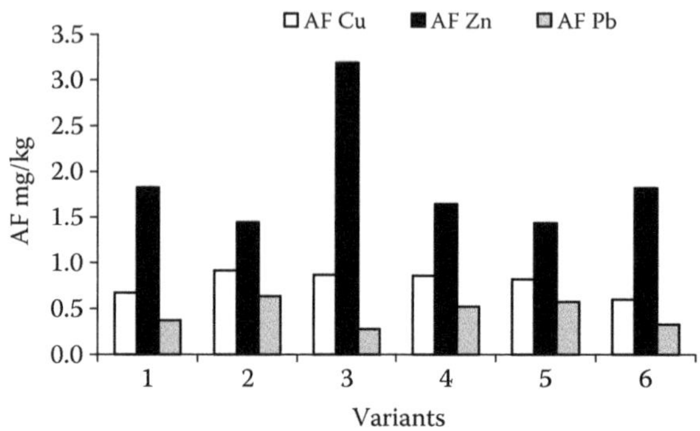

FIGURE 11.7
Accumulation factor of Cu, Zn, and Pb in ryegrass shoots (AF) calculated as the concentration in plants/concentration in 0.43 M HNO_3 extractant.

For the Pirdop soil, there were significant differences ($p < 0.05$) between the means of Zn and Cu in the variants. Lower Cu content for the limed variant V2 was observed, while for the rest of the variants, it ranged from 22.2 to 25.5 mg/kg and corresponded to normal and below toxic levels of trace metals in plant tissue (Kabata-Pendias and Pendias 1992). Higher bioavailability of Zn was registered when organic ameliorants peat and coal were included in the amendments with variants 5 and 6. Speciation showed that 0.2%–0.3% of Zn was present as free ions, but it is also possible that low molecular weight $FAZn^{+}$ organic complexes were bioavailable (Chaudry and Loneragan 1972). The mobility of Zn could be influenced by the presence of P, Ca, Fe oxides, and organic matter in the soil. Zn is displaced by other metals (e.g., Cu and Pb) competing for adsorption sites (Atanassova 1999; Cao et al. 2004). Other authors obtained similar results. Brown et al. (2005) tested different ameliorants (lime, coal dust, etc.), resulting in a decrease in exchangeable Zn, without influencing plant uptake. Hartley et al. (2004) reported a fivefold increase in the Zn mobility in the treatment of the soil with iron sulfate and lime material. The use of different organic compounds in the soil also significantly increased the part of the available forms of Zn (Sanchez-Monedero et al. 2004).

Relationships between Metal Solubility and Uptake by the Test Plants

In order to determine the effects of the ameliorants on solution parameters and plant yields, relationships were determined among plant biomass, pH of soil solution following ameliorant application, metal solution concentrations (mg/L), and metal contents in plant biomass. No significant relationships were observed, either between Pb in plants/solution or plant biomass yields and Pb plant/solution, and values of the correlation coefficients are presented for Cu and Zn only (Tables 11.13 and 11.14).

Statistically significant correlations ($P < 0.05$) were obtained between pH values of the soil solution and the following parameters: biomass ($r = 0.64$),

TABLE 11.13

Correlation Coefficient (r) of the Relationships between Ryegrass Biomass, pH of the Soil Solution, Cu and Zn Solution Concentrations, and Cu and Zn Contents in Biomass at Kremikovtsi Site 1 Soil

	pH	Cu (mg/L)	Zn (mg/L)	Biomass (g)	Cu (mg/kg)	Zn (mg/kg)
pH	1					
Cu (mg/L)	–0.5588	1				
	0.0304					
Zn (mg/L)	–0.6122	0.9357	1			
	0.0153	0.0001				
Plant biomass	0.6364	–0.9101	–0.8542	1		
(g)	0.0108	0.0001	0.0001			
Cu (mg/kg)	0.5819	–0.4826	–0.5634	0.4984	1	
plant	0.0229	0.0684	0.0287	0.0587		
Zn (mg/kg)	–0.1307	–0.2278	–0.2042	0.1278	0.1255	1
plant	0.6424	0.4142	0.4654	0.6499	0.6559	

TABLE 11.14

Correlation Coefficient (r) of the Relationships between Ryegrass Biomass, pH of the Soil Solution, Cu and Zn Solution Concentrations, and Cu and Zn Contents in Biomass at Pirdop Site 2 Soil

	pH	Cu (mg/L)	Zn (mg/L)	Biomass (g)	Cu (mg/kg)	Zn (mg/kg)
pH	1					
Cu (mg/L)	–0.9781	1				
	0.0000					
Zn (mg/L)	0.4035	–0.5638	1			
	0.1358	0.0286				
Plant biomass	0.8762	–0.9296	0.3932	1		
(g)	0.0000	0.0001	0.147			
Cu (mg/kg)	0.9225	–0.9861	0.5514	0.9271	1	
	0.0000	0.0001	0.0331	0.0000		
Zn (mg/kg)	0.6415	–0.7451	0.4658	0.5903	0.7178	1
	0.0099	0.0014	0.0801	0.0205	0.0026	

copper and zinc concentrations in soil solution ($r = -0.56$ and $r = -0.61$), and copper content in ryegrass ($r = 0.58$).

For the Pirdop soil, a stronger correlation was observed between pH in solution and plant biomass, and copper concentration in solution ($r = -0.98$) and copper and zinc contents in ryegrass ($r = 0.92$ and $r = 0.64$). It is obvious that the applied ameliorants, lime, phosphorite–glauconite, peat, coal, and Fe hydroxides may neutralize toxic-exchangeable acidity to plants. High negative correlation between plant biomass and Cu and Zn soil solution concentration was observed with Kremikovtsi soil ($r = -0.91$ and $r = -0.85$), implying *in situ* stabilization and immobilization of the soluble metals. For the two soils, the biomass was positively correlated with Cu and Zn contents, but relationships for the less acidic Kremikovtsi soil were not significant (Tables 11.13 and 11.14). A negative correlation was observed between Cu in plants and the dissolved metal in solution, implying nonbioavailability of the Cu–organic complexes in solution. Zinc relationships were not significant.

Summary

Based on this study, it could be summarized that the organomineral amelioration containing lime, glauconite–phosphorite, and iron hydroxides as inorganic additives and peat, coal, and manure as organic amendments led to a significant stabilization of both the solid and solution phases of the studied acid soils. The effectiveness of the ameliorants was due to neutralization of soil acidity, *in situ* stabilization of the heavy metals, and decrease in the reactive and bioavailable fractions. In all the ameliorated variants, a considerable decrease in Cu, Zn, and Pb concentrations in the soil solution was observed. The major species of Cu, Zn, and Pb in the pH 5.7–7.4 range were organic complexes with DOC. The organomineral amendment effectiveness to the ryegrass was proved by significant increases in the biomass yields and >70% for Zn and >90% for Cu and Pb decrease in metal solution concentration compared to the control variants. A decrease in the percentage of free M^{2+} ions in the soil solution was observed in response to an increasing pH, and the percentage of organo metallic complexes rose to >90% in the soil solution pH range 5.8–7.5. The degree of soil metal contamination and soil organic carbon contents had almost no influence on the relative distribution of metal forms in the soil solution. No significant difference in metal uptake among treated soils and control soils could be discerned. The mineral nutrition and biogeochemical processes in the rhizosphere of ryegrass plants could also influence the pH, ionic composition, metal speciation in solution, and bioavailability.

References

Ahmad, W., B. Singh, F.A. Dijkstra, and R.C. Dalal. 2013. Inorganic and organic carbon dynamics in a limed acid soil are mediated by plants. *Soil Biology and Biochemistry* 57:549–555.

Andersson, S., I. Valeur, and I. Nilsson. 1994. Influence of lime on soil respiration, leaching of DOC, and C/S relationships in the mor humus of a haplic podsol. *Environment International* 20:81–88.

Arsova, A., and T. Raychev. 1999. Bioecological effect of waste materials used for amelioration of copper and arsenic polluted soil located in Kremikovtsi region of Bulgaria. *Journal of Balkan Ecology* 2:47–52.

Arsova, A., and T. Raytchev. 2000. Ameliorating copper polluted soil with carbonate deposit and coal powder. *Journal of Balkan Ecology* 3:85–89.

Arsova, A., and T. Raytchev. 2001. Effect of organo-mineral treatment on productivity of alfalfa grown on copper and zinc polluted soils. *Bulgarian Journal of Plant Physiology* 27:25–35.

Atanassova, I. 1999. Competitive effect of copper, zinc, cadmium and nickel on ion adsorption and desorption by soil clays. *Water, Air, & Soil Pollution* 113:115–125.

Atanassova, I., and M. Okazaki. 1997. Adsorption-desorption characteristics of high levels of copper in soil clay fractions. *Water, Air, & Soil Pollution* 98:213–228.

Basta, N.T., and S.L. McGowen. 2004. Evaluation of chemical immobilization treatments for reducing heavy metal transport in smelter contaminated soil. *Environmental Pollution* 127:73–82.

Benkova, M. 2005. Colloid-chemical interactions at organo-mineral melioration of heavy metal polluted soil (in Bulgarian). PhD thesis, Institute of Soil Science "N. Poushkarov," Sofia, Bulgaria, p. 140.

Benkova, M., T. Raytchev, N. Dinev, M. Hajnos, and G. Jozefaciuk. 2005. Alfalfa growth and heavy metal uptake after addition of lime, peat and ferrous material to acid soil. In: *Physicochemical Management of Acid Soils Polluted with Heavy Metals*, T. Raytchev, G. Józefaciuk, Z. Sokolovska, and M. Hajnos (eds.), Centre of Excellence for Applied Physics in Sustainable Agriculture Agrophysics, Nikola Poushkarov ISS, Institute of Agrophysics PAS, Lublin-Sofia, Bulgaria, pp. 59–74.

Biasi, C., S.E. Lind, N.M. Pekkarinen, J.T. Huttunen, N.J. Shurpali, N.P. Hyvönen, M.E. Repo, and P.J. Martikainen. 2008. Direct experimental evidence for the contribution of lime to CO_2 release from managed peat soil. *Soil Biology and Biochemistry* 40:2660–2669.

Boisson, J., A. Ruttens, M. Mench, and J. Vangronsveld. 1999. Evaluation of hydroxyapatite as a metal immobilizing soil additive for the remediation of polluted soils. Part 1. Influence of hydroxyapatite on metal exchangeability in soil, plant growth and plant metal accumulation. *Environmental Pollution* 104:225–233.

Bolan, N.S., and V.P. Duraisamy. 2003. Role of inorganic and organic soil amendments on immobilisation and phytoavailability of heavy metals: A review involving specific case studies. *Australian Journal of Soil Research* 41:533–555.

Bolan, N.S., D.C. Adriano, P. Duraisamy, A. Mani, and K. Arulmozhiselvan. 2003. Immobilization and phytoavailability of cadmium in variable charge soils. I. Effect of phosphate addition. *Plant Soil* 250:83–94.

Boldirev, A.I. 1983. *Physical and Colloid Chemistry*, Moscow, Vishaya Shkola, p. 408 (in Russian).

Brown, S., B. Christensen, E. Lombi, M. McLaughlin, S. McGrath, J. Colpaert, and J. Vangronsveld. 2005. An inter-laboratory study to test the ability of amendments to reduce the availability of Cd, Pb, and Zn in situ. *Environmental Pollution* 138:34–45.

Cances, B., M. Ponthieu, M. Castrec-Rouelle, E. Aubry, and M.F. Benedetti. 2003. Metal ion speciation in a soil and its solution: Experimental data and model results. *Geoderma* 113:341–355.

Cao, X., L.Q. Ma, M. Chen, S.P. Singh, and W.G. Harris. 2003. Phosphate induced metal immobilization in a contaminated site. *Environmental Pollution* 122:19–28.

Cao, X., L.Q. Ma, D.R. Rhue, and C.S. Appel. 2004. Mechanisms of lead, copper, and zinc retention by phosphate rock. *Environmental Pollution* 131(3):435–444.

Chaudry, F.M., and J.F. Loneragan. 1972. Zinc absorption by wheat seedlings. 1. Inhibition by macronutrient ions in short term experiments and its relevance to long term zinc nutrition. *Soil Society of America Proceedings* 36:323–327.

Cheng, W. 2009. Rhizosphere priming effect: Its functional relationships with microbial turnover, evapotranspiration, and CeN budgets. *Soil Biology and Biochemistry* 41:1795–1801.

Chlopecka, A., and D.C. Adriano. 1997. Influence of zeolite, apatite and Fe-oxide on Cd and Pb uptake by crops. *Science of the Total Environment* 207:195–206.

de Vries, W., S. Lofts, E. Tipping, M. Meili, J.E. Groenenberg, and G. Schutze. 2007. Impact of soil properties on critical concentrations of cadmium, lead, copper, zinc and mercury in soil and soil solution in view of ecotoxicological effects. *Reviews of Environmental Contamination and Toxicology* 191:47–89.

Dumale, W.A. Jr., T. Miyazaki, K. Hirai, and T. Nishimura. 2011. SOC turnover and lime-CO_2 evolution during liming of an acid Andisol and Ultisol. *Open Journal of Soil Science* 1(2):49–53.

Edwards, R., I. Rebedea, N.W. Lepp, and A. Lovell. 1999. An investigation into the mechanism by which synthetic zeolites reduce labile metal concentrations in soils. *Environmental Geochemistry & Health* 21:157–173.

Ganev, S. 1987. Strongly acidic and weakly acidic nature of soil acidity and determining the liming rate of acid soils. *Forestry* 1:19–25 (in Bulgarian).

Ganev, S. 1990. *Modern Soil Chemistry*. Science and Art, Sofia, Bulgaria, p. 371 (in Bulgarian).

Ganev, S., and A. Arsova. 1980. Methods for determination of strongly acidic and weakly acidic cation exchange in soils. *Soil Science and Agrochemistry*, 15:22–33 (in Bulgarian).

Gooddy, D.C., P. Shand, D.G. Kinniburgh, and W.H. van Riemsdijk. 1995. Field-based partition coefficients for trace elements in soil solutions. *European Journal of Soil Science* 46:265–285.

Gworek, B. 1992. Inactivation of cadmium in contaminated soils usingsynthetic zeolites. *Environmental Pollution* 75:269–271.

Hartley, W., R. Edwards, and N.W. Lepp. 2004. Arsenic and heavy metal mobility in iron oxide-amended contaminated soils as evaluated by short- and long-term leaching tests. *Environmental Pollution* 131:495–504.

Haynes, R.J., and R. Naidu. 1998. Influence of lime, fertiliser and manure applications on soil organic matter content and soil physical conditions: A review. *Nutrient Cycling in Agroecosystems* 51:127–137.

He, M., H. Shia, X. Zhao, Y. Yu, and B. Qu. 2013. Immobilization of Pb and Cd in contaminated soil using nanocrystallite hydroxyapatite. *Procedia Environmental Sciences* 18:657–665.

Hooda, P.S., and B.J. Alloway. 1996. The effect of liming on heavy metal concentrations in wheat, carrots and spinach grown on previously sludge-applied soils. *Journal of Agricultural Science* 127:289–294.
Hashimoto, Y., T. Taki, and T. Sato. 2009. Sorption of dissolved lead from shooting range soils using hydroxyapatite amendments synthesized from industrial byproducts as affected by varying pH conditions, *Journal of Environmental Management* 90:1782–1789.
Houba, V.J.G., E.J.M. Temminghoff, G.A. Gaikhorst, and W. van Vark. 1999. Soil analysis procedures. Extraction with 0.01 M $CaCl_2$. Soil Reference Handbook No. 06173011. Wageningen University, Wageningen, Netherlands.
Houba, V.J.G., J.J. Van der Lee, I. Novozamsky, and I. Walinga. 1988. Soil and plant analysis, part 5: Soil analysis procedures, Wageningen University, Wageningen, Netherlands.
Huang, C.L. and E.E. Schulte. 1985. Digestion of plant tissue for analysis by ICP emission spectroscopy. *Communications in Soil Science and Plant Analysis* 16:943–958.
ISO 11466:1995, International Standard. Soil quality—Extraction of trace elements soluble in aqua regia, 1995.
Iyamuremye, F., R.P. Dick, and J. Baham. 1996. Organic amendmentsand phosphorus dynamics: I. Phosphorus chemistry and sorption. *Soil Science* 161:426–435.
Kabata-Pendias, A., and H. Pendias. 1992. *Trace Elements in Soils and Plants,* 2nd Edition, CRC Press, Boca Raton, FL.
Kononova, M.M. 1966. *Soil Organic Matter: Its Nature, Its Role in Soil Formation and Soil Fertility,* 2nd English Edition, Pergamon Press Inc., New York.
Marschner, H. 1991. Root-induced changes in the availability of micronutrients in the rhizosphere. In: *Plant Roots,* J.J. Waisel, A. Eshel, and U. Kafkafi (eds.), Marcel Dekker, New York, pp. 503–528.
McBride, M.B. 1994. *Environmental Chemistry of Soils.* Oxford University Press, New York.
McBride, M.B., and Martinez, C.E. 2000. Copper phytotoxicity in a contaminated soil: Remediation tests with adsorptive materials. *Environmental Science & Technology* 34:4386–4391.
McGowen, S.L., N.T. Basta, and G.O. Brown. 2001. Use of diammonium phosphate to reduce heavy metal solubility and transport in smelter-contaminated soil. *Journal of Environmental Quality* 30:493–500.
McLaren, R.G., D.S. Hogg, and R.S. Swift. 1990. Some factors, affecting the availability of native and applied soil copper in New Zealand soils. *Forest Ecology and Management* 37:131–142.
Mench, M., J. VanGronsveld, V. Didier, and H. Clijsters. 1994. Evaluation of metal mobility, plant availability and immobilization by chemical agents in a limed-silty soil. *Environmental Pollution* 86:279–286.
Miretzky, P. and A. Fernandez-Cirelli. 2008. Phosphates for Pb immobilization in soils: A review. *Environmental Chemistry Letters* 6:121–133.
Naidu, R., N.S. Bolan, R.S. Kookana, and K.G. Tiller, 1994. Ionic strength and pH effects on the adsorption of cadmium and the surface charge of soils. *European Journal of Soil Science* 45:419–442.
Oliver, B.G., E.M. Thurman, and R.L. Malcolm. 1983. The contribution of humic substances to the acidity of colored natural waters. *Geochimica et Cosmochimica Acta* 47:2031–2035.
Orlov, D. 1985. *Soil Chemistry.* University of Moscow, Moscow (in Russian).

Oste, L., T. Lexmond, and W. Van Riemsdijk. 2002. Metal immobilization in soils using synthetic zeolites. *Journal of Environmental Quality* 31:813–821.

Petkova, D., S. Popandova, T. Popova, and K. Chakalov. 2000. Testing of zeolite-glauconite-phosphorite mixtures for decreasing lead uptake by plants. *Bulgarian Journal of Agricultural Science* 6:321–326.

Pierzynski, G.M., and A.P. Schwab. 1993. Bioavailability of zinc, cadmium, and lead in a metal-contaminated alluvial soil. *Journal of Environmental Quality* 22:247–254.

Raichev, T., A. Arsova, S. Popandova, and G. Jozefaciuk, 2000. Effect of application of coal powder and lime on alfalfa growth on copper polluted acidic soil. *International Agrophysics* 14:121–126.

Raychev, T., A. Arsova, and M. Benkova, 2002. Estimation of quantitative rates for organo-mineral liming heavy metal polluted acid soils. *Bulgarian Journal of Agricultural Science* 8:539–550.

Raytchev, T., A. Arsova, and S. Pishmanova. 1999. Meliorant for acid soils polluted with copper and method of application. Patent No. 327 Y1 BG, 30.07.99 (in Bulgarian).

Römkens, P.F.A.M., J.E. Groenenberg, J. Bril, and W. de Vries. 2004. Derivation of partition equations to calculate heavy metal speciation and solubility in soils. ALTERRA report no. 305. Alterra, Wageningen University and Research Center, Netherlands.

Sanchez-Monedero, M.A., C. Mondini, M. de Nobili, L. Leita, and A. Roig. 2004. Land application of biosolids. Soil response to different stabilization degree of the treated organic matter. *Waste Management* 24:325–332.

Shende, A., A.S. Juwarkar, and S.S. Dara. 1994. Use of fly ash in reducing heavy metal toxicity to plants. *Resources Conservation and Recycling* 12:221–228.

Shestopalov, A., and S. Gorbov. 1998. Soil pollution by heavy metals in Rostov-on-Don and the brown coal use as a detoxicant. In: Proceeding Scientific Papers of 16th World Congress of Soil Science, Montpellier, France, August 1998, Vol. 2, pp. 704–705.

Siebers, N., M. Siangliw, and C. Tongcumpou. 2013. Cadmium uptake and subcellular distribution in rice plants as affected by phosphorus: Soil and hydroponic experiments. *Journal of Soil Science and Plant Nutrition* 13(4):833–844.

Smith, S.R. 2009. A critical review of the bioavailability and impacts of heavy metals in municipal solid waste composts compared to sewage sludge. *Environmental International* 35:142–156.

State Gazette, No. 71, August 12, 2008. Decree No. 3 of 01.08.2008 on maximum permissible contents of hazardous substances in soils. Sofia, Bulgaria (in Bulgarian).

State Gazette, No. 87, October 30, 2007. Decree No. 1 of 10.10.2007 on exploration, use and protection of ground waters from pollution and deterioration. Sofia, Bulgaria (in Bulgarian).

Temminghoff, E.J.M., S.E.A.T.M. Van der Zee, and F.A.M. de Haan. 1997. Copper mobility in a copper-contaminated sandy soil as affected by pH and solid and dissolved organic matter. *Environmental Science and Technology* 31:1109–1115.

Waterlot, C., C. Pruvot, H. Ciesielski, and F. Douay. 2011. Effects of a phosphorus amendment and the pH of water used for watering on the mobility and phytoavailability of Cd, Pb, and Zn in highly contaminated kitchen garden soils. *Ecological Engineering* 37:1081–1093.

Wenzel, W.W., and G. Wieshammer. 1995. Suction cup materials and their potential to bias trace metal analyses of soil solution: A review. *International Journal of Environmental Analytical Chemistry* 59:277–290.

12

Influence of Common Ions on Sorption and Mobility of Soil Phosphorus

Sabry M. Shaheen and Jörg Rinklebe

CONTENTS

Phosphorus (P) is an essential macronutrient in biological organisms. Optimizing P use efficiency will deliver agronomic, economic, and environmental benefits as agricultural production systems adjust to meet future global food production targets (Heffer and Prud'homme 2013). Such an optimization will rely on adequate knowledge of the dynamics of soil P pools to enable accurate predictions of the required external P inputs to achieve optimum growth of subsequent crops (Damon et al. 2014).

Bioavailability of soil P is a major research field within soil science. Small concentrations of phosphates in soil solution compared with plant requirement and apparently small recovery of fertilizer phosphate stimulated tremendous research (Delgado et al. 2000; Beauchemin et al. 2003; Wandruszka 2006). Earlier P research focused on P nutrition in plants, while later work concentrated on maximizing P supply for cultivated plants and attempting yield targets (Memon and Fox 1983; Fox 1989). Today, an increasing number of scientists work on reducing the agricultural P inputs and balancing the production with environmental values (Sims et al. 2000; Yuji and Sparks 2001; Bertrand et al. 2001; Zhang et al. 2005). Studies have concentrated on the reactions of added phosphate with soil constituents, common ions, soil amendments, and on mechanisms controlling the amount of phosphate in solution (e.g., Shaheen et al. 2007, 2009; Shaheen and Tsadilas 2013).

The relevance of P for crop production is obvious due to fertilization with P worldwide. The chemistry of P has been investigated in industrial countries on mostly highly weathered and acidic soils (Sinaj et al. 1992; Said and Dakermanji 1993; Frossard et al. 1995). However, parallel studies are missing in developing countries, which are of special importance since P deficiency is widespread and the high cost of inorganic fertilizers limits its application. The serious gap of knowledge hinders improving fertilizer use efficiency while the economies of these countries rely on the progress of the agriculture sector.

Effective soil phosphorus management from both environmental and agronomic points of view requires the knowledge of P bioavailability and sorption characteristics. Several scientists have dealt with P sorption in various soils (e.g., Tsadilas et al. 1996; Hamdi et al. 2014; Yang et al. 2014). However, adequate research focused specifically on P sorption and lability in relation to the effect of common fertilizer ions added to soil with fertilizers and amendments is lacking. The interaction between P and other fertilizer components such as potassium (K^+), ammonium (NH_4^+), calcium (Ca^{2+}), nitrate (NO_3^-), bicarbonate (HCO_3^-), and sulfate (SO_4^{2-}) salts is less understood. Very little literature is available on how these ions affect P behavior in soil. In addition, a clear understanding of P behavior, especially its sorption and mobility as impacted by other fertilizer elements, is necessary to better manage P in these soils.

For a complete understanding of bioavailability, it is essential to elucidate the processes involved in P sorption. An understanding of the chemical parameters required in regulating P uptake and release in natural aquatic systems requires a detailed assessment of the controls on the bioavailability of P. There is a need to indicate the relative sorption association of P with the co-ions, either the cations and/or anions on P sorption, and mobility. In this respect, it is important to understand the nature of sorption reactions, which occur when phosphate fertilizer is added to soils in relation to the other common ions existing in fertilizers. Therefore, this chapter highlights the importance of common ions on sorption and mobility of soil P. This information can help with providing more efficient management of P in relation to other nutrients from both agronomic and environmental points of view.

Deficiency and Excess of Soil Phosphorus

Phosphorus is an essential element for plant growth, and the amount of plant-available P in the soil is often inadequate to meet plant requirements. In soils, plants take up less than 20% of applied P and the remaining is immobilized (Bürkert et al. 2001). The poor mobility of soil inorganic P is thought to be due to (1) low dissolution rate of inorganic P species, (2) sorption of phosphate ions onto iron and aluminum oxides in acidic and onto $CaCO_3$

in calcareous soils, and (3) conversion to relatively insoluble species; thus P becomes fixed in the soil matrix (Hinsinger 2001).

Plant-available P level in soils, especially in calcareous soils, is generally too low for sustained crop yields (Shaheen et al. 2007). Thus, P fertilizer use efficiency is especially low in calcareous soils (Memon et al. 1991). Therefore, application of inorganic and/or organic fertilizer P becomes necessary for commercial crop production (Sharpley et al. 2007). However, intensive fertilization might lead to high concentrations of labile P in the surface soil, which often exceeds the uptake by plants and results in a buildup of P in the soil and increasing the risk of particulate and dissolved P losses in surface runoff (Sharpley and Withers 1994; Turtola and Yli-halla 1999; Sharpley and Tunney 2000; Sharpley et al. 2013; Soinne et al. 2014). Most of the excess P is bound to the soil and some is transformed to less available forms; however, at the same time there is also a risk of some of the excess P being transported from agricultural fields to surface waters. Phosphorus is generally considered to be the limiting nutrient in surface waters, where only minor P loads can have a substantial impact on the ecological status of the recipients, freshwater lakes, and brackish and salty fjords (Søndergaard 2007; Schindler 2012; Baken et al. 2014; Forsmann and Kjaergaard 2014). Phosphorus is often the limiting nutrient in surface water systems, and even small additions of P can have a great influence on primary production in these waters. Thus, excess P can play a major role in the eutrophication of freshwater systems (Schindler 2012; Sharpley et al. 2013). Nutrient losses such as nitrogen and P from agricultural practices continue to contribute to eutrophication of surface waters (Stapleton et al. 2000; Cleneghan 2003; Baken et al. 2014). Eutrophication occurs when surface waters become overenriched with nutrients such as nitrogen (N) and P. Thus, even if losses of P from agriculture are small from an agronomic viewpoint, they constitute a serious environmental problem (Baken et al. 2014; Forsmann and Kjaergaard 2014; Li et al. 2014).

The main processes for P losses from agricultural fields to surface waters are erosion, surface runoff, and subsurface transfer. For a long time, the first two processes were thought to be the most important. Since P is usually strongly bound by the soil, subsurface transfer of P was thought not to be a significant problem. Today it is well known that subsurface transfer from agricultural fields cannot be ignored (Sims et al. 1998; Klimeski et al. 2014). Phosphorus subsurface transfer can occur both in dissolved form or bound to particles that are transported with the water (Heckrath et al. 1995). Dissolved P is directly bioavailable and can therefore affect eutrophication in a short timescale. A significant proportion of P lost from agricultural fields through leaching is transported in dissolved form (Sharpley et al. 2013). Since the concentration of P in the soil solution is critical for losses of dissolved P, processes such as P sorption and release are crucial. Therefore, mobility and availability of P in soils is likely controlled by its sorption behavior.

Sorption of Soil P

Sorption refers to both adsorption on solid surfaces and absorption into solid phases of aluminum (Al) and iron (Fe) oxides and other mineral surfaces (Rhue and Harris 1999). In solution, phosphate reacts quickly with a wide variety of surfaces being taken up by and released from particles through a complex series of sorption reactions. Phosphate sorption has been known to occur both quickly and slowly and be reversible and irreversible depending on the sorption (or buffering) capacity of the solid phase and the residence time in contact with the solution (Barrow 1983; Froelich 1988). The simplest plausible model capable of explaining the wide variety and complexity of observed behavior of phosphate soil–solution reactions involves the uptake and release of dissolved phosphate via a two-step process (Barrow 1983a; Mansell et al. 1995; Chen et al. 1996): a rapid surface adsorption or desorption of phosphate onto or off reactive surfaces, followed by a slow penetration or release by solid-state diffusion of this surface phosphate into or out of the interior of the particles.

An effective soil phosphorus management from both environmental and agronomic points of view requires a proper knowledge of P bioavailability, fractions, and sorption characteristics determining its solubility, and therefore its leaching to groundwater (Shaheen et al. 2007, 2009). Soil P management decisions may differ according to the alkaline or acidic nature of individual soils. Ideally, environmentally sound P management decisions should be made based on a fundamental understanding of soil P chemistry coupled with site-specific information that can provide insights into P availability, retention, and possible loss for individual soils. Phosphorus mobility in soils is controlled by its interaction with soil matrix as well as other nutrients and metal ions.

Ionic Strength of Common Ions and P Sorption

Soil P sorption capacity is correlated with metal oxides, carbonates, soil solution pH, ionic strength, and composition (Tsadilas et al. 1996; Harrell and Wang 2006; Shaheen et al. 2009; Shaheen and Tsadilas 2013). Phosphate sorption has been shown to vary with soil/solution ratios, ionic strength, and cation species of the supporting electrolyte (Barrow et al. 1980; Nair et al. 1984; Bolan et al. 1986; Pardo et al. 1992; Bhadha et al. 2012).

The effect of ionic strength in the solution phase on P mobilization has only received sporadic attention (Koski-Vahala and Hartikainen 2001). In general, sorption increases as the ionic strength of the solution increases (Nair et al. 1984). However, this may not be the case in the presence of anions such as SO_4^{2-}, F^-, and OH^- that compete with phosphate ($H_2PO_4^-$) ions for surface sorption sites, reducing the amount of P sorbed during fast reactions and consequently affecting the rate and mass of phosphate taken up or released during slow reactions (Nair et al. 1984).

The effects of pH, ionic strength, and electrolyte composition on P sorption of several Andisols containing high amounts of variable charge materials were studied by Pardo et al. (1992). They found that in the pH range of 3 to 7, the amount of P sorbed by soil samples decreased as pH increased, but the opposite effect could be observed while raising the pH from 7 to 9. Regarding the effect produced by ionic strength, the results obtained have shown that an increase in the ionic strength clearly produces an increase in the amount of P sorbed, and this effect is maintained along the pH range studied (Figure 12.1).

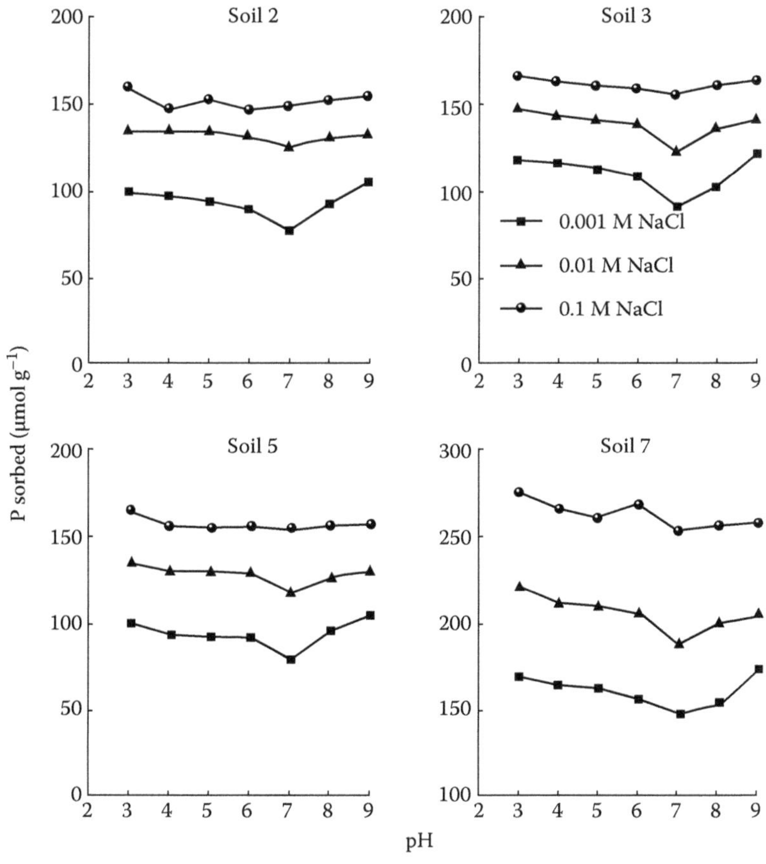

FIGURE 12.1
Effect of changes in pH and ionic strength on the amount of P sorbed. P added: 385 μmol g^{-1}. (Reprinted from *Geoderma*, 54, Pardo, M.T., Guadalix, M.E., Garcia-Gonzalez, M.T., Effect of pH and background electrolyte on P sorption by variable charge soils, 275–284, Copyright 1992, with permission from Elsevier.)

Figure 12.1 shows that the effect of pH on P sorption becomes more pronounced with lower ionic strength of the electrolyte. It is obvious that in simple systems with variable charge, an increase in the ionic strength increases the net surface charge, either positive or negative, depending on the charge on the surface in question. Increasing ionic strength should decrease sorption when the surface is negative and increase sorption when the surface is positive. In contrast, several authors found an opposite effect (Barrow et al. 1980b; Barrow and Ellis 1986). Bolan et al. (1986) suggested that the effect of ionic strength on sorption operates through its influence on electric potential in the plane of sorption rather than through its effect on the surface charge. At low pH values with a surface positively charged, an increase in the ionic strength makes the electric potential less positive as a result of greater concentration of electrolyte anions close to the surface and thus decreases P sorption. At high pH values, with a surface that is negatively charged, increases in salt concentration make the potential in the plane of adsorption less negative and thus increase P sorption. In accordance with this, an intermediate pH would be reached where the ionic strength would not have any effect on P sorption (point of zero salt effect) (Bolan et al. 1986). The results of Pardo et al. (1992) demonstrate that P sorption always increases when the ionic strength increases.

In the study of Pardo et al. (1992), the presence of different electrolytes (sodium chloride [NaCl], potassium chloride [KCl], and calcium chloride [$CaCl_2$]) with the same ionic strength (0.01) and the same pH values (pH 5) also clearly changed the amount of P sorbed by the soil samples (Figure 12.2). Sorption of P was highest in the presence of $CaCl_2$ and markedly higher in the presence of KCl than of NaCl. Thus, the nature of the electrolyte solution had a greater effect on P sorption than either pH or ionic strength. The presence of mono- or divalent cations (Ca^{++}, K^+, and Na^+) as background electrolytes affects P sorption, where the sorption of P in the presence of Ca^{++} was higher than in the presence of K^+ and Na^+.

The presence of mono- or divalent cations affects the charge of the sorbent surfaces and the electric potential. Thus, the greater sorption of P observed in the presence of divalent cations can be explained with the increase in the positive charge near the sorbent surfaces, which reduces electrostatic repulsion between the phosphate anion and the surface. On the other hand, cations such as Ca have a specific affinity for the surface (Barrow et al. 1980a; Marsh et al. 1987; Shaheen et al. 2009). Monovalent cations also formed a sequence, with P sorption increasing as the atomic number increased. For these ions, differences found in the amounts of P sorbed by the soils have to be seen in light of differences in the hydration energy and in polarizability. The lower atomic number of Na compared with K, and therefore its lower hydration energy, causes the effect of the positive cation charge to be smaller, thus reducing the sorption of P by the soils.

The effects of cations and anions on P sorption have been attributed to the formation of surface complexes between P and them (Helyer et al. 1976) and precipitation (Freeman and Rowell 1981). The effect of exchangeable cations

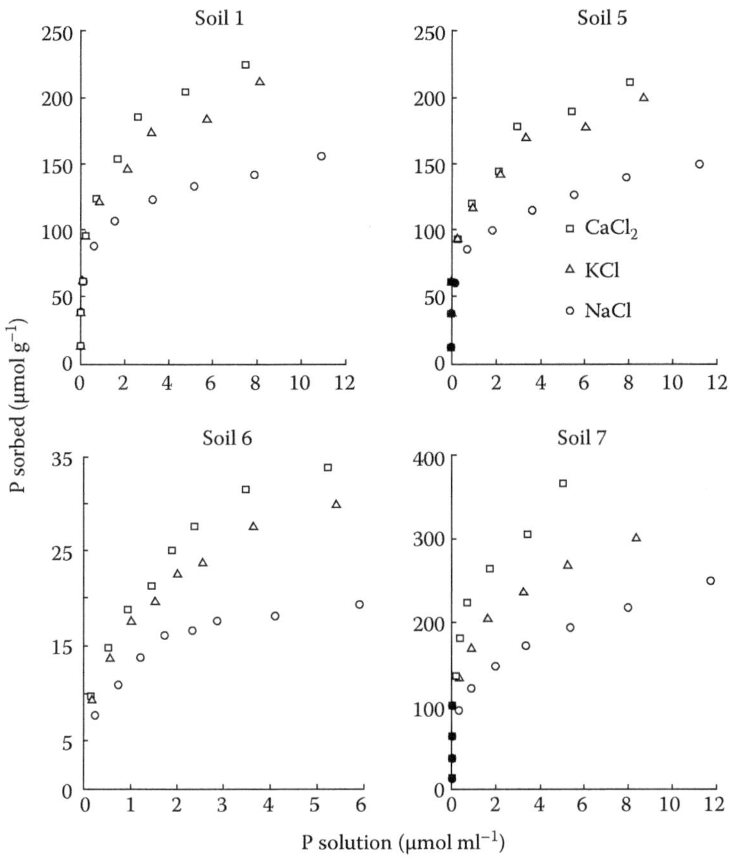

FIGURE 12.2
P sorption curves from different electrolytes of constant pH (pH 5.0) and ionic strength (0.01). (Reprinted from *Geoderma*, 54, Pardo, M.T., Guadalix, M.E., Garcia-Gonzalez, M.T., Effect of pH and background electrolyte on P sorption by variable charge soils, 275–284, Copyright 1992, with permission from Elsevier.)

and anions on the retention of P is likely to be soil pH dependent (Barrow 1984). The relationship between the electrostatic potential and pH varies depending on which anions have previously reacted with the P-retaining materials, and thus, this relationship would be expected to differ between soils and to be affected by previous reactions with P. These effects were attributed to the charge variations and the electric potential of the reacting surfaces (Barrow et al. 1980a, 1980b; Barrow 1984; Pardo et al. 1992; Barrow 2002).

Shaheen et al. (2009) tested the hypothesis that sorption of P by soils is seriously affected by the common cations and anions added to soils with fertilizers. Thus, they investigated the effect of K^+, NH_4^+, Ca^{2+}, NO_3^-, HCO_3^-, and SO_4^{2-} on P sorption in selected acid, and neutral and alkaline Alfisols. Five electrolyte background solutions were used in the study: 50 mM KNO_3,

50 mM NH_4NO_3, 20 mM Ca $(NO_3)_2{\cdot}4H_2O$, 50 mM NH_4HCO_3, and 20 mM $(NH_4)_2SO_4$ corresponding to ionic strength of 0.05 for all the examined cations and anions. The cation effect was evaluated by comparing P sorption in nitrate salts of Ca^{2+}, K^+, and NH_4^+, whereas the anion effect was evaluated in NH_4^+ salts of NO_3^-, HCO_3^-, and SO_4^{2-} backgrounds. The concentrations of 50 mM for the monovalent and 20 mM for the divalent ion background solutions were used to ensure that the effects of cations and anions were evaluated at a dominant and constant ionic strength. They found that both cations and anions had large effects on P sorption (Table 12.1 and Figure 12.3). Of the three cations, the greatest P sorption was in the Ca^{2+} background, followed by NH_4^+ and K^+ backgrounds. This trend was true for all the soils tested (i.e., acid, neutral, and alkaline).

The differences in total P sorption caused by different cations were generally reflected in the distribution (K_d) values for all soils (Table 12.1). The greater P sorption in the presence of divalent cation (Ca^{2+}) in the background solution compared to those containing monovalent cations either K^+ or NH_4^+ can be explained by the increase in the positive charge near the sorbent surfaces in the case of divalent cation, which reduces the electrostatic repulsion between the phosphate anion and the surface (Pardo et al. 1992; Garoma 1996). Similarly, Sample et al. (1980) reported that the positively charged edge

TABLE 12.1

Sorption Parameters Calculated from the Freundlich Equation and Distribution Coefficient (K_d) of P under the Background Electrolytes Used

		Cations and Anions Background				
Coefficient	**Soil pH**	**Ca^{2+}**	**K^+**	**NH_4NO_3**	**HCO_3^-**	**SO_4^{2-}**
K_f, lkg^{-1}	5.1	86.2ab	83.0a	73.4a	94.4a	54.7a
	6.8	78.8a	80.3a	71.5a	86.7a	73.1ab
	8.0	176.1b	159.8b	167.8b	278.5b	137.1b
n	5.1	0.39a	0.35a	0.40a	0.46a	0.45a
	6.8	0.42a	0.37a	0.43a	0.72b	0.35a
	8.0	0.43a	0.46a	0.45a	0.46a	0.45a
K_d, lkg^{-1}[a]	5.1	13.2a	9.8a	10.9a	21.3a	8.6a
	6.8	13.6a	10.5a	12.1a	49.9b	8.8a
	8.0	55.6b	53.4b	53.6b	134.3c	37.6b
RMSE[c]		0.23	0.31	0.53	0.55	0.59
R^2[d]		0.96	0.96	0.89	0.89	0.84

Source: Reproduced from Shaheen, S.M., Tsadilas, C.D., Eskridge, K.M., Effect of common ions on phosphorus sorption and lability in Greek Alfisols with different pH. *Soil Sci.* 174:21–26, 2009. With permission from Lippincott Williams & Wilkins.

[a] Distribution coefficient (K_d) for P concentration 50 mg l^{-1}.

[b] For each one parameter (K_f, n, K_d), numbers in the same column with different letters differ significantly at the probability level ($P < 0.05$) according to the LSD test.

[c] *RMSE* = Root mean square error.

[d] R^2 = Coefficient of determination.

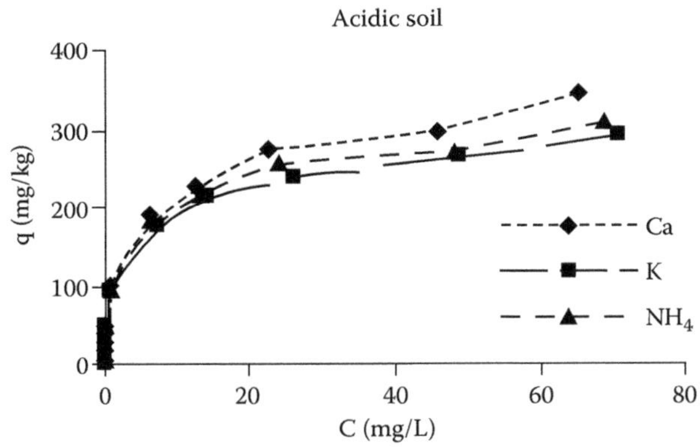

FIGURE 12.3
Phosphorus sorption by the soils as affected by differing cationic and anionic background solutions. (Reproduced from Shaheen, S.M., Tsadilas, C.D., Eskridge, K.M., 2009. Effect of common ions on phosphorus sorption and lability in Greek Alfisols with different pH. *Soil Science* 174:21–26, with permission from Lippincott Williams & Wilkins.)

sites of clay minerals are more accessible to P anions and that divalent cations such as Ca^{2+} enhance P sorption compared to monovalent cations such as Na^+ and K^+. In this respect, Devau et al. (2011) found that the sorption of Ca ions onto soil minerals also promoted sorption of P ions through electrostatic interactions. The increased removal of phosphate from solution with increasing pH in the presence of montmorillonite might be due to the release of exchangeable Ca^{2+} from the montmorillonite and subsequent formation of a calcium phosphate precipitate at pH > 7. Alternatively, phosphate sorption to montmorillonite may have increased at alkaline pH as a result of electrostatic interaction between phosphate and Ca^{2+} bound to deprotonated mineral surfaces (Rietra et al. 2001; Devau et al. 2009, 2011; Klimeski et al. 2014).

Total P sorption differs slightly in NH_4^+ and K^+ background solutions (Figure 12.3 and Table 12.1; Shaheen et al. 2009). The NH_4^+ was thought to resemble K^+ in cation exchange reactions in soils due to their similar ionic size and charge (Sposito 1989). Recent evidence, however, suggests that NH_4^+ and K^+ have different impacts on the interlayer surfaces of certain clay minerals such as vermiculites (Lumbanraja and Evangelou 1990). The ammonium ion induced expansion, rather than collapse, of the interlayer of vermiculite in contrast to K^+ (Lumbanraja and Evangelou 1990). The expansion of the clay mineral interlayer by NH_4^+ would suggest the relatively higher sorption in the presence of NH_4^+ compared to K^+ background (Wang and Harrell 2005).

In order to eliminate some of the inconsistencies associated with the influence of ions on P sorption procedures, Nair et al. (1984) proposed a standard P sorption protocol for soils and sediment based on the ability of multiple laboratories to produce consistent results over a wide range of soil samples. They concluded that P sorption was generally lower (up to 40%) with K^+ as the supporting electrolyte cation compared with Ca^{++}, consistent with the findings of Barrow et al. (1980). The effect was most pronounced at the higher initial P concentration. They also found that changes of the soil/solution ratio (SSR) had an effect on equilibrium phosphorus concentration values, and even though this effect was generally smaller than the effect of supporting electrolyte, they concluded that it was large enough to warrant standardizing both factors in the procedure (Nair et al. 1984).

In order to minimize artifacts associated with solution composition, equilibrium phosphorus concentration (EPC) experiments were performed by Bhadha et al. (2012) using a buffering electrolyte such as 0.01 M $CaCl_2$ or KCl (Figure 12.4). In this study, they evaluated the effect that experimental parameters such as equilibration time, SSR, and solution electrolyte may have on EPC values for two distinctly different soil types: (1) organic muck soil, and (2) loamy mineral soil. Results indicate that there is a significant difference in emerging EPC values with changing equilibrium time, SSR, and electrolyte. For the muck soil, the authors observed up to 97% decrease in EPC while increasing equilibration time from 1 to 8 h using 1:10 SSR. For the same soil, the most profound effect of SSR was observed for 1 h equilibration

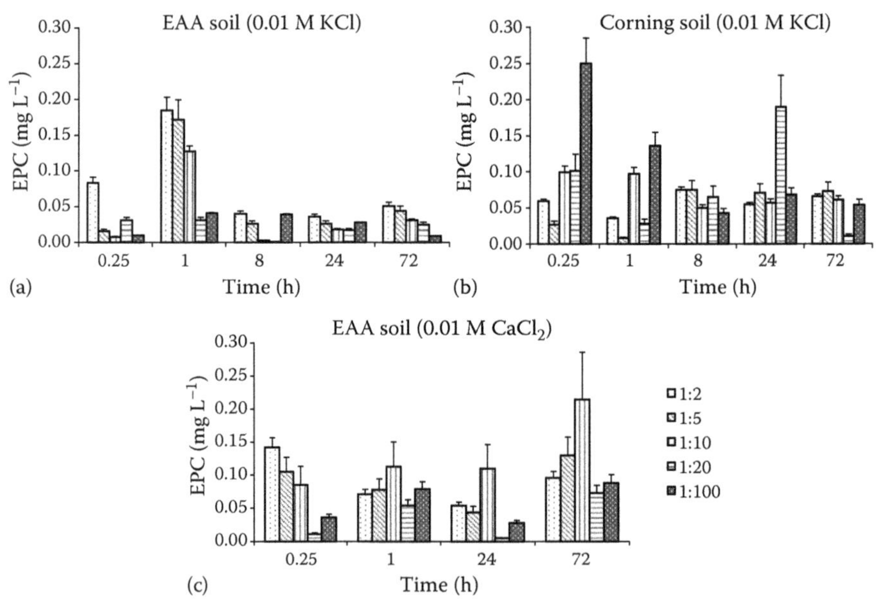

FIGURE 12.4
Changes in equilibrium phosphorus concentration values by varying equilibration time at five different soil/solution ratios for (a) Everglades Agricultural Area (EAA), (b) corning soils using 0.01 M KCl as the solution electrolyte, and (c) EAA soil using 0.01 M $CaCl_2$ as the solution electrolyte. Error bars represent the relative percent difference. (Reprinted from *Geoderma*, 173–174, Bhadha, J.H., Daroub, S.H., Lang, T.A., Effect of kinetic control, soil:solution ratio, electrolyte cation, and others, on equilibrium phosphorus concentration, 209–214, Copyright 2012 with permission from Elsevier.)

where the EPC value decreased from 0.19 mg L^{-1} using 1:2 SSR to 0.03 mg L^{-1} using 1:20 SSR, corresponding to an 84% change. The EPC of the loamy soil decreased by 94% while increasing equilibration time from 24 to 72 h using 1:20 SSR, whereas the same soil change in SSR was highly variable (Figure 12.4). Overall there was a 38% increase in ESP values using 0.01 M $CaCl_2$ compared to KCl as the solution electrolyte in the experiment (Figure 12.5). One possible explanation for this is that divalent cations increase the P sorption because they can fit into cavities of adsorbed phosphate ions (Bhadha et al. 2012). Bolster and Sistani (2009) showed that important differences between soils and $CaCl_2$ versus KCl solutions can also be attributed to factors such as organic matter content and pH.

In soils, the bioavailability of P is also controlled by the interfacial reactions of sorption, desorption, and precipitation with Fe/Al (oxyhydr)oxides (Yan et al. 2014).

The results of Yan et al.'s (2014) research demonstrated that the molecular structure and size of organic P as well as the crystallinity and crystal structure of Al (oxyhydr)oxides are the key factors affecting the sorption of organic P. In acid soils with a pH < 5.0, Al minerals hydrolyze to form octahedron hexahydrate (Al^{3+}) and mononuclear hydroxides ($Al(OH)^{2+}$ and $Al(OH)_2^+$), which are responsible for P sorption (Arias et al. 2006).

High levels of exchangeable Al^{3+} and Fe^{3+} in tropical acid soils lead to P sorption through their reaction with phosphate ions to form insoluble compounds. Additionally, iron oxides play an important role in the sorption and

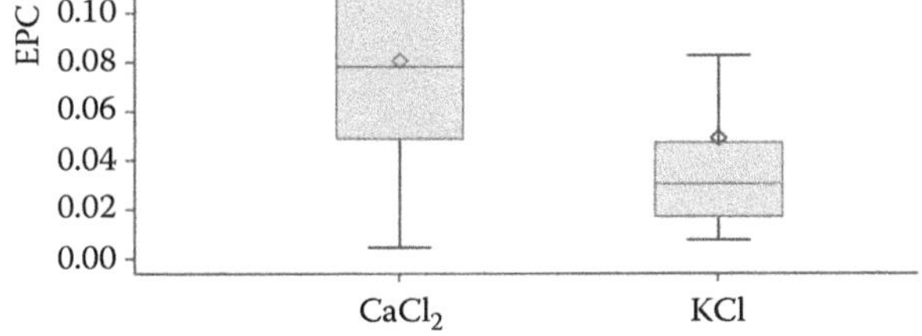

FIGURE 12.5
Box-plot comparison using $CaCl_2$ versus KCl as the solution electrolyte in the EPC experiment. The ◊ signs in the box are the mean, the horizontal lines in the box are the median, the top and bottom of the box represents the 75th and 25th percentiles, the whiskers define the 5th and 95th percentile observations, and the ○ signs outside the boxes are outliers. $n = 20$. (Reprinted from *Geoderma*, 173–174, Bhadha, J.H., Daroub, S.H., Lang, T.A., Effect of kinetic control, soil:solution ratio, electrolyte cation, and others, on equilibrium phosphorus concentration, 209–214, Copyright 2012 with permission from Elsevier.)

mobility of P. Ferrihydrite ($5Fe_2O_3{\cdot}9H_2O$), a naturally occurring nanocrystalline iron hydroxide, is common near surface environments, such as aquatic systems, soils, sediments, and organisms. Its large surface area, high reactivity, and corresponding large adsorption capacity allow nanocrystalline ferrihydrite to act as a sink for various contaminants and nutrient elements such as phosphate, and thus it plays an important role in controlling the bioavailability of P through adsorption and coprecipitation (Wang et al. 2013a,b).

The influence of Fe and Al on sorption of P was studied by Yaghi and Hartikainen (2013). In their study, light expanded clay aggregates (LECAs) and LECAs coated with Al oxide (Al-LECA) or iron (Fe) oxide (Fe-LECA) were tested for their efficiency as P sorbents in the pH range 3–8 (Figure 12.6). The oxide coatings duplicated the actual sorption capacity calculated from the sorption isotherms at the P concentration in the equilibrium solution of 20 μg L^{-1}, assumed to be the allowable P level in purified water. In the oxide-coated LECAs, the sorption was fast while in LECA, sorption was markedly slower (Figure 12.6). This might mean that at higher pH values the competition by hydroxyl ions diminished the sorption in LECA relatively more than that in the coated sorbents. In agreement with the acidity of Al^{3+} being 100 times lower than that of Fe^{3+}, at elevated pH, the sorption by Al-LECA proved to be less reversible than that by Fe-LECA. The results provide evidence that in constructed wetlands, Al-coated sorbents are superior to Fe-coated sorbents, which are redox-sensitive and may lose their sorption properties in anoxic conditions (Yaghi and Hartikainen 2013; Baken et al. 2014).

The influence of Al on sorption and mobility of P in soils is pH dependent (Shaheen et al. 2007). The increase in P sorption with the pH increase in soils with high Al content may be attributed to the increased activity of the amorphous hydroxy-aluminum species resulting from the increased pH (Shaheen

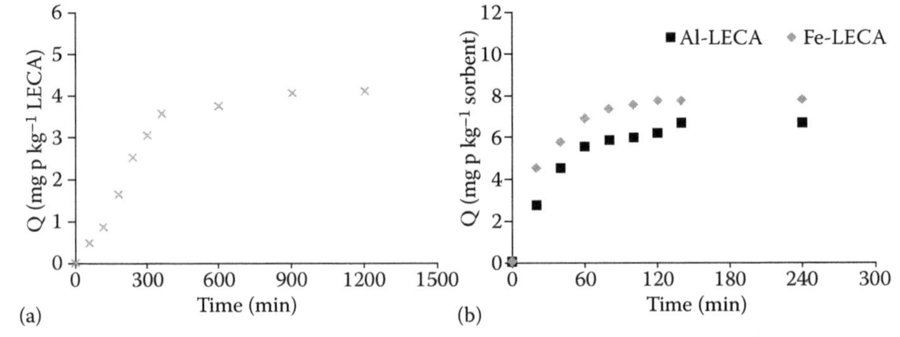

FIGURE 12.6
Phosphorus sorption by (a) LECA and (b) Al-LECA and Fe-LECA as a function of reaction time. Reaction conditions: initial conc. 19.6 μg P L^{-1}, pH = 4.0, sorbent dose 1.1 g L^{-1}, room temperature. Note the different scales. (Reproduced from Yaghi, N., Hartikainen, H., 2013. Enhancement of phosphorus sorption onto light expanded clay aggregates by means of aluminum and iron oxide coatings. *Chemosphere* 93, 1879–1886, with permission from Elsevier.)

et al. 2009). The strong affinity of the trivalent aluminum ion for OH^- ions accounts for its hydrolysis in aqueous solutions. The hydrolyzed Al ion species is then polymerized through a hydroxyl bridge. Polymerization is favored by pH increase, providing the OH^- units needed for the bridges. However, as the polymers increase in size, the average net positive charge for each aluminum ion decreases, and therefore its attraction for additional OH^- ion units becomes weaker. Consequently, the last OH^- ion to the hydroxy-aluminum polymer is weakly attached. Once attached, the last OH^- ion could be readily displaced by other anions. Phosphate anions are adsorbed on the surface when the hydroxyl–aluminum–phosphate attraction is strong enough to displace these weakly held surface hydroxyl ions (Yan et al. 2014).

In the study of Shaheen et al. (2009), of the three used anions, the greatest P sorption was recorded in the HCO_3^- background, followed by NO_3^- and SO_4^{2-} backgrounds (Figure 12.3, Table 12.1). This trend was true for all the tested soils, especially for the alkaline one. The differences in P sorption caused by the anions tested were reflected in K_d values (Table 12.1, Figure 12.3). The HCO_3^- background showed the highest K_d values compared to NO_3^- or SO_4^{2-} backgrounds. The greater P sorption in the HCO_3^- background compared to that in either NO_3^- or SO_4^{2-} backgrounds can be explained by the pH difference that occurred between them. The pH values in HCO_3^- background solutions were higher (ranging between 8.0 and 8.8) than the other two anions, which were lower than 8.0 even in the alkaline soil (except in one case). The increase in the total amount of P sorbed in the alkaline conditions in the presence of HCO_3^- may be due to the higher soil pH (Shaheen et al. 2009). Sorption processes are time-dependent and pH-controlled, and process rates decrease with time (Barrow 1984; Bolan et al. 1985). The explanation of this is that at pH > 7, phosphorus in soil solution exists mainly in the form of HPO_4^{2-}. This divalent ion with two nucleophilic centers and the potential to act as a bidentate ligand may have greater affinity for the adsorbent surfaces (Pardo et al. 1992; Garoma 1996). Morgan (1997) also reported that under alkaline conditions in the presence of free calcium carbonate, sorption of $H_2PO_4^-/HPO_4^{2-}$ on calcite could occur by replacing water, HCO_3^-, or OH^- ions present on the calcite particles.

The presence of SO_4^{2-} in soil solution might decrease P sorption. In this respect, Shaheen et al. (2009) found that SO_4^{2-} background decreased sorption of P especially in alkaline soils (Figure 12.3). This may be explained by the formation of complexes of sulfates with iron and aluminum oxides and the subsequent reduction of their activity in soil solution. In addition, sulfate may form complexes with exchangeable aluminum, therefore decreasing P sorption as suggested by Andrade et al. (2002), who studied the effect of organic residue, limestone, and gypsum on P sorption in lowland soils. Release of sulfate from organic residue decomposition and gypsum to soil solution decreases exchangeable Al and reduces P sorption by the treated soils. In the same study, a positive correlation between P sorption and exchangeable Al was reported. Furthermore, Alva and Sumner (1989) verified that exchangeable Al reduction may be due to its complexation with sulfate ions.

In the study by Zhu et al. (2007), the batch technique was used to investigate mutual effects of fluoride (F) and phosphorus (P) on their cosorption/desorption in an acidic red soil with high contents of Fe and Al (hydro)oxides. Results indicate that in a F–P coexisting system, a decrease in pH enhances the sorption of both F and P. An increase in F concentration suppresses P sorption due to a competitive effect. However, F sorption can be improved in the presence of P due to surface precipitation of (Al, Fe)–F–P. The presence of F has no measurable effect on P desorption, while the stability of F in the presence of P can be significantly diminished in comparison with that in the absence of P, which would lead to an improvement of F mobility (Zhu et al. 2007).

Influence of Common Ions on P Mobility

The mobility of P is strongly affected by its reactions with soil constituents and common ions. Among the principal experimental variables that affect the results of P desorption and mobility studies are the ionic composition in addition to both species and concentrations of the contacting solution. In spite of the realization that ionic species, concentrations, and their compositions effect P sorption and/or desorption, most of the salt-related studies are confined to Cl^- (anion) in association with different cations (Ahmad et al. 2008). Only a small number of previous studies were found to compare the results of more P release by SO_4^{2-} than Cl^- salt. Most of the previous studies focused on anion sorption capacities, but the mechanisms for their sorption are not fully understood. Most of the authors suggested that the mechanisms of SO_4^{2-} and PO_4^{3-} sorption are similar and that both ions compete for the same sorption sites (Pasricha and Fox 1993). Although adsorbed SO_4^{2-} does not compete strongly with PO_4^{3-}, there is likely some competition for the sorption between these anions, which may cause comparatively more P release by SO_4^{2-} than by Cl^- salts. While knowledge about the comparative response of P to Cl^- and SO_4^{2-} ions was lacking, the study by Ahmad et al. (2008) was conducted to evaluate the comparative effects of anions (in association with cations) on inorganic P release and P fractions in the soil. The test soil was amended with livestock compost manure (OP); KH_2PO_4 (IPk), or $Ca(H_2PO_4)_2$ (IPc) at a rate of 1 ppm. Soil was subjected to one salt and nine subsequent water extractions and different P fractions were measured. Four salt types, NaCl, Na_2SO_4, KCl, and K_2SO_4, were used at levels of 0.5 M. Irrespective of P sources, P release was substantially increased in the salt-pretreated soil as compared to the nonsaline soil. Sulfate salts released more P in subsequent water extractions than Cl^-. Phosphorus release decreased for salt types with Na_2SO_4 > NaCl > K_2SO_4 > KCl and for P sources with OP = IPK > control (without P application) > IPc, respectively. Thus, this study (Ahmad et al. 2008) clearly showed that not only cations species differ in P desorption capacity, but associated anions also play a vital role in the fate of P under saline environments. Synergetic effects exist between sodium (Na) and SO_4^{2-} ions, which enhanced the P release. This study has also confirmed the fact that P from organic sources is available from inorganic P sources as well. However, P release depends more on the type of P source applied than on total P. Higher P release by Na saturation could be due to the release of P associated with oxide surfaces or due to dissolution of Ca–P phases (Curtin et al. 1987). The study conducted by Ahmad et al. (2008) highly recommends that more than one anion species must be used in research approaches for evaluating the P response in a saline environment. These results have important implications for future studies, as most of the researchers focus on different cations only for evaluating P response to salts from an environmental point of view. However, Ahmad et al.'s (2008) study has made it clear that anions in association with cations differed for their effects on P release.

Shaheen et al. (2009) tested the influence of the common cations and anions on the mobility of the sorbed P. At the end of the sorption experiment, sorbed P was partitioned into labile (extracted by $NaHCO_3$) and nonlabile (the difference between total P sorbed and the labile fraction). The amount of labile P differed among the tested soils and certain ionic backgrounds (Table 12.2). On average, only 34% to 63% of the total sorbed P was found in labile form in the studied soils with all examined ionic backgrounds. The amount of labile P was smaller in alkaline soil than in neutral and acid soils. This result could be explained by the high pH values and an increase in exchangeable Ca in the alkaline soil compared to the neutral and acid soils (Table 12.1). This relation, together with findings from previous studies in an Entisol and Aridisol of Egypt (Shaheen et al. 2007), verifies that the $NaHCO_3$–P fraction, which is a reliable estimation of labile P, is

TABLE 12.2

Range and Average Percentage of Labile Pool[a] of the Total Sorbed P in the Presence of Different Cations and Anions

		Cations and Anions Background				
Soils		Ca^{2+}	K^+	NH_4NO_3	HCO_3^-	SO_4^{2-}
Acid soil	Range	35.0–50.2	42.2–78.1	48.6–76.4	40.8–64.3	46.5–80.9
	Average	43.5a[b]	59.0a	60.9a	53.7a	62.1a
Neutral soil	Range	33.2–59.3	45.1–68.6	52.7–65.6	45.6–65.0	47.6–83.4
	Average	46.6a	60.4a	59.8a	56.8a	62.9a
Alkaline soil	Range	23.8–52.5	42.2–67.4	40.6–65.3	6.6–63.3	43.6–58.2
	Average	38.8a	53.3a	52.2a	33.7b	48.9b

Source: Reproduced from Shaheen, S.M., Tsadilas, C.D., Eskridge, K.M., Effect of common ions on phosphorus sorption and lability in Greek Alfisols with different pH. *Soil Sci.* 174:21–26, 2009. With permission from Lippincott Williams & Wilkins.

[a] Phosphorus extracted by $NaHCO_3$ at the end of sorption experiment. Average percentage of labile P was calculated from all concentration levels of sorption.

[b] Numbers in the same columns with different letters differ significantly at the probability level $P < 0.05$) according to the LSD test.

negatively correlated with pH. In agreement to this, Rupa et al. (2001) reported that soils with high exchangeable Ca and magnesium (Mg) level decreased P desorption at high pH values due to precipitation of P as a Ca-phosphate.

Both cations and anions had a large effect on the lability of the sorbed P that was more pronounced as the amount of P sorbed increased. Of the three cations tested, the lowest amount of labile P was found in the Ca^{2+} background for the soils studied (Rupa et al. 2001). The average percentage of the labile P ranged from 39% to 47% in Ca^{2+}, 53% to 60% in K^+, and from 52% to 61% with NH_4^+ background. The decrease in labile P in Ca^{2+} background may be due to the precipitation of P as a Ca-phosphate (Rupa et al. 2001). In relation to this, McCullum (1996) reported that Ca-phosphates are quite stable and very insoluble at higher pH, thus becoming the major forms of inorganic P in neutral and alkaline soils. The effect of NH_4^+ and K^+ on the lability of the sorbed P was similar in the soils studied (Table 12.2). However, in the alkaline soil, the percentage of labile P retrieved was lower than in acid and neutral soils (Table 12.2). The average percentage of the total P sorbed being in labile form ranged in the three soils from 34% to 57% in HCO_3^-, 52% to 61% in NO_3^-, and 49% to 63% in SO_4^{2-} backgrounds. The lower percentage of labile P in HCO_3^- background may be due to the increase in pH values in the soil solution mixture, which might lead to precipitation of P in a nonlabile form or the increased carbonate ions in the soil solution mixture, possibly reacted with P to form precipitates (Morgan 1997). On the other hand, the higher percentage of labile P in NO_3^- and SO_4^{2-} may be explained by the decrease in pH values in the soil solution mixture, as shown in Table 12.2. Furthermore, SO_4^{2-} might cause a decrease in exchangeable Al that may precipitate P in a nonlabile form (Alva and Sumner 1989). Another possible explanation might be what McBride (1989) has suggested: that nonlability of nutrients and metal ions in soil is attributed to the entrapment of these ions into small pores of particle aggregates or clay mineral structures, the strong surface sorption by oxides, and/or the formation of insoluble ion precipitates.

The release and mobility of P in wetland soils in relation to the redox chemistry of Fe were also studied. Although several studies have demonstrated the risk of P release following rewetting of lowland soils (Aldous et al. 2007; Kjaergaard et al. 2012; Abit et al. 2013), the ability to actually predict P release rates following wetland restoration is extremely challenging due to complex interactions of soil biogeochemistry and wetland hydrology (Young and Ross 2001; Kjaergaard et al. 2012). Reductive dissolution of Fe(III) causing oxides to partly or fully dissolve and resulting in the release of Fe(II) and sorbed P is generally considered as the major cause of P release following rewetting (Shenker et al. 2005; Geurts et al. 2008; Zak et al. 2008; Kjaergaard et al. 2012). Readsorption of P to nonreduced Fe(III) and/or redox-stable Al oxides (Heiberg et al. 2010), as well as precipitation products such as Fe(II) phosphates (Heiberg et al. 2012; Walpersdorf et al. 2013) or calcium phosphates (Shenker et al. 2005), may, however, complicate the prediction of P release. Forsmann and Kjaergaard (2014) indicated that concurrent Fe and P release was controlled by reductive Fe(III) dissolution.

Actual P release rates following restoration of wetlands on peat/organic lowland soils depend on a range of local variables such as the supply of terminal electron acceptor on redox dynamics (Shenker et al. 2005; Kjaergaard et al. 2012), the groundwater supply of sulfate facilitating precipitation of ironsulfides in anoxic environments increasing the risk for P release (e.g., Lucassen et al. 2004), flowrate (Kjaergaard et al. 2012), and transport across an aerobic soil–water interface with reoxidation/precipitation of Fe(III) and resorption of soluble P (Zak et al. 2004).

Summary and Future Directions

An effective soil phosphorus management from both an environmental and agronomic point of view requires the knowledge of P bioavailability and sorption characteristics in relation to the effect of common ions added to soil with fertilizers and/or amendments. Phosphorus sorption has been shown to vary with soil/solution ratios, solution pH, ionic strength, and cation species of the supporting electrolyte.

Sorption of P increases as the ionic strength of the solution increases (Pardo et al. 1992). However, this may not be the case in the presence of anions such as SO_4^{2-}, F^-, and OH^- that compete with phosphate ($H_2PO_4^-$) ions (Nair et al. 1984). Increasing ionic strength should decrease sorption when the surface is negative and should increase sorption when the surface is positive. In contrast, several authors found the opposite effect (Barrow et al. 1980b; Barrow and Ellis 1986).

The presence of mono- or divalent cations as background electrolytes affects P sorption. The greater sorption of P is observed in the presence of divalent cations (i.e., Ca^{++}) in the background solution compared to those containing monovalent cations (NH_4^+, Na^+, and K^+). The greatest P sorption i was recorded in the presence of HCO_3^- in the background solution, followed by NO_3^- and SO_4^{2-} backgrounds. High levels of exchangeable Al^{3+} and Fe^{3+} in tropical acid soils lead to P sorption through their reaction with phosphate ions to form insoluble compounds. The influence of Al on sorption and mobility of P in soils is pH dependent.

The mobility of P is strongly affected by its reactions with soil constituents and common ions. A few studies (e.g., Ahmad et al. 2008) showed that P release was substantially increased in salt-pretreated soil compared to nonsaline soil. Sulfate salts released more P in subsequent water extractions than Cl^-. Phosphorus release decreased for salt types with $Na_2SO_4 > NaCl > K_2SO_4 > KCl$. This clearly showed that not only do cation species differ in P desorption capacity, but associated anions also play a vital role in the fate of P in saline environments. Synergetic effects exist between Na and SO_4^{2-} ions that enhanced P release. The lability of the sorbed P is affected by the common ions and these effects are pH dependent. Some studies (e.g., Shaheen et al. 2009) showed that the amount of labile P was smaller in alkaline soil than in neutral and acid soils. Both NH_4^+ and K^+ increased P lability equally, while Ca^{2+} decreased it in all soils. The anions NO_3^- and SO_4^{2-} increased the labile P in contrast to HCO_3^-, which decreased it at all pH levels. Sulfate, however, maintained slightly more P in the labile pool compared to NO_3^- in the acid and neutral soils, while NO_3^- favored the highest labile P in alkaline soil.

We hope that the information in this chapter will contribute to a more efficient management of P in relation to other nutrients, and in terms of environmental relevance, the chapter has also provided valuable information that is relevant for operational predictions on potential P release rates especially following restoration of wetlands on peat/organic lowland soils.

References

Abit, S.M., Vepraskas, M.J., Duckworth, O.W., Amoozegar, A. 2013. Dissolution of phosphorus into pore-water flowing through an organic soil. *Geoderma* 197:51–58.

Ahmad, Z., Faridullah, El-Sharkawi, H., Irshad, M., Honna, T., Yamamoto, S., Al-Busaidi, A.S. 2008. Changes in water-extractability of soil inorganic phosphate induced by chloride and sulfate salts. *Environ. Sci. Pollut. Res.* 15:23–26.

Aldous, A.R., Craft, C.B., Stevens, C.J., Barry, M.J., Bach, L.B. 2007. Soil phosphorus release from a restoration wetland, Upper Klamath Lake, Oregon. *Wetlands* 27: 1025–1035.

Alva, A.K., Sumner, M.E. 1989. Alleviation of aluminum toxicity to soybeans by phosphogypsium or calcium sulfate in dilute nutrient solutions. *Soil Sci.* 147:278–285.

Andrade, A.T., Fernandes, L.A., Faquin, V. 2002. Organic residues, limestone, gypsum, and phosphorous adsorption by lowland soils. *Sci. Agric.* 59:349–355.

Arias, M., Da Silva-Carballal, J., Garcia-Rio, L., Mejuto, J., Nunez, A. 2006. Retention of phosphorus by iron and aluminium-oxides-coated quartz particles. *J. Colloid Interface Sci.* 295:65–70.

Baken, S., Nawara, S., Van Moorleghem, C., Smolders, E. 2014. Iron colloids reduce the bioavailability of phosphorus to the green alga Raphidocelis subcapitata. *Water Res.* 59:198–206.

Barrow, N.J. 1983a. A mechanistic model for describing the sorption and desorption of phosphate by soil. *J. Soil Sci.* 34:733–750.

Barrow, N.J. 1983b. On the reversibility of phosphate sorption by soils. *J. Soil Sci.* 34:751–758.

Barrow, N.J., Bowden, J.W., Posner, A.M., Quirk, J.P. 1980a. An objective method for fitting models of ion adsorption on variable charge surfaces. *Aust. J. Soil Res.* 18:37–47.

Barrow, N.J., Bowden, J.W., Posner, A.M., Quirk, J.P. 1980b. Describing the effects of electrolyte on adsorption of phosphate by a variable charge surface. *Aust. J. Soil Res.* 18:395–404.

Beauchemin, S., Hesterberg, D., Chou, J., Beauchemin, M., Simrad, R.R., Sayers, D.E. 2003. Speciation of phosphorus in phosphorus enriched agricultural soils using X-ray absorption near-edge structure spectroscopy and chemical fractionation. *J. Environ. Qual.* 32:1809–1819.

Bertrand, I., Grignon, N., Hinsinger, P., Souche, G., Jaillard, B. 2001. The use of secondary ion mass spectroscopy coupled with image analysis to identify and locate chemical element in soil minerals: The example of phosphorus. *Scanning* 23:279–291.

Bhadha, J.H., Daroub, S.H., Lang, T.A. 2012. Effect of kinetic control, soil:solution ratio, electrolyte cation, and others, on equilibrium phosphorus concentration. *Geoderma* 173-174:209–214.

Bhadha, J.H., Harris, W.G., Jawitz, J.W. 2010. Soil phosphorus release and storage capacity from an impacted subtropical wetland. *Soil Sci. Soc. Am. J.* 74:1816–1825.

Bolan, N.S., Syers, J.K. Tillman, R.W. 1986. Ionic strength effects on surface charge and adsorption of phosphate and sulphate by soils. *J. Soil Sci.* 37:379–388.

Bolster, C.H., Sistani, K.R. 2009. Sorption of phosphorus from swine, dairy and poultry manures. *Communications in Soil Science and Plant Analysis* 40:1106–1123.

Bürkert, A., Bagayoko, M., Alvey, S., Bationo, A. 2001. Causes of legume rotation effects in increasing cereal yields across the Sudanian, Sahelian and Guinean zone of West Africa. In: *Plant Nutrition. Food Security and Sustainability of Agro-Ecosystem through Basic and Applied Research,* W. Horst. et al. (eds.). Kluwer Academic Publishers, Dordrecht, Netherlands, 972–973.

Chen, J.S., Mansell, R.S., Nkedi-Kizza, P., Burgao, B.A. 1996. Phosphorus transport during transient, unsaturated water flow in an acid sandy soil. *Soil Sci. Soc. Am. J.* 60:42–48.

Cleneghan, C. 2003. *Phosphorus Regulations National Implementation Report, 2003.* Environmental Protection Agency, Wexford, Ireland.

Curtin, D., Syers, J.K., Smillie, G.W. 1987. The importance of exchangeable cations and resin-sink characteristics in the release of soil phosphorus. *J. Soil Sci.* 38:711–716.

Damon, P.M., Bowden, B., Rose, T., Rengel, Z. 2014. Crop residue contributions to phosphorus pools in agricultural soils: A review. *Soil Biol. Biochem.* 74:127–137.

Delgado, A., Torrent, J. 2000. Phosphorus forms and desorption patterns in heavily fertilized calcareous and limed acid soils. *Soil Sci. Soc. Am. J.* 64:2031–2037.

Devau, N., Hinsinger, P., Le Cadre, E., Colomb, B., Gerard, F. 2011. Fertilization and pH effects on processes and mechanisms controlling dissolved organic phosphorus in soils. *Geochim. Cosmochim. Acta* 75:2980–2996.

Devau, N., Le Cadre, E., Hinsinger, P., Jaillard, B., Gerard, F. 2009. Soil pH controls the environmental availability of phosphorus: Experimental and mechanistic modelling approaches. *Appl. Geochem.* 24:2163–2174.

Forsmann, D.M., Kjaergaard, C. 2014. Phosphorus release from anaerobic peat soils during convective discharge–effect of soil Fe:P molar ratio and preferential flow. *Geoderma* 223–225:21–32.

Fox, L.E. 1989. A model for inorganic control of phosphate concentrations in river waters. *Geochim. Cosmochim. Acta* 53:417–428.

Froelich, P.N. 1988. Kinetic control of dissolved phosphate in natural rivers and estuaries: A primer on the phosphate buffer mechanism. *Limnol. Oceanogr.* 33:649–668.

Frossard, E., Brossard, M., Hedley, J., Metherell, A. 1995. Reactions controlling the cycling of P in soils. In: *Phosphorus in the Global Environment,* H. Tiessen (ed.). John Wiley & Sons, New York, 107–138.

Garoma, M.D. 1996. Biogeochemistry of phosphorus in the highland plateau soils of Ethiopia. Ph.D. Theses. North Carolina state university. 256 pp. http://books.google.com.eg/books/about/Biogeochemistry_of_Phosphorus_in_the_Hig.html?id=_mqDtgAACAAJ&redir_esc=y.

Geurts, J.J.M., Smolders, A.J.P., Verhoeven, J.T.A., Roelofs, J.G.M., Lamers, L.P.M. 2008. Sediment Fe: PO(4) ratio as a diagnostic and prognostic tool for the restoration of macrophyte biodiversity in fen waters. *Freshw. Biol.* 53:2101–2116.

Goyne, K.W., Jun, H.-J., Anderson, S.H., Motavalli, P.P. 2008. Phosphorus and nitrogen sorption to soils in the presence of poultry litter-derived dissolved organic matter. *J. Environ. Qual.* 37:154–163.

Hamdi, W., Pelster, D., Seffen, M. 2014. Phosphorus sorption kinetics in different types of alkaline soils. *Arch. Agron. Soil Sci.* 60:577–586.

Harrell, D.L., J.J. Wang. 2006. Fractionation and sorption of inorganic phosphorus in Louisiana calcareous soils. *Soil Sci.* 171:39–51.

Heckrath, G., Brookes, P.C., Poulton, P.R., Goulding, K.W.T. 1995. Phosphorus leaching from soils containing different phosphorus concentrations in the Broadbalk Experiment. *J. Environ. Qual.* 24:904–910.

Heffer, P., Prud'homme, M. 2013. Nutrients as limited resources: Global trends in fertilizer production and use. In: *Improving Water and Nutrient-use Efficiency in Food Production Systems,* Z. Rengel (ed.), John Wiley & Sons, Ames, IA, 57–78.

Heiberg, L., Koch, C.B., Kjaergaard, C., Jensen, H.S., Hansen, H.C.B. 2012. Vivianite precipitation and phosphate sorption following iron reduction in anoxic soils. *J. Environ. Qual.* 41:938–949.

Heiberg, L., Pedersen, T.V., Jensen, H.S., Kjaergaard, C., Bruun Hansen, H.C. 2010. A comparative study of phosphate sorption in lowland soils under oxic and anoxic conditions. *J. Environ. Qual.* 39:734–743.

Hinsinger, P. 2001. Bioavailability of soil inorganic P in the rhizosphere as affected by root-induced chemical changes: A review. *Plant Soil* 237:173–195.

Kinraide, T.B. 1991. Identification of rhizotoxic aluminum species. *Plant Soil* 134: 167–178.

Kjaergaard, C., Heiberg, L., Jensen, H.S., Hansen, H.C.B. 2012. Phosphorus mobilization in rewetted peat and sand at variable flow rate and redox regimes. *Geoderma* 173:311–321.

Klimeski, A., Risto Uusitalo, R., Eila Turtola, E. 2014. Screening of Ca- and Fe-rich materials for their applicability as phosphate-retaining filters. *Ecol. Eng.* 68: 143–154.

Kochian, L.V. 1995. Cellular mechanism of aluminium toxicity and resistance in plants. *Annu. Rev. Plant Physiol.* 46:237–260.

Koski-Vahala, J., Hartikainen, H. 2001. Assessment of risk of phosphorus loading due to resuspended sediment. *J. Environ. Qual.* 30:960–966.

Li, H., Sun, A.-H., Hou, J., Chen, X., Bai, Y.-Q., Cao, X.-Y., Song, C.-L., Zhou, Y.-Y. 2014. Phosphorus fractions, sorption characteristics, and its release in sediments of a Chinese eutrophic river (Nanfei River). *Fresen. Environ. Bull.* 23:1222–1231.

Lucassen, E., Smolders, A.J.P., van de Crommenacker, J., Roelofs, J.G.M. 2004. Effects of stagnating sulphate-rich groundwater on the mobility of phosphate in freshwater wetlands: A field experiment. *Arch. Hydrobiol.* 160:117–131.

Mansell, R.S., Bloom, S.A., Nkedi-Kizza, P. 1995. Phosphorus transport in spodosols impacted by dairy waste. *Ecol. Eng.* 5:281–299.

McBride, M.B. 1989. Reactions controlling heavy metals solubility in soils. *Adv. Soil Sci.* 10:1–55.
McCullum, R.E. 1996. Soil phosphorus and potassium. In: *The Nature and Properties of Soils,* 2nd Edition, N.C. Brady, R.R. Weil (eds.). Prentice-Hall International, Inc., Upper Saddle River, NJ, 445–487.
Memon, K.S., Fox, R.L. 1983. Utility of phosphate sorption curves in estimating P requirements of cereal crops: Wheat (*Triticum aestivum*). Proceeding of the Third International Congress on phosphorus compounds. Brussels. Belgium Oct 4–6.
Memon, K.S., Puno, H.K., Fox, R.L. 1991. Phosphate sorption approach for determining phosphorus requirements of wheat in calcareous soils. *Fert. Res.* 28:67–72.
Morgan, M.A. 1997. The behavior of soil and fertilizer phosphorus. In: *Phosphorus Loss from Soil to Water.* H. Tunney, O.T. Carton, P.C. Brookes, A.E. Johnston (eds.). Cab International, Wallingford, Oxon, UK, 137–150.
Nair, P.S., Logan, T.J., Sharpley, A.N., Sommers, L.E., Tabatabai, M.A., Yuan, T.L. 1984. Interlaboratory comparison of a standard phosphorus adsorption procedure. *J. Environ. Qual.* 13:591–595.
Pardo, M.T., Guadalix, M.E., Garcia-Gonzalez, M.T. 1992. Effect of pH and background electrolyte on P sorption by variable charge soils. *Geoderma* 54:275–284.
Pasricha, N.S., Fox, R.L. 1993. Plant nutrient sulphur in the tropics and the sub-tropics. *Adv. Agron.* 50:209–269.
Rhue, R.D., Harris, R.G. 1999. Phosphorus sorption/desorption reactions in soils and sediments. In: *Phosphorus Biogeochemistry in Subtropical Ecosystems,* K.R. Reddy, G.A. O'Connor, C.L. Schleske (eds.). Lewis Publishers, Boca Raton, FL, 187–206.
Rietra, R.P.J.J., Hiemstra, T., Van Riemsdijk, W.H. 2001. Interaction between calcium and phosphate adsorption on goethite. *Environ. Sci. Technol.* 35:3369–3374.
Rupa, T.R., Tomar, K.P., Srinivasa Rao, C.H., Subba Rao, A. 2001. Kinetics of phosphate sorption-desorption as influenced by soil pH and electrolyte. *Agrochimica* 45:124–133.
Said, M.B., Dakermanji, A. 1993. Phosphate adsorption and desorption by calcareous soils of Syria. *Commun. Soil Sci. Plant Anal. J.* 24:197–210.
Schindler, D.W. 2012. The dilemma of controlling culturalc eutrophication of lakes. *Proc. Biol. Sci.* 279:4322–4333.
Shaheen, S.M. 2009. Sorption and lability of cadmium and lead in different soils from Egypt and Greece. *Geoderma* 153:61–68.
Shaheen, S.M., Tsadilas, C.D. 2013. Phosphorus sorption and availability to canola grown on an Alfisol amended with various soil amendments. *Commun. Soil Sci. Plant Anal.* 44:89–103.
Shaheen, S.M., Tsadilas, C.D., Eskridge, K.M. 2009. Effect of common ions on phosphorus sorption and lability in Greek Alfisols with different pH. *Soil Sci.* 174:21–26.
Shaheen, S.M., Tsadilas, C.D., Stamatiades, S. 2007. Inorganic phosphorus forms in some Entisol and Aridisols of Egypt. *Geoderma* 142:217–225.
Sharpley, A., Jarvie, H.P., Buda, A., May, L., Spears, B., Kleinman, P. 2013. Phosphorus legacy: Practices to mitigate future water quality impairment. *J. Environ. Qual.* 42:1308–1326.
Sharpley, A., Tunney, H. 2000. Phosphorus research strategies to meet agricultural and environmental challenges in the 21st century. *J. Environ. Qual.* 29:176–181.

Sharpley, A.N., Daniel, T.C., Wright, B., Kleinman, P., Sobecki, T., Parry, R., Joern, B. 1999. National research project to identify sources of agricultural phosphorus loss. *Better Crops* 83:12–15.

Sharpley, A.N., Herron, S., Daniel, T. 2007. Overcoming the challenges of phosphorus-based management in poultry farming. *J. Soil Water Conserv.* 62:375–389.

Sharpley, A.N., Withers, P.J.A. 1994. The environmentally-sound management of agricultural phosphorus. *Fertil. Res.* 39:133–146.

Shenker, M., Seitelbach, S., Brand, S., Haim, A., Litaor, M.I. 2005. Redox reactions and phosphorus release in re-flooded soils of an altered wetland. *Eur. J. Soil Sci.* 56:515–525.

Sims, J.T., Edwards, A.C., Schoumans, O.F., Simard, R.R. 2000. Integrating soil phosphorus testing into environmentally based agricultural management practices. *J. Environ. Qual.* 29:60–71.

Sims, J.T., Simard, R.R., Joern, B.C. 1998. Phosphorus loss in agricultural drainage: Historical perspective and current research. *J. Environ. Qual.* 27:277–293.

Sinaj, S., Frossard, E., Morel, J.L. 1992. Phosphate availability in Albanian soils. In: *Proceedings of Second European Society of Agronomy Congress,* A. Scaife (eds.), Warwick, UK, August 23–28, 1992. ESA, HRI, Wellesbourne, UK, 306–307.

Soinne, H., Hovi, J., Tammeorg, P., Eila Turtola, E. 2014. Effect of biochar on phosphorus sorption and clay soil aggregate stability. *Geoderma* 219–220:162–167.

Tisdale, S.L., Nelson, W.L., Beaton, J.D. 1990. *Soil Fertility and Fertilizers.* Macmillan, New York.

Tsadilas, C.D., Samaras, V., Dimoyiannis, D. 1996. Phosphate sorption by red Mediterranean soils from Greece. *Commun. Soil Sci. Plant Anal.* 27:2279–2293.

Turtola, E., Yli-Halla, M. 1999. Fate of phosphorus applied in slurry and mineral fertilizer: Accumulation in soil and release into surface runoff water. *Nutr. Cycl. Agroecosyst.* 55:165–174.

Walpersdorf, E., Koch, C.B., Heiberg, L., O'Connell, D.W., Kjaergaard, C., Hansen, H.C.B. 2013. Does vivianite control phosphate solubility in anoxic meadow soils? *Geoderma* 193:189–199.

Wandruszka, R. 2006. Phosphorus retention in calcareous soils and the effect of organic matter on its mobility. *Geochem. Trans.* 7:6, doi:10.1186/1467-4866-7-6.

Wang, X., Li, W., Harrington, R., Liu, F., Parise, J.B., Feng, X., Sparks, D.L. 2013a. Effect of ferrihydrite crystallite size on phosphate adsorption reactivity. *Environ. Sci. Technol.* 47:10322–10331.

Wang, X., Liu, F., Tan, W., Li, W., Feng, X., Sparks, D.L. 2013b. Characteristics of phosphate adsorption-desorption onto ferrihydrite: Comparison with well-crystalline Fe (hydr)oxides. *Soil Sci.* 178:1–11.

Yaghi, N., Hartikainen, H. 2013. Enhancement of phosphorus sorption onto light expanded clay aggregates by means of aluminum and iron oxide coatings. *Chemosphere* 93:1879–1886.

Yan, Y.P., Liu, F. Jr., Li, W., Liu, F., Feng, X.H., Sparks, D.L. 2014. Sorption and desorption characteristics of organic phosphates of different structures on aluminium (oxyhydr)oxides. *Eur. J. Soil Sci.* 65:308–317.

Yang, G.-R., Hao, X.-Y., Li, C.-L., Li, Y.-M. 2014. Effect of land use on soil phosphorus sorption-desorption under intensive agricultural practices in plastic-film greenhouses. *Pedosphere* 24:367–377.

Young, E.O., Ross, D.S. 2001. Phosphate release from seasonally flooded soils: A laboratory microcosm study. *J. Environ. Qual.* 30:91–101.

Yuji, A., Sparks, D.L. 2001. ATR-FTIR Spectroscopic investigation on phosphate adsorption mechanism at the ferrihydrite-water interface. *J. Colloid Interface Sci.* 241:317–326.

Zak, D., Gelbrecht, J. 2007. The mobilisation of phosphorus, organic carbon and ammonium in the initial stage of fen rewetting (a case study from NE Germany). *Biogeochemistry* 85:141–151.

Zhang, H., Schroder, J.L., Fuhrman, J.K., Basta, N.T., Storm, D.E., Payton, M.E. 2005. Path and multiple regression analysis of phosphorus sorption capacity. *Soil Sci. Soc. Am. J.* 69:96–106.

Zhu, M.-X., Ding, K.-Y., Jiang, X., Wang, H.-H. 2007. Investigation on co-sorption and desorption of fluoride and phosphate in a red soil of China. *Water Air Soil Pollut.* 183:455–465.

Index

Page numbers followed by f and t indicate figures and tables, respectively.

B

D

N

O

P

T

For Product Safety Concerns and Information please contact our EU
representative GPSR@taylorandfrancis.com
Taylor & Francis Verlag GmbH, Kaufingerstraße 24, 80331 München, Germany

www.ingramcontent.com/pod-product-compliance
Ingram Content Group UK Ltd.
Pitfield, Milton Keynes, MK11 3LW, UK
UKHW020025200926
475702UK00014B/2073

* 9 7 8 1 1 3 8 8 0 3 1 8 3 *